Engineering Materials

This series provides topical information on innovative, structural and functional materials and composites with applications in optical, electrical, mechanical, civil, aeronautical, medical, bio- and nano-engineering. The individual volumes are complete, comprehensive monographs covering the structure, properties, manufacturing process and applications of these materials. This multidisciplinary series is devoted to professionals, students and all those interested in the latest developments in the Materials Science field, that look for a carefully selected collection of high quality review articles on their respective field of expertise.

Indexed at Compendex (2021) and Scopus (2022)

Ali Dad Chandio · Iftikhar Ahmed Channa
Editors

Thin Films and Coatings

Engineering Applications

Springer

Editors
Ali Dad Chandio
Advanced Materials and Sustainable Environment (AMSE) Research Group
Department of Metallurgical Engineering and Technology
NED University of Engineering and Technology
Karachi, Pakistan

Iftikhar Ahmed Channa
Advanced Materials and Sustainable Environment (AMSE) Research Group
Department of Metallurgical Engineering and Technology
NED University of Engineering and Technology
Karachi, Pakistan

ISSN 1612-1317 ISSN 1868-1212 (electronic)
Engineering Materials
ISBN 978-981-96-0004-5 ISBN 978-981-96-0005-2 (eBook)
https://doi.org/10.1007/978-981-96-0005-2

This Springer imprint is published by the registered company Springer Nature Singapore Pte Ltd.
The registered company address is: 152 Beach Road, #21-01/04 Gateway East, Singapore 189721, Singapore

Contents

Introduction to Thin Films and Coatings

Jaweria Ashfaq, Mubashir Hassan, Talal Saeed, and Muhammad Ahsan khan

Abstract This first chapter provides an overview of thin films and coatings, reviewing their timeline and increasing relevance in the engineering domain. Thin films that are a number of layers deposited in very small thicknesses have been in use for many years and have undergone various changes. Starting from their initial application in relatively coarse ideas, they successfully evolved and are now critical elements of many important fields. Thin films are applied in electronics, optics and material engineering as these materials enhance the performance, reliability and efficacy of machines and systems. The chapter also provides the state of art on thin films and how this technology has enhanced the development of more accurate as well as efficient coatings. Analyzing the development of thin films, it is possible to understand how these materials evolved from simple tools to the key components of technologies. Although it is the first chapter in this book, it provides the fundamental knowledge about thin films and coatings to help prepare the reader for the remaining chapters that focus on the more detailed aspects of these materials and their applications in engineering.

1 Definition and Overview

Thin-film technology is a microfilm that has grown significantly to be an important faction of the present society because of its thin layers that range from the nanometer range of between 0.001 and 0.01 mm and the micrometer range of between 0.0001 and 0.01 mm. It is important to note that these films have different applications in numerous fields including electronics, optic, energy, and health care. They provide variation in material qualities depending on the function, as well as thickness control, which broadens potential applications and innovations in a number of sectors [1].

J. Ashfaq (✉) · M. Hassan · T. Saeed · M. A. khan
Thin-Film Lab As Part of Advanced Materials and Sustainable Environment (AMSE) Research Group, Department of Metallurgical Engineering, NED University of Engineering and Technology, Karachi 75270, Sindh, Pakistan
e-mail: Jaweriaashfaq72@gmail.com

A. D. Chandio and I. A. Channa (eds.), *Thin Films and Coatings*, Engineering Materials,
https://doi.org/10.1007/978-981-96-0005-2_1

2 Composition of Thin Films

Thin film is an extremely useful feature because it is available in a variety of composition types. It is possible to prepare these films from metals, semiconductors, dielectrics, polymers, and related composite materials. This is because when each material type is incorporated into the thin film, the resultant thin film has attributes that are different from other types of thin films, and this makes it possible to thin film it to have the desired functionality in many sectors of technology. Metals are often used in thin films due to their enhanced electrical conductivity as compared to other materials. Metal thin films are widely used in forming the electrical conduction paths and connections in circuits to be incorporated in the electronic devices. On the other hand, semiconductors are useful in electrical applications for instance as transistors due to their mid-range electrical conductivity. Dielectric thin films are applied in the fabrication of insulating coatings for pieces of electronic equipment and for dielectrics due to their high insulating property [2].

The applications of flexible electronics and biomedical engineering can be brought by the use of polymeric thin films which are flexible and biocompatible [3]. Composite materials are liberal methods to enable specific performance for using the characteristics of several materials in integration. These many material options contribute in enhancing flexibility of the thin films to fit any need in technological applications [4] (Fig. 1 and Table 1).

3 Deposition Techniques

There are various techniques of depositing thin films on substrate surfaces, and each has its own characters as well as applications [9]. These tendering strategies encompass the following (Fig. 2).

3.1 Physical Vapor Deposition (PVD)

Deposition of resistive thin films relies mainly on PVD which is known to be a versatile technique required in various applications. The ability to slow down the material arrival from the solid source at the indicated substrate is one of the distinguished features of the PVD procedures. This control helps to achieve thin films having specific characteristics and features that meet the requirements of specifications being targeted [10].

One of the most widely used form of PVD technique is named evaporation and is based on turning a solid material into vapor following application of heat energy. This then forms a thin layer of condenses on top of it from the vapor that has been produced. This method is effective for application which demand accurate properties

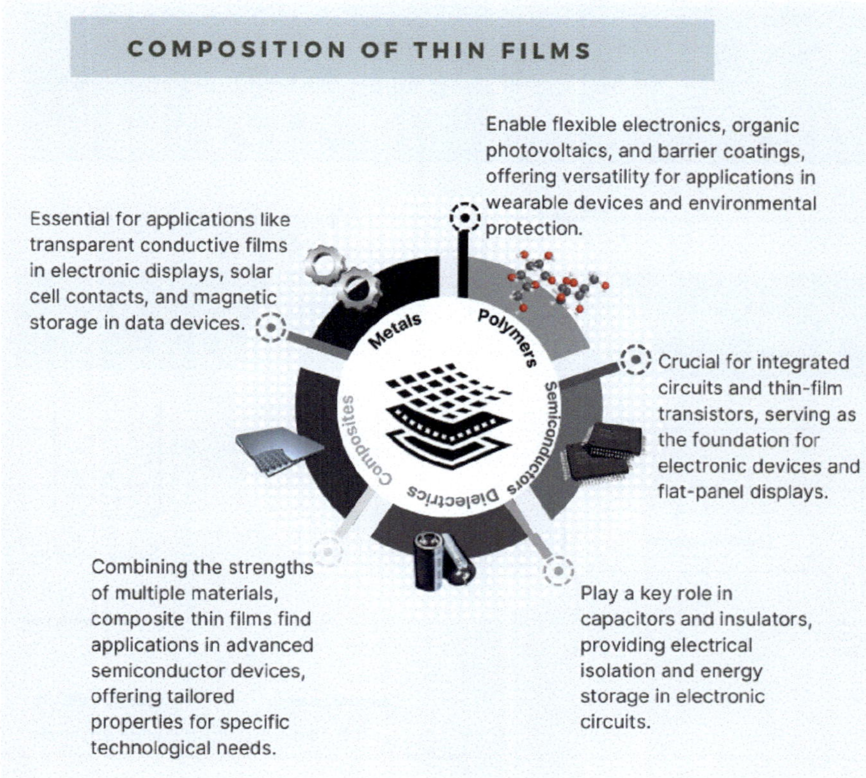

Fig. 1 Composition of thin films

Table 1 Materials used in thin-film technology and applications

Material	Application	References
Metals	Electronics, Coatings, Sensors	[5]
Semiconductors	Integrated circuits, Solar cells	[6]
Dielectrics	Capacitors, Insulating layers	[7]
Polymers	Biomedical science protective coatings and electronic components that are adapt	[3]
Composites	Specializations for certain uses	[8]

of thin films through the deposition of high purity and uniformity. On the other hand, sputtering is also a common category of PVD technique, in which ion impact is used to displace atoms from target surface. These are knocked off the target by this impact and stick to the substrate to form a thin film. Attributes such as tremendous, layer adhesion, regulation of the rate of deposition, and capability of depositing both metal, alloy as well as ceramic material are other advantages of sputtering. Sputtering

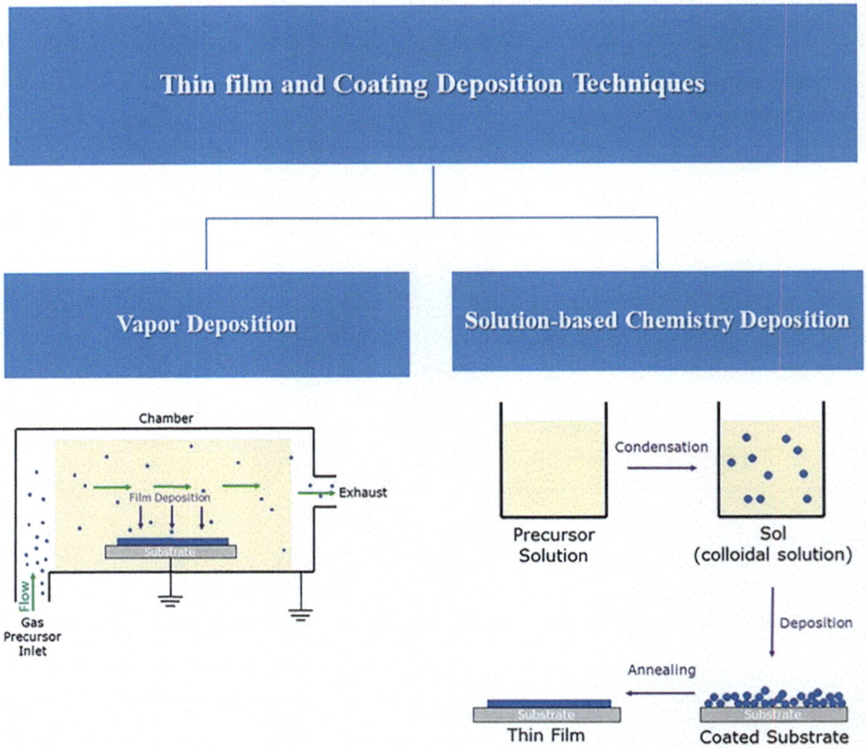

Fig. 2 A thin-film coating on a substrate is represented in this diagram

and evaporation: though similar in some ways, possess their own characteristics that determine their suitability in certain processes involving thin-film deposition. In selecting these techniques, there are often factors such as the properties that one wants to achieve in the final film, the type of substrate, deposition conditions, and scale of production that influence their choice. Taking all these factors into consideration, it can be concluded that PVD techniques are critical to development of such things as semiconductors, optical coatings, and even more [11].

3.2 Chemical Vapor Deposition (CVD)

One of the most widely used techniques for thin-film formation, chemical vapor deposition CVD is a popular choice among the semiconductor industry due to the inline control of composition, film thickness, and uniformity needed to produce very high-quality films. In CVD processes, a substrate surface is directly coated with a thin layer through the subsequent chemical reaction of gaseous precursors, in contrast to PVD processes where the physical sputtering of the material is involved [12].

Another noteworthy feature of CVD is its versatility to potentially deposit a large number of materials of metals and non-metals including semiconductors, ceramics, and even some complicated compounds like oxides and nitrides. It also allows the production of thin-film material with desirable properties including mechanical, optical, and even electrical properties. Another essential aspect of achieving CVD is maintaining the temperature, pressure, flow rate of the gas, and the concentration of the precursor gases [13]. With such delicate control, scientists and engineers are able to exert foregiven control over several film characteristics such as shape, crystallinity, and stoichiometry of the film. For businesses in the thin-film industry delivering products that are stable and consistent, especially in the semiconductor manufacturing and other technology-intensive industries, such control is important. In addition, at CVD process is generalized, it can be employed both for mass production and for different experiments in a laboratory. Different CVD systems offer the capability of accommodating large and complex substrate sizes and shapes at fixed film deposition and operating rates by enhancing the reactor structures and operation conditions. Furthermore, CVD techniques can also be controlled to provide thin films with high uniformity even if on complex 3D shapes, thus allowing for the development of complication of innovative thin-films material into complex devices. This feature is particularly applicable for manufacturing complex optics, optically well-laid coatings, microelectrical mechanical devices, and electronic devices [14].

3.3 Solution-Based Chemistry (SBC) Deposition

Chemical deposition techniques are solution-based and are suitable in most of the techniques of thin-film deposition with a substrate. Here, a solution with the precursor molecules is utilized, and by the chemical reactions occurring at the substrate surface, the thin film is produced. One of the widely used techniques is Chemical Bath Deposition or CBD. The general procedure of CBD is that the substrate is exposed to the solution containing metal ions and one or more reducing agents. Reactions on the substrate surface that adhere and solidify on surface may cause the formation of a thin layer. It is especially preferred because of its ease of use and, more importantly, its affordability and ability to be used for large-scale coating. CBD can be utilized throughout the diverse sectors because of its low-temperature processing which enables the make an excellent film deposit on temperature-sensitive materials [15].

The contribution of the chemical deposition techniques is founded in its capability of producing solution-based thin films at a relatively affordable price and on a large-scale production. They are relatively easy to implement, which makes them suitable for use in research work as well as in business entities. They offer a method in applying thin, uniform layers over substrates which is relevant in industries like electronics, optics, and renewable energies. In addition, the application of a low-temperature solution process makes it compatible with various substrates and lower-cost materials, such as plastic and flexible substrates for applications in flexible electronics and wearable electronic devices [16].

Another solution is coating, which involves applying liquid to the substrate, which rotates and spreads the film due to centrifugal force. Spin coating is known for its simplicity and speed in producing smooth, smooth films. This method is used in semiconductor manufacturing, photovoltaics, and the production of optical coatings. The solution based on the chemical release process has the advantage of being used for a variety of materials and allows controlling the layer thickness and composition. These methods can be adapted to different deposition environments, making them suitable for many applications. Especially in large-scale business processes, the scalability of this process is one of the most important factors in their development. In addition, solution-based solutions play an important role in thin-film production because they are simple, practical, and versatile. This process is important in many industries focused on electronics, optics, and renewable energy technologies. The solution-based technology has unique characteristics such as low performance and robustness, making it an important tool for the production of new paper and machinery [17].

4 Advances in Thin-Film Deposition Techniques

Significant progress has been made in the development of the thin-film process, which opens the door to controlling the quality of the film.

4.1 Atomic Layer Deposition (ALD)

ALD (Atomic Layer Deposition) is considered a very effective technique. In ALD, atomic level control of film thickness and uniformity is achieved by continuous self-deposition surface reactions. Applications requiring precision such as the fabrication of nanoscale devices and microelectronic devices benefit greatly from this technology as shown in figure in which the process of ALD is defined for thin-film deposition techniques. For non-uniform control of tape material, ALD is the preferred method due to its small resolution in the output video [18] (Fig. 3).

4.2 Molecular Beam Epitaxy (MBE)

One way to create thin films at the atomic scale is called Molecular Beam Epitaxy (MBE). By evaporating the material in vacuum, MBE can produce a fine crystalline

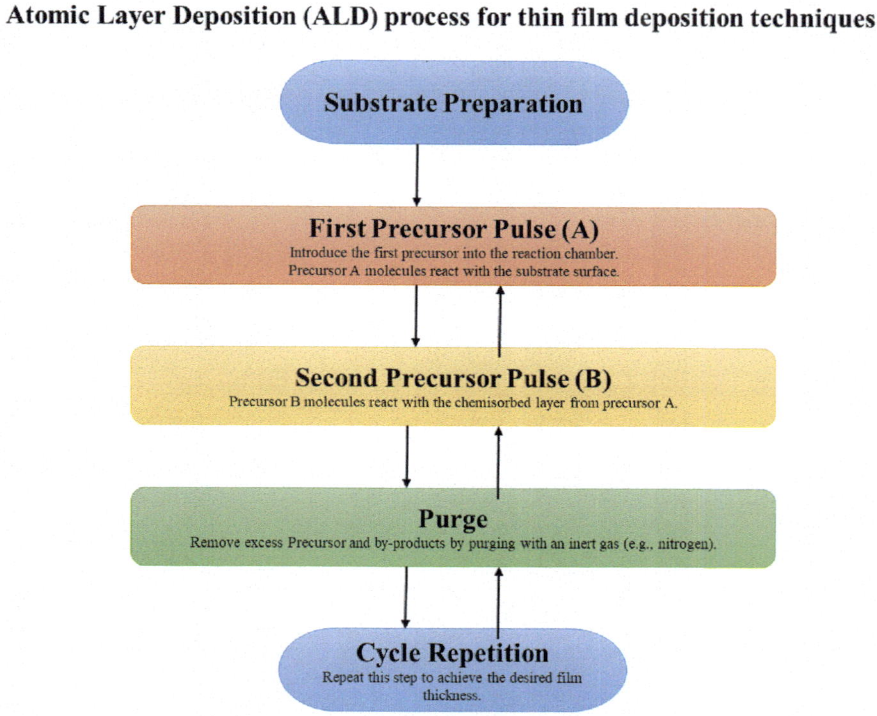

Fig. 3 Graphical representation of thin-film coating on a substrate through ALD

film. In Fig. 4, it shows that diagram illustrating the typical process of Molecular Beam Epitaxy (MBE) that is widely used in thin-film deposition. Some other structures that are also part of the diagram are ultra-high vacuum chamber, effusion cells, molecular beams, substrate holder, substrate, and RHEED system. Precise control of the film and structure is important in the design and fabrication of semiconductor devices, so this method is often used. MBE enables epitaxial growth of thin films, ensuring high product purity and unique crystal structure. Due to the good electrical properties of the film, MBE is important for the advancement of semiconductor technology [19].

4.3 Laser Ablation

Since its introduction in the 1980s, laser ablation has evolved into a much improved thin-film deposition method. A laser is used to selectively remove material from the target to create a thin film on the substrate. The ability to accurately adjust the release process to allow patterns and patterns to be created is one of the important features of laser ablation. The resulting films often have special surface properties

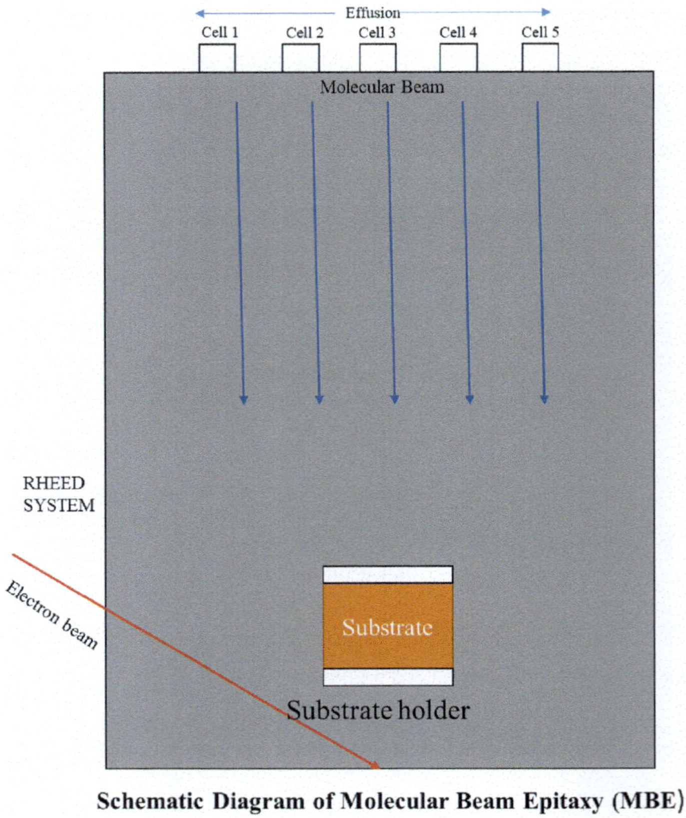

Fig. 4 Schematic diagram of MBE for thin-film deposition

such as modified microstructure and controlled roughness. Applications requiring high performance, such as optical systems, medical devices, and microelectronics, greatly benefit from laser ablation [20, 21] (Fig. 5).

4.4 Pulsed Laser Deposition (PLD)

Introduced in the 1980s, Pulsed Laser Deposition (PLD) is a unique and effective method for producing thin films. PLD involves removing a target material with a laser and then depositing a cloud of material onto the substrate as shown in Fig. 6. Film quality, stoichiometric control, and the ability to stack the mixture together are just some of the advantages of this process. PLD is especially used in applications such as electronics and electronics where composition and film quality are important. Increases the efficiency of thin films, in part due to the ability to arrange particles into a precise stoichiometry [22].

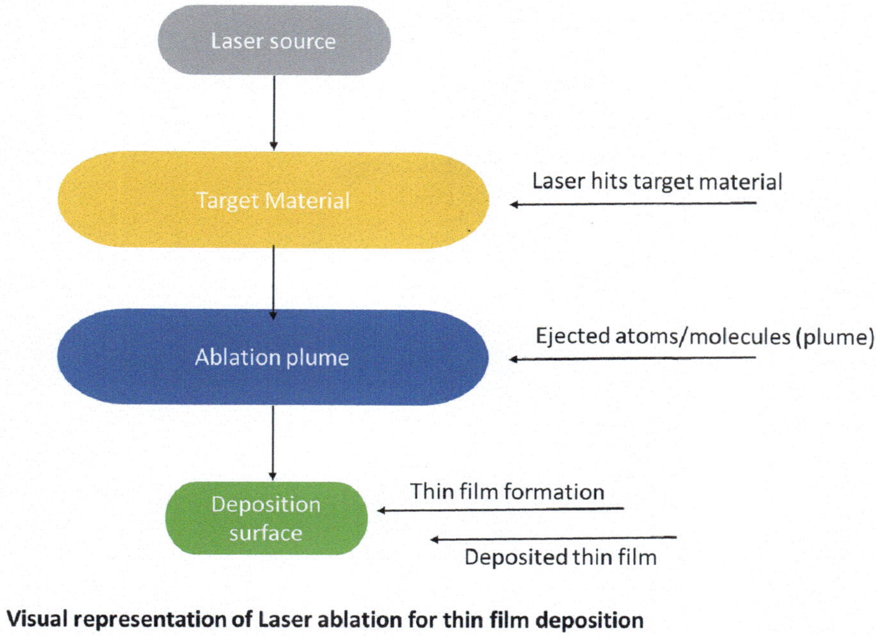

Fig. 5 Visual representation of laser ablation for thin-film deposition technique

Fig. 6 Schematic diagram of pulsed laser deposition

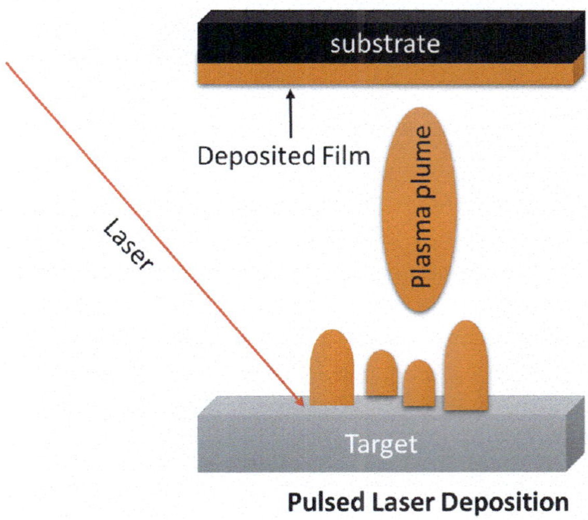

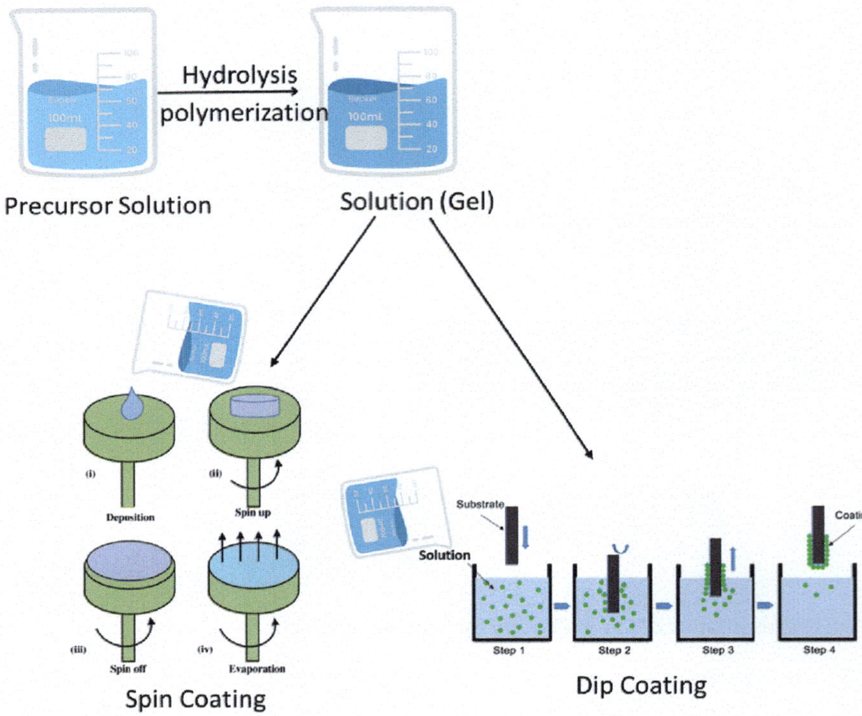

Fig. 7 Step of sol–gel method

4.5 *Sol–Gel Methods*

Sol–gel technology is a revolutionary thin-film separation method that first turns the colloidal solution into a gel and then forms it into a thin layer as shown in Fig. 7. Adaptive technology is particularly important in making glass and glass with special properties. Transparency, refractive index, and thermal stability are just some of the properties that can be controlled by adjusting the composition and structure of the film using sol–gel technology. The resulting films are designed for applications requiring optical and protective treatments, as they often exhibit good consistency and adhesion to the substrate [23].

4.6 *Spray Coating*

Spray coating is a simple, commercial film production process that requires first spraying a liquid onto the surface of the substrate as shown in Fig. 8. This method is particularly suitable for large layers and can be used for many materials such as

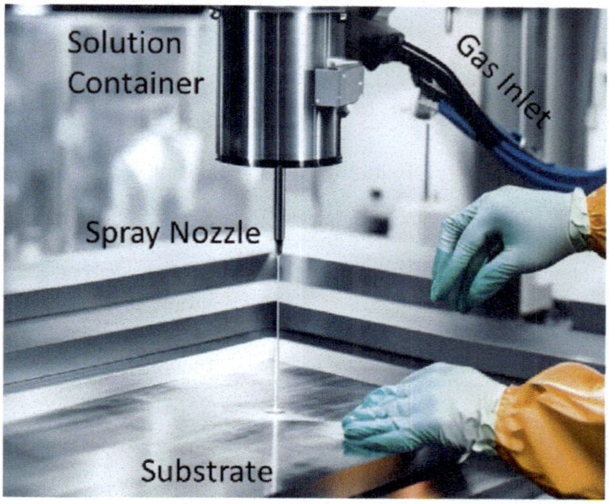

Fig. 8 Spray coating for thin-film deposition

metals, ceramics, and polymers. Spray coating is a popular choice for industries that require extensive coverage, such as construction and automotive, because it is inexpensive and easy to use. Due to its versatility in processing a variety of materials, spray coating is a suitable choice for many applications, from mass production to protective treatments [24].

5 Influence of Thickness on Thin-Film Properties

Thin-film strength is closely related to its thickness, material composition, and deposition process. The optical, electrical, and mechanical properties of the film depend on many factors and its thickness. This section examines how thickness affects the properties of films and the various applications resulting from this control [25].

5.1 Thickness Classification

The thickness of the film varies greatly, from a few nanometers to several micrometers. This is generally classified according to the type of product and its use, as shown in Table 2.

This classification demonstrates the importance of controlling thickness while maximizing the width of film application.

Figure 9 shows how the optical properties of a film, including transparency and reflection, can be changed by varying the film thickness. This control is necessary in cases where limiting noise is important for visual clarity, such as with anti-noise

Table 2 Thin-film classification according to thickness as well as material

Material	Thickness range	Typical applications	References
Metals	10 nm–10 μm	Electronics, Conductive coatings, Optical filters	[26]
Semiconductors	10 nm–1 μm	Integrated circuits, Transistors, Solar cells	[27]
Dielectrics	10 nm–10 μm	Insulating layers, Capacitors, Optical coatings	[28]
Polymers	10 nm–100 μm	Organic electronics, Protective coatings, Sensors	[29]
Composites	10 nm–10 μm	Tailored properties, Enhanced functionalities	[30]

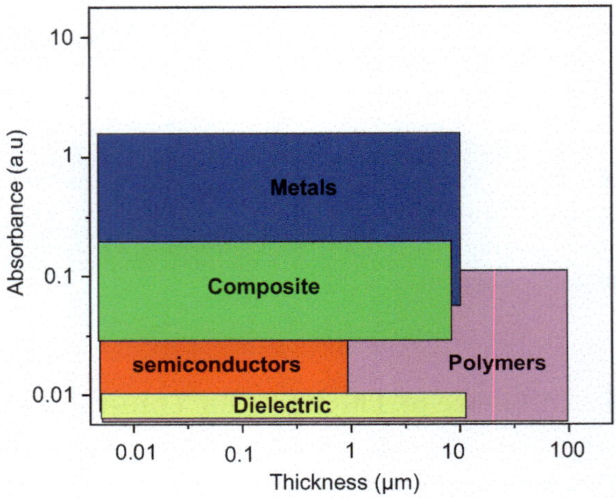

Fig. 9 Give a graphical illustration of the connection between the optical qualities and film thickness

layers. This figure shows how flexible film products vary and highlights the need for thickness control to meet performance requirements in a variety of applications [31].

6 Historical Perspective

Thin films and coatings have a long and varied history that has developed over millennia in tandem with advances in materials science and technology. The application of thin oxide coatings to metal surfaces to stop corrosion and the use of gold leaf as decoration are two early uses of thin films. But the twentieth century witnessed the discipline reach important turning points because of advancements in deposition and vacuum technologies [32].

Figure 10 depicts major milestones that shaped the development of thin-film technology. The first patent for a vacuum deposition technique was filed in the early twentieth century, establishing the framework for future advances. The sputter gun, a critical tool for the deposition of metallic thin films, was invented in the 1920s. The Chemical Vapor Deposition (CVD) technology was invented in the 1930s, and it broadened the spectrum of materials that could be deposited as thin films.

The transistor revolutionized the world of electronics throughout the 1950s, marking a watershed point. Thin films were crucial in the development and production of transistors, paving the way for the downsizing of electronic components. As integrated circuits gained popularity in the 1970s, the demand for high-quality thin films grew.

The emergence of novel deposition techniques such as laser ablation and pulsed laser deposition marked the beginning of a new era in the 1980s. These techniques enabled the accurate deposition of thin films, broadening their applications across multiple industries. As we entered the twenty-first century, thin films remained at the

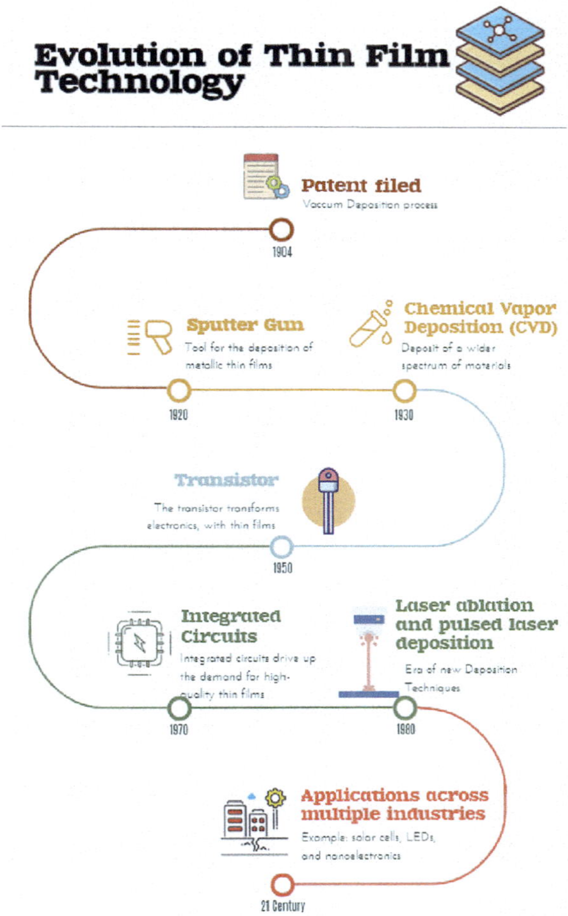

Fig. 10 Timeline of key milestones in thin-film technology development

vanguard of technical breakthroughs, helping to produce solar cells, light-emitting diodes (LEDs), and nanoelectronics [33].

7 Importance in Engineering

Thin films are critical for increasing material performance in a variety of engineering applications. The capacity to carefully manage the thickness and composition of thin films enables the increase of important qualities, such as:

7.1 Electrical Conductivity

In the field of electronics, thin films play an important role in improving electrical conductivity. They are essential components in transistors, capacitors, and integrated circuits. The capacity to precisely deposit small layers of conductive materials allows for the fabrication of high-performance and efficient electrical devices [34].

7.2 Optical Transmission

Thin films are extensively used in optics to control light transmission. Anti-reflective coatings, for example, are used on lenses and optical surfaces to reduce reflections and improve optical clarity. This quality is essential in many optical instruments, including cameras, microscopes, and eyewear [35].

7.3 Mechanical Strength

Coatings with customized mechanical characteristics improve the strength and durability of materials. In aerospace applications, for example, thin films are used as protective coatings on structural components to provide resistance to mechanical loads, wear, and abrasion [36].

7.4 Chemical Resistance

Thin films improve the chemical resistance of materials, making them acceptable for harsh conditions. Coatings protect components against corrosive substances in

Table 3 Applications of thin films in electronics

Application	Thin-film component	References
Transistors	Thin-film transistors (TFTs)	[39]
Capacitors	Thin-film capacitors	[40]
Integrated Circuits (ICs)	Thin-film deposition on semiconductor substrates	[41]
Displays	Thin-film technology in LCDs, OLEDs, and other display technologies	[42]

industries such as chemical processing and manufacturing, increasing equipment longevity and lowering maintenance costs [37].

8 Applications of Thin Films in Various Engineering Fields

Thin films have a wide range of applications across engineering disciplines, providing personalized solutions to improve material performance and functionality. The subsequent sections go into specific applications in various fields:

8.1 Electronics

Thin films are essential to the operation of electronic devices, acting as critical components in transistors, capacitors, integrated circuits, and displays. The downsizing of electrical components via thin-film technology has been a driving force in the evolution of modern electronics. This shrinking enables higher performance, smaller size, and greater efficiency in electrical devices [38] (Table 3).

The curve in Fig. 11 depicts the influence of electronics miniaturization accomplished through thin-film technology. As electronic components get smaller, the overall performance and efficiency of electronic devices improve.

8.2 Optics

Thin films serve an important role in optical applications where precise control of optical characteristics is required. Optics applications include anti-reflective coatings, mirrors, filters, and lenses. Thin films provide precise control, which is critical in the production of high-quality optical components for cameras, microscopes, telescopes, and other optical equipment [43] (Table 4).

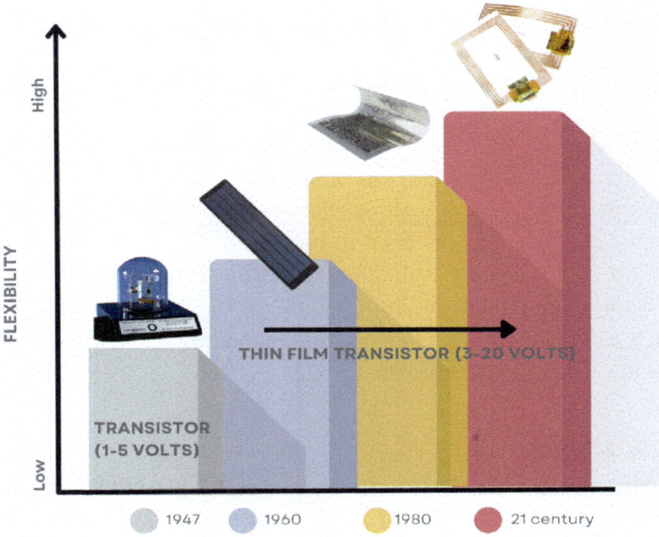

Fig. 11 Miniaturization impact in electronics

Table 4 Applications of thin films in optics

Application	Thin-film component	References
Cameras	Anti-reflective coatings on lenses	[44]
Microscopes	Coatings on mirrors and lenses	[45]
Telescopes	Protective coatings on optical elements	[46]
Displays	Thin-film coatings in display technologies	[47]

Figure 12 depicts thin film-enhanced optical components. Anti-reflective coatings, protective layers, and other thin-film treatments help optical equipment function better.

8.3 Energy

Thin films serve an important part in a variety of energy applications, such as solar cells, fuel cells, and thermoelectric devices. The capacity to deposit thin films on multiple substrates has expanded the possibilities for efficient energy harvesting and storage systems. Thin-film solar cells, in particular, have gained popularity because of their flexibility and lightweight design [48] (Table 5).

Figure 13 depicts the evolution of thin-film solar cells, demonstrating their growing popularity as an effective and adaptable alternative for capturing solar energy.

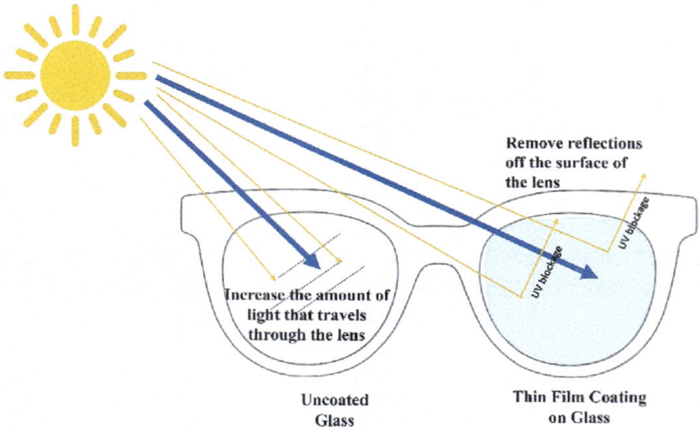

Fig. 12 Optical components enhanced by thin films

Table 5 Applications of thin films in energy

Application	Thin-film Component	References
Solar cells	Thin-film photovoltaic cells	[49]
Fuel cells	Catalyst coatings on electrodes	[50]
Thermoelectric devices	Thin-film materials for efficient heat conversion	[51]

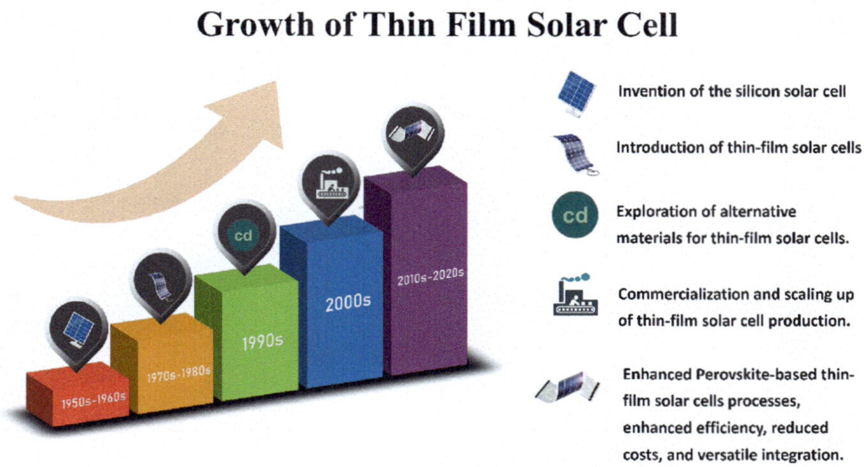

Fig. 13 Growth of thin-film solar cells

8.4 Biomedicine

Thin-film technology has valuable applications in biomedicine, including biosensors, implants, and drug delivery systems. Thin films are useful in medical applications because they are biocompatible and have controlled release capabilities. Thin-film coatings on medical implants enhance integration with biological tissues, lowering the chances of adverse reactions [52] (Table 6).

Figure 14 depicts the usage of thin films in biomedical applications, namely how they improve the functionality and compatibility of medical devices.

Table 6 Applications of thin films in biomedicine

Application	Thin-film component	References
Biosensors	Thin-film sensor layers for detection	[53]
Implants	Biocompatible coatings on implant materials	[54]
Drug delivery systems	Controlled-release thin-film coatings	[55]

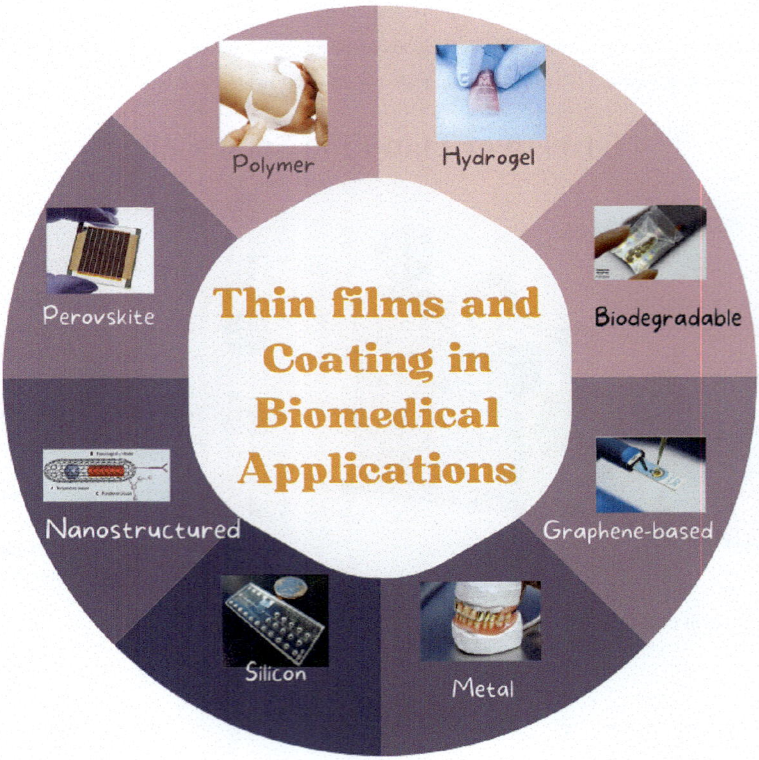

Fig. 14 Thin films in biomedical applications

8.5 *Aerospace*

Thin films play an important role in temperature control coatings and protective coatings for aerospace components exposed to extreme environmental conditions. These coatings improve the durability and performance of materials used in aeronautical applications [56] (Table 7).

Figure 15 depicts the use of thin films in aerospace, emphasizing their significance in temperature control, protection, and the development of lightweight structural materials.

9 Future Directions and Challenges

As thin-film technology advances, various new paths and difficulties emerge that will determine its trajectory:

Table 7 Applications of thin films in aerospace

Application	Thin-film component	References
Thermal control coatings	Thin-film coatings for temperature regulation	[57]
Protective coatings	Corrosion-resistant coatings for structural components	[58]
Lightweight structural materials	Thin-film composites for enhanced strength-to-weight ratio	[38]

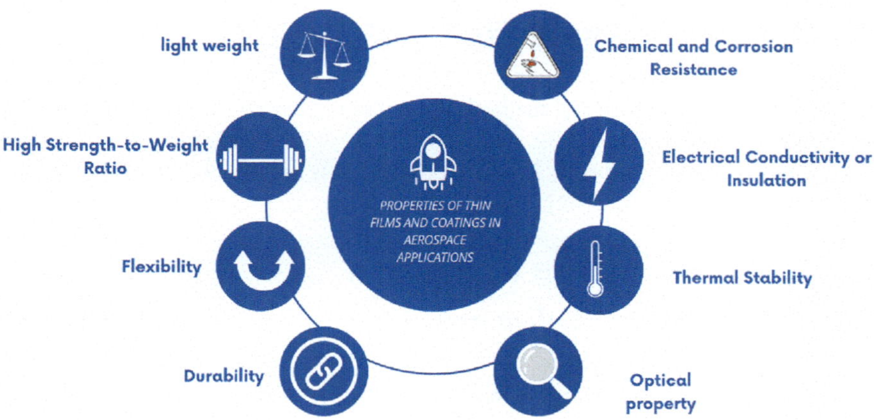

Fig. 15 Properties of thin films in aerospace applications

9.1 Emerging Materials

Advances in materials research, such as the creation of two-dimensional materials and metamaterials, open up new possibilities for thin-film applications with distinct properties and functionalities. These materials have the potential to transform the properties of thin films in a variety of technical fields [59].

9.2 Sustainable Technologies

There is a rising emphasis on producing environmentally friendly thin-film technology. This includes environmentally friendly deposition processes, the use of recyclable components, and energy-saving manufacturing techniques. Sustainable practices are consistent with broader global initiatives promoting environmentally friendly technologies [60].

9.3 Integration with Emerging Technologies

Thin films are expected to play an important role in new technologies such as flexible electronics, wearable devices and the Internet of Things (IoT). The film's lightness, flexibility and many properties make it ideal for integration into technologies that require high-quality materials [61].

9.4 Manufacturing Scalability

Scaling up thin-film production methods remains a challenge. Advances in automation processes, material handling, and quality control are needed to meet demands for new products and equipment. Scalable production is important for the widespread use of thin-film technology [62].

9.5 Multidisciplinary Collaboration

Solving complex engineering problems in thin-film systems requires a collaborative effort. Materials scientists, chemists, physicists, and engineers must work together to create integrated solutions for practical applications [63].

10 Conclusion

In summary, the examination of thin films and coatings in this chapter provides a better understanding of their importance in engineering. Analysis of their content and context shows that films and techniques play an important role in many applications in the industry. Views on history have many changes that reflect the constant progress and development that has led to its current state. This information plays an important role in improving the performance, durability, and performance of various materials and models. From microelectronics to optics, the impact of thin films and coatings is present in all technologies and is driving growth. Basic concept. The properties of many of these materials require complex knowledge, and evaluation of their content, historical context, and technology forms the basis for future research. In the ever-changing information science and engineering environment, thin films and coatings are important and have the potential for innovation and growth.

References

1. Frey, H., Khan, H.R.: Handbook of Thin-Film Technology
2. Wasa, K., Kitabatake, M., Adachi, H.: Chapter 2 thin film materials technology. Thin Film Mater. Technol. 17–69 (2004) [Online]. Available: http://www.sciencedirect.com/science/article/pii/B9780815514831500034
3. R.P.M.: A review on nano-indentation of thin polymeric films. IJERT-Int. J. Eng. Res. Technol. (2019)
4. Petersen, R.J.: Composite reverse osmosis and nanofiltration membranes. J. Membr. Sci. **83**(1), 81–150 (1993). https://doi.org/10.1016/0376-7388(93)80014-O
5. Rim, Y.: Review of metal oxide semiconductors-based thin-film transistors for point-of-care sensor applications. J. Inf. Disp. **21**, 1–8 (2020). https://doi.org/10.1080/15980316.2020.1714762
6. Lee, S.Y.: Comprehensive review on amorphous oxide semiconductor thin film transistor. Trans. Electr. Electron. Mater. **21**(3), 235–248 (2020). https://doi.org/10.1007/s42341-020-00197-w
7. Salvatore, G.A., et al.: Wafer-scale design of lightweight and transparent electronics that wraps around hairs. Nat. Commun. **5**(1), 2982 (2014). https://doi.org/10.1038/ncomms3982
8. Wang, Q., Chen, W., Zhu, W., McClements, D.J., Liu, X., Liu, F.: A review of multilayer and composite films and coatings for active biodegradable packaging. NPJ Sci. Food **6**(1) (2022). https://doi.org/10.1038/s41538-022-00132-8
9. Yu, Y., Gong, M., Dong, C., Xu, X.: Thin-film deposition techniques in surface and interface engineering of solid-state lithium batteries. Next Nanotechnol. **3–4**, 100028 (2023). https://doi.org/10.1016/j.nxnano.2023.100028
10. Liu, F., Hashim, N.A., Liu, Y., Abed, M.R.M., Li, K.: Progress in the production and modification of PVDF membranes. J. Membr. Sci. **375**(1–2), 1–27 (2011). https://doi.org/10.1016/j.memsci.2011.03.014
11. Gutpa, J., Shaik, H., Naveen Kumar, K., Sattar, S.A.: PVD techniques proffering avenues for fabrication of porous tungsten oxide (WO3) thin films: a review. Mater. Sci. Semicond. Process. **143**, 106534 (2022). https://doi.org/10.1016/j.mssp.2022.106534
12. Xu, S., Zhang, L., Wang, B., Ruoff, R.S.: Chemical vapor deposition of graphene on thin-metal films. Cell Rep. Phys. Sci. **2**(3), 100372 (2021). https://doi.org/10.1016/j.xcrp.2021.100372
13. Carlsson, J.O., Martin, P.M.: Chemical vapor deposition. In: Handbook of Deposition Technologies for Films and Coatings, pp. 314–363 (2010). https://doi.org/10.1016/B978-0-8155-2031-3.00007-7

14. Abro, I.A., Chandio, A.D.: Analysis and evolution on diffusional stability of nickel aluminide bond coat via nickel electro-plating. Eur. Phys. J. Plus. **138**(3) (2023). https://doi.org/10.1140/epjp/s13360-023-03816-6
15. Rauf, A., Sabri, A., Kanwal, Z., Naseem, S., Raza, M., Riaz, S.: Size-and shape-dependent antibacterial studies of silver nanoparticles synthesized by wet chemical routes. Nanomaterials **6**(4), 74 (2016). https://doi.org/10.3390/nano6040074
16. Das, A., Uppaluri, R., Das, C.: Feasibility of poly-vinyl alcohol/starch/glycerol/citric acid composite films for wound dressing applications. Int. J. Biol. Macromol. **131**, 998–1007 (2019). https://doi.org/10.1016/j.ijbiomac.2019.03.160
17. Thakur, D., Kumar, Y., Sharanagat, V.S., Srivastava, T., Saxena, D.C.: Development of pH-sensitive films based on buckwheat starch, critic acid and rose petal extract for active food packaging. Sustain. Chem. Pharm. **36**, 101236 (2023). https://doi.org/10.1016/j.scp.2023.101236
18. Gao, L., et al.: TiO_2-coated ZnO nanowire arrays: a photocatalyst with enhanced chemical corrosion resistance. Catalysts **11**(11) (2021). https://doi.org/10.3390/catal11111289
19. Voigt, C., Wagner, B., Vogel, E.: Molecular beam epitaxy of in 2 Se 3 thin films. ECS Meet. Abstr. **MA2022–02**, 816 (2022). https://doi.org/10.1149/MA2022-0215816mtgabs
20. Turkevich, J., Stevenson, P.C., Hiller, J.: Synthesis of gold nanoparticles. Discuss. Faraday Soc. **11**, 55–75 (1951). https://doi.org/10.13140/RG.2.2.15238.37449
21. Mao, X.L., Ciocan, A.C., Borisov, O.V., Russo, R.E.: Laser ablation processes investigated using inductively coupled plasma-atomic emission spectroscopy (ICP-AES). Appl. Surf. Sci. **127–129**, 262–268 (1998). https://doi.org/10.1016/S0169-4332(97)00787-3
22. Obata, Y., Karateev, I., Pavlov, I., Vasiliev, A., Haindl, S.: Challenges for pulsed laser deposition of FeSe thin films. Micromachines **12**, 1224 (2021). https://doi.org/10.3390/mi12101224
23. Khan, R.: Hierarchical Nanostructures of Titanium Dioxide: Synthesis and Applications. In: Javed, S. (ed.), p. Ch. 1. IntechOpen, Rijeka (2018). https://doi.org/10.5772/intechopen.74525
24. Liao, T.-Y., et al.: Multifunctional cold spray coatings for biological and biomedical applications: a review. Prog. Surf. Sci. **97**(2), 100654 (2022). https://doi.org/10.1016/j.progsurf.2022.100654
25. Manev, E.D., Nguyen, A.V.: Critical thickness of microscopic thin liquid films. Adv. Colloid Interface Sci. **114–115**, 133–146 (2005). https://doi.org/10.1016/j.cis.2004.07.013
26. Gordon, A.P., McDowell, D.L.: Numerical simulation of time-dependent fracture of graded bimaterial metallic interfaces. Int. J. Fract. **126**(4), 321–344 (2004). https://doi.org/10.1023/B:FRAC.0000031093.85061.c9
27. Losurdo, M., et al.: Spectroscopic ellipsometry and polarimetry for materials and systems analysis at the nanometer scale: state-of-the-art, potential, and perspectives. J. Nanopart. Res. **11**(7), 1521–1554 (2009). https://doi.org/10.1007/s11051-009-9662-6
28. Wu, X., Chen, X., Zhang, Q.M., Tan, D.Q.: Advanced dielectric polymers for energy storage. Energy Storage Mater. **44**, 29–47 (2022). https://doi.org/10.1016/j.ensm.2021.10.010
29. Mary, S.K., Koshy, R.R., Daniel, J., Koshy, J.T., Pothen, L.A., Thomas, S.: Development of starch based intelligent films by incorporating anthocyanins of butterfly pea flower and TiO_2 and their applicability as freshness sensors for prawns during storage. RSC Adv. **10**(65), 39822–39830 (2020). https://doi.org/10.1039/d0ra05986b
30. Zuo, G., Song, X., Chen, F., Shen, Z.: Physical and structural characterization of edible bilayer films made with zein and corn-wheat starch. J. Saudi Soc. Agric. Sci. **18**(3), 324–331 (2019). https://doi.org/10.1016/j.jssas.2017.09.005
31. Karsandık, Ö., Özdal, T., Kavak, H.: Influence of thickness and annealing temperature on properties of solution processed bismuth sulfide thin films. J. Mater. Sci. Mater. Electron. **33** (2022). https://doi.org/10.1007/s10854-022-08662-3
32. Greene, J.E.: Review Article: Tracing the recorded history of thin-film sputter deposition: from the 1800s to 2017. J. Vac. Sci. Technol. A Vacuum, Surfaces, Film. **35**(5) (2017). https://doi.org/10.1116/1.4998940
33. Sponsors, S., News, L.N.: The history and working principle of the scanning electron microscope (SEM), pp. 1–8 (2015)

34. Liang, J., Luo, J., Sun, Q., Yang, X., Li, R., Sun, X.: Recent progress on solid-state hybrid electrolytes for solid-state lithium batteries. Energy Storage Mater. **21**, 308–334 (2019). https://doi.org/10.1016/j.ensm.2019.06.021
35. Kurt, H., et al.: Nanoplasmonic biosensors: theory, structure, design, and review of recent applications. Anal. Chim. Acta **1185**, 338842 (2021). https://doi.org/10.1016/j.aca.2021.338842
36. Cardoso, G.C., et al.: Preparation and characterization of novel as-cast Ti–Mo–Nb alloys for biomedical applications. Sci. Rep. **12**(1), 1–9 (2022). https://doi.org/10.1038/s41598-022-14820-8
37. Keller, C.-P., Hosoda, T.: 21—Functional Fluoropolymers. In: Plastics Design Library, S.B.T.-I. to F. (Second E. Ebnesajjad, Ed.), pp. 379–388. William Andrew Publishing, Oxford (2021). https://doi.org/10.1016/B978-0-12-819123-1.00021-5
38. Krishnapriya, R., Laishram, D., Shidhin, M., Sharma, R.K.: Chapter Four—Multifunctional materials for clean energy conversion. In: Costa, P., Costa, C.M., Lanceros-Mendez, S.B.T.-A.L.M.M. (eds.) Woodhead Publishing in Materials, pp. 131–152. Woodhead Publishing (2021). https://doi.org/10.1016/B978-0-12-818501-8.00007-X
39. Lu, C., et al.: Progress in flexible organic thin-film transistors and integrated circuits. Sci. Bull. **61**(14), 1081–1096 (2016). https://doi.org/10.1007/s11434-016-1115-x
40. Al Shaqsi, A.Z., Sopian, K., Al-Hinai, A.: Review of energy storage services, applications, limitations, and benefits. Energy Rep. **6**, 288–306 (2020). https://doi.org/10.1016/j.egyr.2020.07.028
41. Kwon, J., Takeda, Y., Shiwaku, R., Tokito, S., Cho, K., Jung, S.: Three-dimensional monolithic integration in flexible printed organic transistors. Nat. Commun. **10**(1), 54 (2019). https://doi.org/10.1038/s41467-018-07904-5
42. Shar, A.M., Mahesar, A.A., Chandio, A.D., Memon, K.R.: Impact of confining stress on permeability of tight gas sands: an experimental study. J. Pet. Explor. Prod. Technol. **7**(3), 717–726 (2017). https://doi.org/10.1007/s13202-016-0296-9
43. Baldini, G., Rigaldi, L.: Thin Films in Optics (1976). https://doi.org/10.1007/978-1-4684-0847-8_13
44. Choi, J.Y., Yun, B.Y., Kim, Y.U., Kang, Y., Lee, S.C., Kim, S.: Evaluation of thermal/acoustic performance to confirm the possibility of coffee waste in building materials in using bio-based microencapsulated PCM. Environ. Pollut. **294**, 118616 (2022). https://doi.org/10.1016/j.envpol.2021.118616
45. Fouda, A., EL-Din Hassan, S., Salem, S.S., Shaheen, T.I.: In-vitro cytotoxicity, antibacterial, and UV protection properties of the biosynthesized Zinc oxide nanoparticles for medical textile applications. Microb Pathog **125**, 252–261 (2018). https://doi.org/10.1016/j.micpath.2018.09.030
46. Bracco, G., Holst, B.: Surface science techniques. Springer Ser. Surf. Sci. **51**(1), 3–34 (2013). https://doi.org/10.1007/978-3-642-34243-1
47. Du, M., Li, Q., Zhao, Y., Liu, C.-S., Pang, H.: A review of electrochemical energy storage behaviors based on pristine metal–organic frameworks and their composites. Coord. Chem. Rev. **416**, 213341 (2020). https://doi.org/10.1016/j.ccr.2020.213341
48. Uddin, A., Upama, M.B., Yi, H., Duan, L.: Encapsulation of organic and perovskite solar cells: a review. Coatings **9**(2) (2019). https://doi.org/10.3390/coatings9020065
49. Olabi, A.G.: Renewable energy and energy storage systems. Energy **136**, 1–6 (2017). https://doi.org/10.1016/j.energy.2017.07.054
50. Hannan, M.A., et al.: Battery energy-storage system: a review of technologies, optimization objectives, constraints, approaches, and outstanding issues. J. Energy Storage **42**, 103023 (2021). https://doi.org/10.1016/j.est.2021.103023
51. Suresh Khurd, A., Kandasubramanian, B.: A systematic review of cellulosic material for green electronics devices. Carbohydr Polym Technol Appl **4**, 100234 (2022). https://doi.org/10.1016/j.carpta.2022.100234
52. Tavares, M.I.B., da Silva, E.O., da Silva, P.R.C., de Menezes, L.R. (2017). Polymer Nanocomposites. In: Seehra, M.S. (ed.), p. Ch. 7. IntechOpen, Rijeka (2017). https://doi.org/10.5772/intechopen.68142

53. de Oliveira Filho, J.G., et al.: Curcumin: a multifunctional molecule for the development of smart and active biodegradable polymer-based films. Trends Food Sci Technol **118**, 840–849 (2021). https://doi.org/10.1016/j.tifs.2021.11.005
54. Thanka Rajan, S., Subramanian, B., Arockiarajan A.: A comprehensive review on biocompatible thin films for biomedical application. Ceram. Int. **48**(4), 4377–4400 (2022). https://doi.org/10.1016/j.ceramint.2021.10.243
55. Memon, A.G., et al.: Citrate-capped AuNP fabrication, characterization and comparison with commercially produced nanoparticles. Crystals **12**(12) (2022). https://doi.org/10.3390/cryst12121747
56. Giglio, M., Manes, A.: Crack propagation on helicopter panel: experimental test and analysis. Eng. Fract. Mech. **75**(3), 866–879 (2008). https://doi.org/10.1016/j.engfracmech.2007.02.024
57. Tirado-Gallegos J.M., et al.: Elaboration and characterization of active apple starch films incorporated with ellagic acid. Coatings **8**(11) (2018). https://doi.org/10.3390/coatings8110384
58. Wang, D., Bierwagen, G.P.: Sol-gel coatings on metals for corrosion protection. Prog. Org. Coat. **64**(4), 327–338 (2009). https://doi.org/10.1016/j.porgcoat.2008.08.010
59. Madhusudanan, S.P., Caroline, S.C., Batabyal, S.K.: Chapter 2—Sustainable energy harvesting technologies. In: Dalapati, G., Shun Wong, T., Kundu, S., Chakraborty, A., S.B.T.-S., S.B.M. for Zhuk, E.A. (eds.), pp. 15–33. Elsevier (2022). https://doi.org/10.1016/B978-0-323-99860-4.00015-0
60. Wang, Z., Ma, J., Tang, C.Y., Kimura, K., Wang, Q., Han, X.: Membrane cleaning in membrane bioreactors: a review. J. Membr. Sci. **468**, 276–307 (2014). https://doi.org/10.1016/j.memsci.2014.05.060
61. Choudhury, S.: Review of energy storage system technologies integration to microgrid: types, control strategies, issues, and future prospects. J. Energy Storage **48**, 103966 (2022). https://doi.org/10.1016/j.est.2022.103966
62. Xu, R., et al.: Influences of support layer hydrophilicity on morphology and performances of polyamide thin-film composite membrane. Sep. Purif. Technol. **281**, 119884 (2022). https://doi.org/10.1016/j.seppur.2021.119884
63. Ajakaye, J., Ogunniyi, S.: 21st-Century Multidisciplinary Collaboration in Research in Library

Fundamentals of Thin Films

Nida Naveed

Abstract The essential parts of thin films are described in the chapter entitled "Fundamentals of Thin Films." Consequently, this chapter focuses on the different deposition techniques, the criteria for selecting the material, the growth mechanism, and the morphology of the film. Studying these factors helps to uncover how thin films work and how they can be designed for certain uses. The discussion involves an analysis of methods of film depositions—physical and chemical, impact of different materials on the film characteristics, and complex processes that define the growth of films and their structure and properties. In a way which details the broad nature of the field, this chapter seeks to lay down strong conceptual basis for comprehending the imperative role and the versatility of thin films in modern technology and diverse industrial applications.

Introduction

This chapter serves as a comprehensive exploration of thin films, investigating their fundamental principles, deposition techniques, material selection considerations, and the elaborate interplay between growth mechanisms and film morphology. This chapter aims to provide a detailed overview, offering insights into thin films and helping readers better understand their wide range of uses and capabilities.

1 Fundamentals of Thin Films

Thin film is a very comprehensive branch known under many names and studied in many branches of science. It lays the platform from which many technological advancements are announced. In essence, thin films are made by applying a very thin amount of material—at the nano- or microscale—to a surface, usually known as a substrate [1]. When it comes to scientists, engineers, or researchers doing their work

N. Naveed (✉)
University of Sunderland, Sunderland, UK
e-mail: nida.naveed@sunderland.ac.uk

A. D. Chandio and I. A. Channa (eds.), *Thin Films and Coatings*, Engineering Materials,
https://doi.org/10.1007/978-981-96-0005-2_2

in disciplines such as material science, optics, electronics, or surface engineering the understanding of what thin films are, how they are formed and their characterization is an inevitable part of the way how they develop their work [1]. Thus, deposition is a vital aspect of thin films and its fundamental process is demonstrated in the sections below. The method of layering is mainly dependent on techniques like Chemical Vapor Deposition (CVD) and Physical Vapor Deposition (PVD). Indeed, all that CVD does is use chemical reactions to do what PVD does by mechanically transferring material from a source to a substrate [2]. However, to control the properties of the created thin films to meet specific requirements for protective coatings, semiconductors or optical coatings, the mastery of these deposition methods is mandatory [1, 3].

Generally, the thickness plays a paramount role in thin film both in the design and usage since it exercises a strong influence over the disposition of the substance. The physical properties of the thin films in the optical, electrical, and mechanical regime often vary greatly from those displayed by the bulk system [3]. Consequently, when working with such micro- and ultra-thin layers, the engineers have to take into account items such as the quantum effects of interference phenomena [4]. The fundament of new technologies such as sensors, anti-reflective coatings, or thinner-film solar cells is built up by such knowledge [4].

In conclusion, thin-film principles are the basis of most technologies (semiconductors, solar cells, coatings, sensors, optical devices, and memory storage) in that they act as the starting point. The elaborate knowledge of thin films' characteristics and properties is essential for the development of numerous fields from electronic and optical to others. This is because deposition processes are somewhat complicated, especially when it comes to deposition procedures and the arch features at the nano-level are also unique [3, 4].

2 Thin-Film Deposition Techniques

Many sciences and technologies rely on thin-film deposition techniques to make films with specified qualities. Metal vapor vacuum deposition is a method used to create electrical equipment, optical coatings, novel materials, and other products. To make use of the qualities of thin films, scientists, engineers, and researchers must understand the fundamental principles of thin-film deposition.

2.1 Physical Vapor Deposition (PVD)

2.1.1 Introduction

Some coatings that are made on surface by a method of vacuum deposition are called Physical Vapor Deposition (PVD) coatings. These coatings increase the hardness, the wear, and the corrosion properties of the material due to changes in the surface characteristics of the material. Such coatings as PVD are applied in areas that require very high standards of surface textures like electronics, automotive, and aerospace. The focus of this chapter is to clearly shed light on the basics, types, and application of PVD coatings and stress the importance of PVD coatings in the modern world of engineering and technology.

2.1.2 Working Principle

The initial PVD step illustrated in Fig. 1 involves the preparation of the substrate, which mainly involves washing off impurities that might interfere with the bonding of the film to the substrate. The source material is then evaporated in a vacuum chamber using one of three methods: such as sputtering, resistive heating, or electron beam heating. To ensure that the gas molecules do not collide with one another, this makes a vapor phase in the chamber with a low pressure. After that, the vaporized material condenses on the substrate which is at a lower temperature to control the film properties and for better adhesion. The process of film growth was explained by the manner in which atoms or molecules in the vapor state stick together in small clusters that expand to form the last film. Doping can be effectively controlled since the characteristics of the film can be easily manipulated such as thickness, composition, and pressure through the deposition parameters such as rate, temperature, and pressure structure [5].

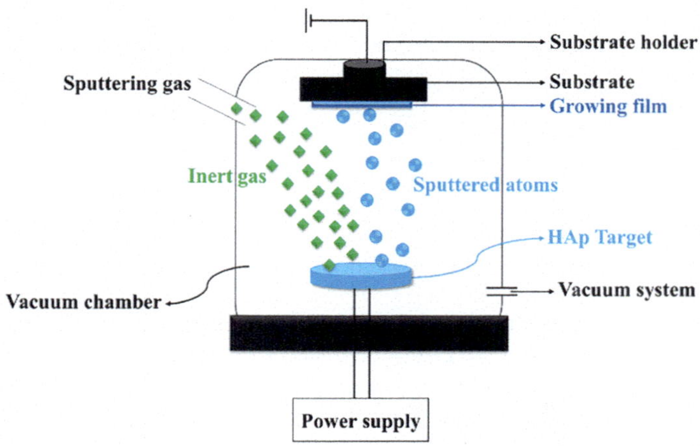

Fig. 1 Schematic diagram showing that how PVD coating is done on a substrate Copied/modified/redrawn from [https://images.app.goo.gl/Aj9g2RZ5fRwUn9XUA] licensed by CC BY 4.0

2.1.3 Advantages

- They improve the hardness of the base material to an unparalleled level thus making it more intense with abrasion and wear.
- Thus, in corrosive conditions, PVD coatings can help to increase the durability of materials due to protection against corrosion.
- It should be mentioned that PVD coatings can be utilized in severe conditions as it might preserve the features at high temperatures.
- PVD coatings are widely used for decorative purposes since they can be easy to be deposited in various colors and finishes.
- One might observe that PVD procedures yield less hazardous wastes than the hitherto mentioned plating techniques and hence are considered environmentally friendly.

2.1.4 Disadvantages

For this reason, PVD coatings are most likely to be less successful in applications that call for thick coatings.

- The flexibility of PVD coatings could be restricted by the kind of substrata and surface treatments that need to be used to gain proper adhesion on the substrate material.
- PVD equipment can also be expensive and highly technical and may require special skills for services.
- A disadvantage of dip coating is that the thickness of the formed coating is not consistent sometimes because of its dependence on the angle of inclination of the plate or the irregularities of the outer side of the sample.
- PVD coatings to some of the applications may be limited to some extent on the basis of materials that mainly dictate the PVD coatings.

2.1.5 Importance

Thin films are crucial in enhancing the surface features and this is the reason PVD is important in the electronics, optical, and aerospace industries. PVD coatings, for example, are widely used in the electronics sector to increase the performance and lifetime of semiconductors and microelectronic devices by adding protective layers that promote conductivity and reduce wear.

PVD processes are used in the optical sector to generate anti-reflective coatings on lenses and mirrors, thereby increasing efficiency and durability. A recent study found that PVD coatings dramatically enhanced the performance of optical fibers by lowering signal loss and enhancing data transmission speeds.

PVD coatings are used to protect turbine blades and other aerospace components against oxidation and wear at high temperatures. For example, a case study on aircraft engines found that PVD-coated turbine blades had a longer lifespan and greater performance than uncoated blades, resulting in lower maintenance costs and increased engine efficiency.

Understanding and applying PVD principles can help companies increase product performance and durability significantly [5, 6].

2.1.6 Applications

PVD is used for:

- Providing a protective and improving layer to the semiconductor devices.
- Production of anti-reflective coatings for glasses and other optical parts.
- Development of anti-wear coating on cutting tools [5, 6].

Case Study: A 2021 study showed that TiN-coated cutting tools in automotive manufacturing plants exhibited up to a 400% increase in lifespan compared to uncoated tools, reducing downtime and maintenance costs significantly.

2.1.7 Sputter Deposition

Introduction

Sputter coatings known as thin film depositions are formed out of sputtering, a technique where energetic particles erode a solid target material and produce atoms to be emitted out The consequent sputter coatings are often used to alter the superficial conductivity, wear and visibility of materials commonly in electronics, optics, and for solar energy uses Thin-film depositions are useful in numerous arenas due to the precision and versatility of the process and constitute the subject.

Working Principle

The substrate, which is the surface to be coated and the target or material to be deposited are loaded into the vacuum chamber at the early stages of sputter coating as explained in Fig. 2. A familiar material is a solid metal or an alloy compounded from two or more metals. An electric field at a high voltage is applied to expose the target and produce the ion plasma. The electric field then accelerates these ions towards the target and upon impact they hit and expel other target material atoms. The ejected atoms stick to the substrate and move in straight line making a thin film. In most cases, a revolving stage is created so that the substrate can be placed in a proper manner to guarantee a uniform layer of the coating. This way, the target material, voltage, gas pressure, and mesh thickness as well as the constitution of the deposited layer can be controlled. To prevent such flow of contaminants into the growth chamber and ensure that a substance with the required characteristics of a deposited film is produced, the process is done in a vacuum. Sputter coating is generally used in coating semiconductors and mirrors in electronics industry and cutting tools in automobile industry [7].

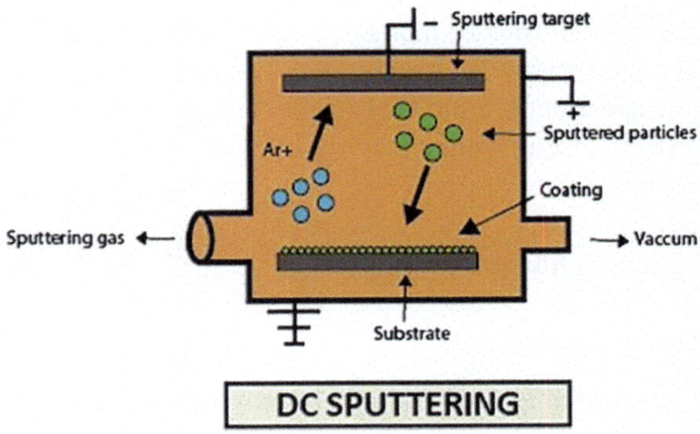

Fig. 2 The diagram illustrating the Sputter coating mechanism

Advantages

- Sputter coatings are ideal for depositing thin films on substrates with complex shapes for the reason that the process delivers excellent thickness uniformity over large areas.
- Sputter coatings make it possible to regulate the deposited film composition and establish multilayer formations with the required characteristics of the material.
- The film can be less contaminated and have fewer impurities if it is appreciated that sputter coatings having high purity can be deposited.
- Sputter coatings can be put on metals, alloys, oxides, nitrides, etc.
- Sputter coatings often exhibit good substrate adhesion, which means good longevity of the coating.

Disadvantages

- deposition also has some drawbacks Sputter coatings cannot be applied to a non-conductive substrate, thereby limiting the choice of substrate a little.
- A disadvantage of sputter coatings is that they are generally thin and this may not serve the intended purpose of a thick coating.
- Sputter deposition equipment, especially for industrial uses, could be expensive to procure as well as to use.
- The sputter deposition process can be somewhat time-consuming, especially for thick coatings; this could be a drawback on production.
- Sputtering, one of which is that heat produced on the substrate could be detrimental to some of the materials.

Importance

Sputter coating is important in electronics and optic industries due to ability to create thin layers of high-purity deposit and homogeneity.

In the electronics sector, sputter coating is used to deposit thin layers of conductive materials on semiconductor wafers, therefore boosting integrated circuit performance and reliability. For example, a recent study found that sputter-coated copper sheets improved conductivity and lowered resistance in semiconductor devices, resulting in speedier and more efficient electronic components.

In the optics sector, sputter coating is used to create anti-reflective coatings on lenses and mirrors, considerably improving optical performance. A case study shown that sputter-coated lenses enhanced light transmission and reduced glare, which is critical for high-precision optical devices like telescopes and microscopes.

Industries can significantly improve product performance and reliability by taking advantage of the precision and quality of sputter coating [7].

Applications

Sputter coating is used for:

- Applying thin layers on the microstructures of mirrors and lenses.
- Engaging in the production of thin film solar cells.
- Coating a glass to make it more harder to scratch [7].

Case Study: In a 2022 study, sputter-coated indium tin oxide (ITO) layers on touch-screens enhanced electrical conductivity and durability, leading to more responsive and longer lasting devices.

2.2 *Chemical Vapor Deposition (CVD)*

Introduction

The approach of thin-film deposition commonly used is chemical vapor deposition or CVD which makes the coatings with precision control over thickness, composition and structures. In CVD, a solid film is formed when a volatile precursor material is chemically transformed on a substrate surface. The application of materials with specific properties such as purity, homogeneity, and adhesion of the coating is achieved through this method. It is apparent that a wide variety of fields rely on CVD as it can be used for the synthesis of novel materials and in semiconductor fabrication. This work describes the general and specific characteristics of CVD coatings with focus on the importance of these coatings in today's materials science and technology.

Working Principle

The CVD process is initiated with an initial substrate in a reaction chamber, as illustrated in Fig. 3. The precursor gases of the required elements enter the chamber and chemically react there often with the help of heat, plasma, or catalyst. In these processes, the precursor molecules are broken and release species such as radicals, ions, and atoms. After getting on to the surface of the substrate, those species undergo the following steps in order to form a solid layer. The film is built up layer by layer as more quantities of the precursor molecules are provided and they get deposited on the substrate surface. This implies that the control of heat, pressure, and flow rates

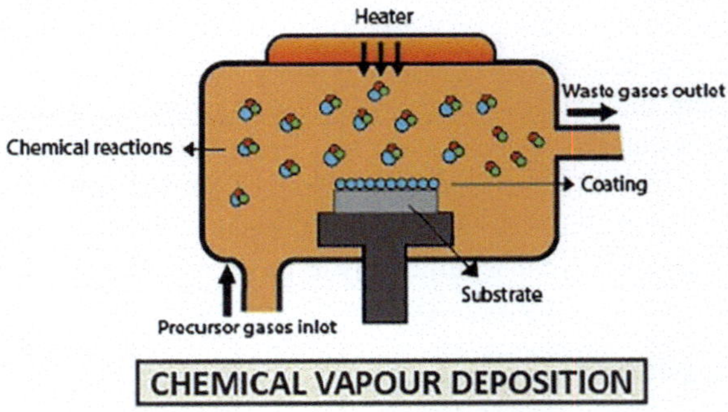

Fig. 3 Diagram illustrating the CVD coating on a substrate

of the gases being used in the film fabrication process is critical in achieving the desired properties of the film such as structure, composition, and thickness. By far, CVD is widely applied in the semiconductor industry for the precise manipulation of film characteristics especially in thin films applied to integrated circuits [8, 9].

Advantages

- Stanford said CVD coatings may form and cover complex shapes and structures with thickness and surface roughness.
- The kind, thickness, and structure of the deposited film are all programmable through CVD.
- Homogeneous coatings can be produced when CVD is used, thus increasing the purity of the final product.
- Many forms of materials for instance metals, ceramics, semiconductors among others can be deposited through CVD.
- However, as CVD may reach high rates of deposition in some cases, then this technique is suitable for industrial application.

Disadvantages

- Buying and maintaining CVD equipment can be expensive, especially when it comes to the expensive hybrid equipment used in higher performance applications.
- Precision can be challenging to control in CVD procedures including temperature, pressure, and gases flowing through the reactors.
- Generally, only materials that are resistant to heat treatment as well as reactive gaseous environment can be used for the CVD coatings.
- Some of the processes in CVD involve the application of poisonous or hazardous gasses, therefore concerning the aspect of safety and the environment.
- It must also be clarified that—and this applies to other processes too—although CVD can deposit quite thick layers of specific materials, it could be comparatively slower with others, and so slow down production rates.

Importance

Countless industries ranging from semiconductor materials production, tool coatings to many others rely on CVD thin-film applications. In the semiconductor industry, CVD is used to deposit high-quality thin films of materials such as silicon, silicon dioxide, and tungsten, all of which are required for integrated circuit production. A recent study found that CVD-deposited silicon sheets considerably increased the performance and efficiency of solar cells.

CVD is used in the tool coating industry to apply hard coatings like titanium nitride (TiN) and diamond-like carbon (DLC) to cutting tools, thereby increasing their hardness, wear resistance, and durability. A case study found that cutting tools coated with CVD TiN had a significant increase in durability and cutting efficiency compared to uncoated tools, resulting in lower tool replacement costs and increased productivity.

By utilizing CVD thin-film applications, industries can achieve superior product performance and extended service life, making CVD an indispensable technology [8, 9].

Applications

CVD is used for:

- Coating thin films for electronic chips for the semiconductor industry.
- Applying anti-corrosive layers on the material.
- Formation of ultra-thin films used in fabrication of solar cells and LEDs [8, 9].

Case Study: A 2022 study found that CVD-deposited silicon films improved the efficiency of photovoltaic cells by 15%, highlighting the technology's impact on renewable energy applications.

2.3 Atomic Layer Deposition (ALD)

Introduction

With atomic-level control over thickness, Atomic Layer Deposition (ALD) is a precise thin-film deposition process that allows for the creation of conformal and homogeneous coatings. A substrate is successively exposed to alternating precursor gases in order for ALD to function and enable the self-limiting surface reactions required for regulated film growth. Coatings with outstanding homogeneity, thickness control, and high aspect ratios are produced by this special technique. Applications for ALD are numerous in the fields of surface engineering, semiconductor fabrication, and nanotechnology. An introduction of ALD's concepts, procedures, and applications is given in this study, with special attention to how important it is for facilitating the creation of nanoscale devices and sophisticated materials [10].

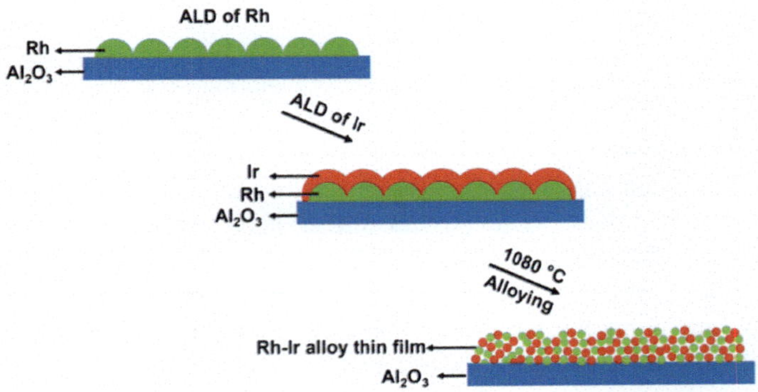

Fig. 4 Illustrating the ALD coating technique in the simplest way[11] Creative common under license 4.0

Working Principle

Atomic Layer Deposition (ALD) is seen in Fig. 4, where the reaction chamber is filled with a substrate. The sequential, self-limiting surface reaction mechanism used by ALD sets it apart from other methods. A precursor gas is first injected into the chamber, where it forms a monolayer of material by reacting in a self-limiting way with the substrate surface. After that, surplus precursor and reaction by-products are purged out of the chamber. Subsequently, another precursor gas is added, and it reacts with the substrate surface monolayer to generate another monolayer of a distinct substance. To increase the required film thickness, this cycle of precursor exposure and purging is repeated several times, creating a precise and consistent layer of material with each cycle. Nanotechnology and semiconductor fabrication, for example, require high precision, and ALD provides excellent control over coating composition and thickness [10].

Advantages

- As a result of this, ALD makes it possible to deposit layers, coatings of extremely thin and homogeneous layers with nearly precise layer thickness at the atomic level.
- High-aspect ratio 3D structures can be coated with ALD films to conformal thickness because of the characteristics of this technique.
- Therefore, ALD coatings are ideal for use in semiconductor and microelectronic devices due to the uniformity that can be achieved over large spaces.
- Films with better step coverage can be deposited by ALD thus ensuring that features with high aspect ratio have better coverage.
- Due to the ability to regulate the composition of the deposited films with a high level of accuracy, ALD opens up the prospects for creating new materials and multilayer structures.

Disadvantages

- Some of the major disadvantages of the ALD are that it's commonly associated with slower deposition rates than other thin-film deposition techniques thus limiting its usage in any high-throughput processes.
- The ALD methods as well as its equipment are often expensive and complex which would require expertise and maintenance.
- Since ALD is commonly applied for thin films, this technique may not be suitable for the applications that require thicker layers.
- The ALD techniques of depositing the films can be affected by varying thickness of the layer and purity and quality of the precursor.
- The physical and chemical properties of the material systems may not permit the use of ALD and this technique is mostly applied in layering thin films of particular materials.

Importance

ALD is at the core of making semiconductors and other electronic applications, such as nanotechnology, surface engineering, and development of ultrathin films to enhance devices' performance. In the semiconductor sector, ALD is used to build precise and conformal coatings of materials such as hafnium oxide and aluminum oxide, which are required for the production of sophisticated microprocessors and memory devices. A recent study found that ALD-deposited hafnium oxide films greatly enhanced gate dielectric characteristics in transistors, resulting in better device performance and energy efficiency.

In nanotechnology, ALD allows for the creation of nanoscale structures with precise thickness and composition controls. For example, ALD has been used to coat nanoparticles with thin films, which improves their catalytic activity and stability in chemical processes.

Surface engineering uses ALD to create ultrathin, high-quality coatings that improve the mechanical, thermal, and chemical aspects of surfaces. A case study showed that ALD coatings on medical implants increased their biocompatibility and corrosion resistance, resulting in more durable and safer medical equipment.

Industries can significantly improve product performance and innovation by exploiting ALD's precision and versatility [10].

Applications

ALD is used for:

- Conveying gate dielectrics and metal electrodes in modern-day state-of-the-art CMOS devices.
- Drug delivery system and catalysis through coating nanoparticles.
- Manufacturing competitive thin-film batteries [10].

Case Study: In a 2023 study, ALD-coated platinum nanoparticles showed a 30% increase in catalytic efficiency and a significant improvement in stability for fuel cell applications.

2.4 *Electrochemical Deposition*

Introduction

Electrodeposition which can also be referred to as electroplating is one of the most versatile processes as far as the placement of metals and alloys on conductive substrates is concerned. Metal plating is defined as the relative process of depositing a metal on the substrate surface by way of a reduction reaction that is triggered by an electric current. As an advanced technique for the deposition of coatings with high control on thickness, composition, and shape, electrochemical deposition is applied in various fields including surface treatment, metalizing, and protection against corrosion. This paper also explains the general ideas about electrochemical deposition and presents the current state of its implementation in the frame of the modern material science and engineering.

Working Principle

This technique of building a layer of metal or an alloy on a conductive material is known as electrochemical deposition or electrodeposition or electroplating. In this process, the substrate to be coated is electrically connected to the negative terminal of the power supply solution, or cathode, while the positive terminal or anode is connected to a metal electrode. The electrolyte solution, in which one can dissolve the metal ions which are to be deposited, is immersed along with the substrate as well as the metal electrode. At the cathode (substrate) when an electrical current is applied, metal ions in the electrolyte deposit on it forming a metal layer on the substrate surface. To keep the concentration of the metal ions at a constant, metal atoms of the anode go into the solution. It is believed that current density, deposition time, and type and concentration of the electrolyte can act as parameters which allow controlling the thickness and properties of the formed layer. It is widely used in electronics, automobiles, and jewelries for coating of parts, manufacturing of circuit boards, and application of show plating [12].

Advantages

- Electrochemical deposition is applicable to practically all types of metals and alloys that are capable of being deposited on different types of substrates.
- Consulting of electrochemical deposition is useful in high-throughput production because it can achieve relatively high deposition speeds.
- As can be observed, on both complex shapes and complex surfaces, electrochemical deposition can provide quite good homogeneity of the coating.
- Electrochemical depolarization allows a precise control of the thickness of the coat that is coated.
- With regard to the methods of depositing metal coatings, electrochemical deposition can be regarded as cheaper.

Disadvantages

- A major limitation of the application of electrochemical deposition is that the materials on which the process is performed have to be conductive; non-conductive materials cannot be used.
- Before using the electrochemical deposition technique to create a solid coat on also the surface of the substrate must be well prepared.
- Electrochemical deposition offers the advantages of solutions and chemicals, and some of the solutions used are electrolytes, and this can cause disposal and environmental problems.
- Most of the metals and alloys that can be plated from, respectively, soluble in aqueous electrolytes are used in electrochemical deposition.
- Some of the electroplated coatings are brittle and they can crack leading to reduction of their life in certain applications.

Importance

Electrochemical deposition has business uses in the formation of anti-corrosive layers as well as in producing electronics. Electrochemical deposition is used to put tiny layers of materials such as zinc, nickel, and chromium on metal surfaces, protecting them against rust and corrosion. A recent study found that electrochemically deposited zinc–nickel alloys provided better corrosion resistance than standard coatings, prolonging the life of industrial components in severe conditions.

Electrochemical deposition is used in electronics to create printed circuit boards (PCBs) and semiconductor devices. This process is used to deposit conductive metals, such as copper, onto surfaces, resulting in the sophisticated circuitry needed for electronic devices. For example, a case study showed that electrochemically formed copper layers improved the electrical conductivity and dependability of PCBs, resulting in more efficient and durable electronic goods.

By utilizing electrochemical deposition, industries can achieve enhanced material properties and improved product performance, making it a vital process in both corrosion protection and electronics manufacturing [12].

Applications

Electrochemical deposition is used for:

- To produce a more appealing appearance of the metal object and enhance the metal's resistance to corrosion.
- Rare uses of electroplating cost include the creation of metal layers in microelectronics and semiconductor devices.
- Promoting the manufacture of fuel cells and other electrochemical systems.

Case Study: A 2021 study showed that electrochemically deposited zinc coatings on steel bridges reduced corrosion rates by 50%, significantly extending the lifespan of the infrastructure.

2.5 *Spin Coating*

Introduction

One of the most common methods of coating a flat substrate with a uniform layer of a thin layer is spin coating. It involves rapidly rotating a substrate on which a liquid coating material has to be spread. The coating material is uniformly distributed by the force of spin and the thickness of the formed film is even. Spin coating is favored by coaters for being simple to use, relatively economical, and for being able to produce high-quality thin-film coatings with uniform thickness in ambient conditions. In this paper, the possible theories, techniques, and application of spin coating are presented with special emphasis towards putting into focus the technological impact of this method at the confines of surface science, optics, and microelectronics.

Working Principle

The process for spin coating is illustrated in Fig. 5 with a substrate; this could be a silicon wafer or any other flat structure placed on an arm of the spin coater in a controlled environment. A small quantity of liquid coating known as the "photoresist" is applied to the core substrate of the product. The liquid then spreads out as a result of centrifugal force when the substrate is spun briefly by the spin coater. The liquid coating substance in this procedure is applied on the substrate's surface uniformly and thinly since the substrate is in a rotary motion. To achieve the correct film thickness, the spinning time as well as the rotational speed of the dance floor are regulated. Further, after the spinning process, the coated substrate is often baked in order to remove the residual solvent and achieve better film to substrate adhesion. Spin coating is often used in application of thin and uniform films for various purposes in diverse vocations such as in the production of semiconductors where the method is embraced to deposit layers of photo resists in the photolithography process [13].

Advantages

- Spin coating is useful when it is necessary to apply the thin and uniform layers on a significant part of the surface with the complex shape.
- Required equipment and facility for the spin coating is very limited and the process is comparatively easy to establish.

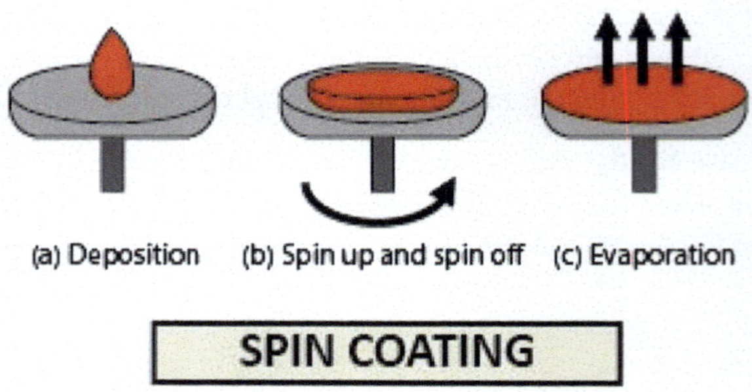

Fig. 5 Illustration of spin coating in simple and easy-to-understand manner

- Spin coating is relatively fast and typically takes only several seconds to a couple of minutes to perform.
- The spin coating is relatively cheap as compared to other thin-film deposition techniques.
- Polymers, nanoparticles, sol–gel precursors, and many other substances can be coated employing the spin coating technique.

Disadvantages

- There is always a somewhat challenge of achieving a very tight control over the amount of coating that is applied, especially when applying thin or thick layers of the coating.
- In spin coating process, there is a substantial amount of material wasted, especially if the substrates have a small size.
- The technique may not accomplish large-area coatings perhaps because spin coating is appropriate for small to medium substrates.
- Spin coating is best suited for flat or mildly rough surfaces; however, this technique may not be very good for rough or uneven surface.
- The ideal coating material for spin coating consists of low viscosity, whereas the high viscosity coating material will not normally spin uniformly.

Importance

Spin coating is widely used in today's industries specifically in microelectronics, optics, and biomaterial industries in the fabrication of thin film. Spin coating is used in the microelectronics industry to apply photoresists and dielectric layers on semiconductor wafers, which are key steps in the manufacture of integrated circuits. A recent study found that spin-coated dielectric layers increased the performance and reliability of microelectronic devices by covering them uniformly and without defects.

Spin coating is used in the optics sector to produce anti-reflective and protective coatings for lenses and other optical components. For example, a case study showed that spin-coated optical films improved the light transmission and durability of lenses, making them more effective and long-lasting.

In the biomaterials sector, spin coating is used to create thin films for medical devices and tissue engineering applications. A recent example is the use of spin-coated biodegradable polymers to produce scaffolds for tissue regeneration, which provided a homogeneous and controlled surface for cell proliferation and tissue development.

By using spin coating's precision and versatility, industries may produce high-quality thin films that improve the performance and usefulness of a wide range of products [13].

Applications

Spin coating is used for:

- Deposition on photoresist in photolithographic operations.

- Organic light-emitting diodes (OLED) fabricating.
- Preparation of thin films applicable for use in biomedical areas, for instance, in the delivery of drugs and in the fabrication of tissues [13].

Case Study: A 2022 case study demonstrated that spin-coated anti-reflective films on eyeglass lenses increased light transmission by 10% and reduced glare, enhancing visual clarity and comfort.

2.6 *Organic Vapor Phase Deposition (OVPD)*

Organic vapor phase deposition (OVPD) is one such procedure for deposition of organic thin films, which is widely incorporated in the fabrication of OEPDs, OPVs, and OLEDs. Organic molecules are vaporized in OVPD, and finally, they are deposited to form a monolayer on a substrate. With this method, it is easy to produce high uniformity large area organic films with controlled thickness.

The available material, required film thickness, substrate, and particular application needs are some factors among several other things that determine the thin-film deposition technique to be used. Due to the fact that every strategy comes with positive and negative aspects, it is the decision of the researcher or the engineer to select the most appropriate one depending on the intended use. In the long run, thin-film deposition methods create a premise for various scientific and technological discoveries. All deposition techniques, from the specialist OVPD technique to the PVD technique such as evaporation and sputtering to the CVD technique such as ALD, are crucial when tailoring thin films for specific applications. Controllability is an essential characteristic of modern technology due to the perspective of controlling thin-film thickness, composition, and structure. This enables innovations in electronics, optic, energy storage, and other sectors and fields. Thin films are an advanced innovative concept in several industries and science is ceaselessly working and experimenting on such processes.

3 Summary

Various industries require coating techniques with a view to increasing the durability and effectiveness of products. Chemical Vapor Deposition (CVD) allows for accurate control of content and structure of thin films as this process uses the chemical reaction of gaseous precursors of the thin films and deposits them on substrates. Giving terrific adherence and homogeneity, PVD is a kind of deposition that entails the use of physical means such as sputtering and evaporation in the laying down of thin layers. Repeating the self-limiting processes of atomic layer deposition, it is possible to achieve coverage and thickness that resides in thin films coating. Since spin coating is user-friendly and is rather economical, it entails spinning of a substrate to spread a uniform thin layer. Thin films are developed through the technique known as electrochemical deposition of electricity current, which offers the facility of high purity and superior controllability of properties. PVD is another technique where, due to use of plasma for sputtering the atoms from the target material onto a substrate, it is possible to deposit multielement materials and very high-purity thin films. These coating techniques impart different utilities and applications and all these are useful in advancing fields such as material science, electronics, and optical engineering.

4 Material Selection in Thin-Film Technology

Another vital aspect that is universally employed when it comes to thin film technology is the material selection which determines the use, efficiency, and durability of thin films. Thin films belong to the field of electronics, optics, coatings, and many other applications. They are in most cases defined as multilayers having a thickness from one nanometer to one micrometer. Thin films are derived from materials which have properties that determine the features of the obtained thin films. Through this broad study, the various ways in which material selection applies to thin films are examined in relation to the major characteristics that are imperative and the various areas that can benefit from precise material selection.

4.1 Significance of Material Selection

Material selection is the most critical process that shapes the subsequent design and production of thin films. The nature of thin film youngster optical, electrical mechanical, and thermal characteristics also depend on the selection of material. When selecting materials appropriate for thin film deposition, there are things like the application that the thin film will be used for, the environment the thin film will be in, and the performance the researcher or engineer wants the thin film to deliver. From enhancing the anti-corrosive properties or developing the transparent conductive layer to enhancing the working of semiconductor devices, the ultimate aim is the ability to perform specific functions.

4.2 Optical Properties

Several thin films use call for specific attention being paid to properties such as transparency, reflectivity, and absorptance. Such materials as indium tin oxide (ITO) are used in optoelectronics as they are conductive and transparent; they are applied in uses such as transparent electrode in solar cell and display. The optical qualities of material can be used in a selective fashion to make thin films for different applications such as optical filters and anti-reflective layers.

4.3 Electrical Properties

Hence, thin-layer electrical conductivity and resistivity are measures of great importance in electronic applications. Because of their distinct electrical properties, materials such as silicon and compound materials such as gallium arsenide are used often. Thus, it is evident that the appropriateness of conductive materials is a critical factor in the creation of thin-film transistors, integrated circuits, and other electronic devices. For example, the material needs to meet specific electrical characteristics whereby a certain conductivity, resistivity, or bandgap is desired.

4.4 Mechanical Properties

Thermal, mechanical, electrical, and chemical properties that include hardness, flexibility, and stickiness affect the thin films in their performance as well as their durability. The selection of materials having appropriate mechanical characteristics is highly relevant in such areas as, for example, flexible electronics or protective layers. For instance, the wear resistance is achieved by using the deposition of the hard layer materials for instance diamond-like carbon (DLC), and equally the bendable and elastic thin film devices can also be realized by the use of flexible substrates and polymers.

4.5 Thermal Properties

Film thickness stability and thermal conductivity are essential where the temperature changes are expected to take place. In the operation of electronic gadgets, one challenge that can greatly affect the performance of the gadgets is usually the heat generated, hence the need to achieve proper thermal control. Thin film heat sinks can be made of copper and/or aluminum that has high thermal conductivity while thermal barriers can be of silicon dioxide which is an insulating material. There is a need to dissipate heat and ensure the reliability of thin film components which affects the materials choice.

4.6 Chemical Properties

Some chemical characteristics that come under hostile condition include chemical reactivity and corrosion that are important when used in thin films. Coatings applied on thin films for automobiles or planes for example should be resistant to corrosive agents. High chemical-resistant material is chosen like tantalum and titanium nitride. The performance of thin-film coatings is critically dependent on the chemical reaction that takes place in an environment, and as a result, specific chemical compatibility must be achieved.

To put in a nutshell, one can relate that the material, which is chosen, forms the basic element in the field of thin films as it governs their properties and consequently, their applicability. Here, it is suggested that thin-film applications should be done according to the mechanical, electrical chemical, optical, and thermal properties of materials. Development of thin film technology is advancing hand in hand with material research, and as both fields evolve, it introduces new possibilities and raises the standard of achievement in electronics, energy, and numerous other fields.

5 Growth Mechanisms and Film Morphology in Thin-Film Deposition

Thin film deposition is central to numerous technological processes, especially in the manufacturing of semiconductors as well as in optics, and sensors. Thus it is important to study the various growth mechanisms and the film morphology when the requirements of the thin films need to be specified. Moving down the line in the classification of the coating techniques, the growth mechanisms and the film morphology further complicate and professionalize the development of functional thin films.

5.1 Growth Mechanisms

Three major modes have been determined in the growth mechanism of thin films: The three types are Frank–van der Merwe, layer-by-layer, Stranski–Krastanov, and Volmer–Weber, Island. If the material is deposited in the Volmer–Weber mode then the first phase to appear on the substrate is in the form of isolated islands. These islands grow and are added to by the deposition process and it is for this reason that there are gaps between them. The Frank–van der Merwe mode is different in that it is layer-by-layer growth; the molecules or atoms first bond to the surface of the substrate and form complete layers before the formation of the following layer begins. Movies created with this technique are free of voids, get continuous, and are produced smoothly. A blend of the first two modes is the Stranski–Krastanov mode.

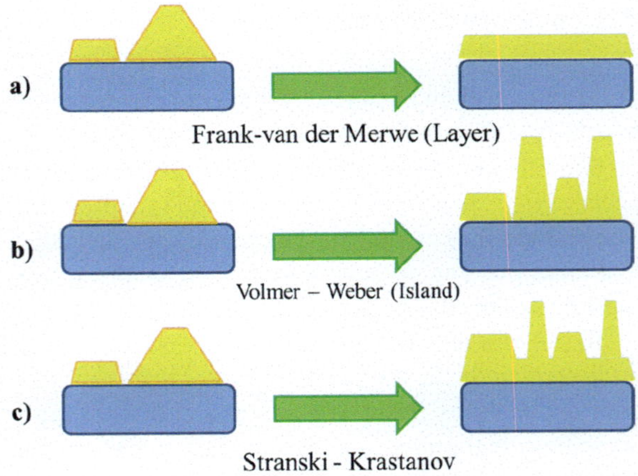

Fig. 6 Three major modes of growth in thin films are illustrated in this figure

Initially, there are a few layers that develop layer by layer (Frank–van der Merwe). However, as the film becomes thicker the strain energy builds up and at some point results in the Volmer–Weber formation of islands on the serried layers [14, 15]. Figure 6 shows how the morphology of the film is influenced by the dominating mode of film growth.

5.1.1 Nucleation and Growth Dynamics

The process of thin film formation: The first step is nucleation in which small clusters of atoms or molecules first appear on the substrate. Nucleation is the same with germination of seeds but in preparing for the formation of a continuous film. Nucleation processes are sensitive to factors like temperature, pressure, as well as the characteristics of a substrate on which the process is administered. Nucleation kinetics is studied to understand the important issues like the critical nuclei size, nucleation rate and how they both are influential for the whole.

Nucleation is followed by the growth mechanisms as nucleation continues to unfold itself. The layer-by-layer growth is frequently observed when the atoms or molecules adhere to the existing layers creating the film's thickness. It is essential, therefore, to understand how and in what way the layers are built so as to end up with films that have the right characteristics. This growth mode is especially typical for technologies such as molecular beam epitaxy (MBE) in which the control for the deposition rate is critical for the formation of intricate structures [16, 17].

5.1.2 Factors Influencing Growth Mechanisms

The decision on the choice of growth mechanisms is sometimes determined by the need of the application, by its specifics. For example, when constructing semiconductor devices, epitaxial growth of crystalline layers is utilized due to the effective control over the crystalline lattice. This is where the conflict between thermodynamics and kinetics comes into play—the crystal structure which is desirable may be thermally favorable but the deposition process in which the material is being subjected to may produce a different structure [18].

Knowledge in growth mechanisms and their control also requires the consideration of the energy impact on deposition. Methods based on the pulsed laser deposition (PLD) involve the use of energy pulses to remove the material from the target and affect the kinetics of nucleation and growth. This control over the energy input is that depending on the required properties of the thin film, the researcher can alter the process in order to receive the necessary characteristics, such as adhesion or crystallographic orientation [18].

5.2 *Film Morphology*

5.2.1 Surface Roughness and Its Implications

Morphology of film deals with the surface coatings of thin films and the roughness of the surface is a basic parameter. Indeed, the features of the shape of the surface of the film can affect the process significantly. The studies of texture roughness incorporate methods such as Atomic Force Microscopy (AFM), Scanning Electron Microscopy (SEM), etc.

In areas like optics and sensors, the surface roughness is usually kept to a minimum to avoid scattering of light or to make good contact with a sensing component. On the other hand, it is a requirement in several circumstances to have strictly controlled roughness in order to improve certain characteristics. For example, in the solar cells, the nanostructures are designed in concentration of light hence enhancing the performance of the structure.

5.2.2 Crystallinity and Grain Structure

The crystallinity of thin films is also another aspect of the film morphology that defines their characteristics, p. 52. The structure of silicate in crystalline films has a distinct geometry affecting electrical conductivity, optical transparency, and mechanical strength. The material often has to be grown under a particular set of conditions of temperature and pressure to achieve a desired crystal structure.

Focused on the structure of films, where various specimens combined with the polycrystalline nature of the films under investigation, namely, the presence of several crystal grains, transfers an added challenge. Grain and grain boundaries' alignment determines aspects such as electrical conductivity and mechanical strength. For instance, sputtering and chemical vapor deposition (CVD) delivered grain size and orientation control, which is essential for the enhancement of the film properties' performance.

5.2.3 Tailoring Grain Boundaries for Functionality

In some cases, grain boundaries can be deliberately controlled in their favorable applications to provide unconventional properties to thin films. These can act as the preferred sites for chemical reactions or as paths of movement of the charge carriers. Therefore, understanding these grain boundaries and being able to engineer them allow new ways of improving the effectiveness of thin films in applications such as catalysis or in the manufacture of electronic devices.

In the complex process of thin film deposition and growth, the underlying processes and the films' microarchitecture tell a tale of order and mastery. Potential of creating films with desired properties suggests the analysis of various opportunities for improvement of a wide variety of industries related to film production. Today, semiconductor devices, microelectronics, photovoltaics, and optical coatings—the processes creating new generations of materials are based on the theory of thin film deposition and the intimate relation with coating techniques. The essence of thin-film engineering is still subject to further discovery and research, the possibilities are vast, and with further theme discoveries more technologies of the future will unfold.

Conclusion

Thus, the chapter "Fundamentals of Thin Films" offers a sufficient background of the key factors in thin-film technologies. We have described deposition techniques as well as the selection of the material, growth mechanisms, and the morphology of the film so that the audience could realize that thin films are created with the help of complex processes. Such knowledge is vital to the promotion of thin-film engineering across various disciplines that include microelectronics, optics, and surface engineering. This chapter shows there still exists the need for the development and improvement of thin-film technology and how the performance and longevity of these materials and devices can be increased.

References

1. Gilmer, G.H., Huang, H., Roland, C.: Thin film deposition: fundamentals and modeling. Comput. Mater. Sci. **12**(4), 354–380 (1998). https://doi.org/10.1016/S0927-0256(98)00022-6
2. Schalk, N., Tkadletz, M., Mitterer, C.: Hard coatings for cutting applications: Physical vs. chemical vapor deposition and future challenges for the coatings community. Surf. Coatings Technol. **429**, 127949 (2022). https://doi.org/10.1016/j.surfcoat.2021.127949
3. Kaiser, N.: Review of the fundamentals of thin-film growth.pdf. **41**(16) (2002)

4. Properties, E.D.: Technical paper. Arab Energy **1**(5) (1984). https://doi.org/10.1017/cbo9780511791376.004
5. Gupta, G., Tyagi, R.K., Rajput, S.K., Saxena, P., Vashisth, A., Mehndiratta, S.: PVD based thin film deposition methods and characterization/property of different compositional coatings—a critical analysis. Mater. Today Proc. **38**, 259–264 (2021). https://doi.org/10.1016/J.MATPR.2020.07.132
6. Haq, I., et al.: Linkage of thermoelectric properties with the structural parameters in zinc indium oxide (ZnInO) thin films grown by physical vapor deposition (PVD). Inorg. Chem. Commun. **149**, 110438 (2023). https://doi.org/10.1016/J.INOCHE.2023.110438
7. Shaji, K., Haviar, S., Zeman, P., Kos, Š, Čerstvý, R., Čapek, J.: Controlled sputter deposition of oxide nanoparticles-based composite thin films. Surf. Coatings Technol. **477**, 130325 (2024). https://doi.org/10.1016/J.SURFCOAT.2023.130325
8. Struppert, T., Heft, A., Grünler, B.: Thin functional films by combustion chemical vapour deposition (C-CVD). Thin Solid Films **520**(12), 4106–4109 (2012). https://doi.org/10.1016/J.TSF.2011.06.048
9. Istanbullu, O.B., Akdogan, G.: Increased body fluid repellency and electrochemical corrosion resistance of intravascular stent materials by ICP-CVD-based DLC thin-film deposition. Diam. Relat. Mater. **138**, 110251 (2023). https://doi.org/10.1016/J.DIAMOND.2023.110251
10. Jolivet, A., et al.: Structural, optical, and electrical properties of TiO_2 thin films deposited by ALD: impact of the substrate, the deposited thickness and the deposition temperature. Appl. Surf. Sci. **608**, 155214 (2023). https://doi.org/10.1016/J.APSUSC.2022.155214
11. Guo, Y., et al.: Noble metal alloy thin films by atomic layer deposition and rapid Joule heating. Sci. Rep. **12**, 2522 (2022). https://doi.org/10.1038/s41598-022-06595-9
12. Bai, H., et al.: Preparation of antimony selenide thin films by electrochemical deposition and application in optoelectronic devices. Mater. Sci. Semicond. Process. **171**, 108027 (2024). https://doi.org/10.1016/J.MSSP.2023.108027
13. Yao, K.J.A., et al.: Sol-gel deposition of TiO_2 thin films by spin coating for photovoltaic applications: effect of acetylacetone stabilizer on structural and optical properties. Mater. Today Proc. (2024). https://doi.org/10.1016/J.MATPR.2024.02.003
14. Lee, W.J., Li, C., Gunning, J., Burke, N., Patel, J.: Is the structure of anisotropic pyrolytic carbon a consequence of growth by the Volmer-Weber island growth mechanism? Carbon N. Y. **50**(13), 4773–4780 (2012). https://doi.org/10.1016/J.CARBON.2012.05.052
15. Lozovoy, K.A., Korotaev, A.G., Kokhanenko, A.P., Dirko, V.V., Voitsekhovskii, A.V.: Kinetics of epitaxial formation of nanostructures by Frank–van der Merwe, Volmer-Weber and Stranski-Krastanow growth modes. Surf. Coatings Technol. **384**, 125289 (2020). https://doi.org/10.1016/J.SURFCOAT.2019.125289
16. Riley, P.R., Joshi, P., Narayan, J., Narayan, R.J.: Enhanced nucleation and large-scale growth of CVD diamond via surface-modification of silicon-incorporated diamond-like carbon thin films. Diam. Relat. Mater. **120**, 108630 (2021). https://doi.org/10.1016/J.DIAMOND.2021.108630
17. Nguyen, V.H., Sekkat, A., Van Bui, H., Muñoz-Rojas, D., Bellet, D.: Nucleation and growth of thin films. Nucleation Growth Appl. Mater. 41–64 (2024). https://doi.org/10.1016/B978-0-323-99537-5.00002-7
18. Wu, S., Xu, R., Guo, B., Ma, Y., Yu, D.: Influence of growth parameters and systematical analysis on 8-inch piezoelectric AlN thin films by magnetron sputtering. Mater. Sci. Semicond. Process. Semicond. Process. **169**, 107895 (2024). https://doi.org/10.1016/J.MSSP.2023.107895

Film Thickness and Its Influence

Muhammad Hasan, Syed Darain Ahmed, Arham Ansar, and Iftikhar Ahmed Channa

Abstract Control of the thin-film thickness is very crucial which is described in the chapter, "Film Thickness and Its Influence." In most of the applications, there is a need for a thin film that requires the thin film to be fabricated with great accuracy of thickness. This chapter focuses on some of the enhanced approaches of measuring thickness that include ellipsometry, profilometry, and atomic force microscopy before concluding. Further, the role of film thickness on other characteristics like optical activities, electrical conductance, and mechanical solidity is also described, which again illustrates how films can be designed for application use. Field of applications ranging from microelectronics and optical coatings to biomedical devices as well as sensors, precise management and measurement of film thickness is necessary for effectiveness and development in these fields. Altogether, it is endorsing the theme of this chapter that film thickness assumes a significant position in studying thin films and applying them to various technologies.

1 Film Thickness and Its Influence

Film thickness is known to be one of the most significant factors that play a pivotal role in controlling the overall performance that affects different industries directly or indirectly. Whether it is in the ultra-miniature world of microelectronics fabrication or in biomedical application, film thickness is a significant factor when it comes to energy harvest or saving a life of patient with implants [1]. In this chapter, the critical necessity implied in the ability to control the thickness of films to overall performance is discussed [2].

This chapter explores and identifies different approaches of characterizing film thickness hence offering a comprehensive literature on the techniques applied across the industrial sectors. From the advanced techniques of semiconductor fabrication

M. Hasan · S. D. Ahmed · A. Ansar · I. A. Channa (✉)
Thin-Film Lab as part of Advanced Materials and Sustainable Environment (AMSE) Research Group, Department of Metallurgical Engineering, NED University of Engineering and Technology, Karachi 75270, Sindh, Pakistan
e-mail: iftikharc@neduet.edu.pk

A. D. Chandio and I. A. Channa (eds.), *Thin Films and Coatings*, Engineering Materials,
https://doi.org/10.1007/978-981-96-0005-2_3

right down to the fine points of medical instruments, the focus that is placed on the precise manufacturing of film thickness becomes in turn the unifying focus that ties together disparate fields. In addition, the analysis extends to variable-thickness films, which requires different degrees of flexibility and customized management [3]. It is important to note that films with varying thicknesses are employed under different situations due to the varying requirements depending upon the applications. These applications take it one step forward in innovation; giving custom solutions in such fields as electronics, optics, and even medical diagnosis. This chapter can be regarded as the general plan, the goal of which is to help the readers orient themselves in the variety of options that can be introduced by variable-thickness films. With each later chapter, the dependency between film thickness to other characteristics becomes evident and signifies how this book unravels the intricate connection between different aspects of the coating material (shown in Fig. 1). It is also not just a technical decision; rather, it is a sophisticated process of choosing a proper solution that implies not only profound analysis of the scientific and engineering approaches but also the perfect understanding of the specific demands of the two industries. Consequently, the analysis extends beyond simple trends in thickness control and extends to the very core of material engineering to where each nanometer can be the key to opening up new possibilities for the end product [4, 5].

In short, the chapter provides users with an opportunity to delve into a world where the control of film thickness is not just a technicality; it is the essence of turning the fate of materials into a work of art. It directs attention to this parameter when this parameter plays a critical role in the magnificent suite of industrial operations. Starting from the minuscule measure of thickness essential for fabricating semiconductor wafers and ending with the broad perspective for the medical implant coatings, film thickness is all-pervading and plays an essential role everywhere. Thus,

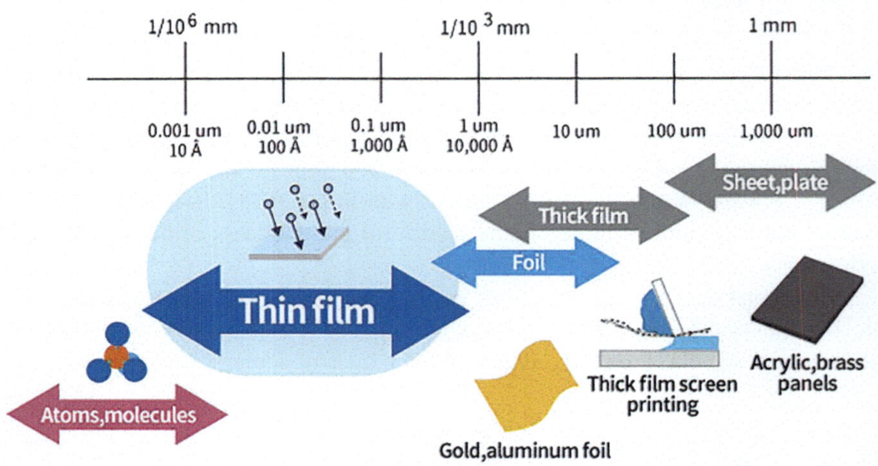

Fig. 1 Thin-film thickness range

in addition to outlining the procedural steps, it is possible to grasp the essence of the change in such thin layers and be inspired by the true potential that lies behind them.

2 Importance of Controlling Thickness

The film thickness is a critical parameter because it directly determines the service characteristics of the materials and products. Another key factor is management of thickness, as it cast a significant impact on the uniformity, the quality as well as the efficiency. Film thickness is controlled by a number of characteristics and significant aspects, including.

2.1 Parameters for Controlling Thickness

2.1.1 Uniformity

Thin films can be described as layers that are usually less than a micrometer in thickness; nonetheless, there are certain considerations that need to be taken into account and one of these is uniformity [6]. A plane distribution without any variation in thickness across the full width of thin films is desirable since it highly impacts the result of operations. This characteristic is important in most of the applications but particularly helpful in applications where precise and reliable materials are needed, for instance in microelectronic, optical coating, and thin-film device as illustrated in Fig. 2 while in which thin-film thickness defines the level of energy absorption and reflection by solar cells. When the thickness of the film applied on the surface is at the required level, as on the left of the picture, the energy is absorbed on the right however shows that when the applied film thickness is increased, the reflection is enhanced and the energy conversion is decreased. This showcases the important factors that determine the overall performance of the solar cells and the other photonic devices on the thickness of the film. These include matters of deposition process, monitoring as well as the preparation of the substrates needed for the formation of a consistent layer of the thin film [7]. To ensure that the coating accurately replicates the desired design and is invariant across the substrate surface, sophisticated deposition methods like Chemical Vapor Deposition (CVD) and Physical Vapor Deposition (PVD) are used [8, 9]. Accuracy is important for deposition process because real-time monitoring systems are the primary means of ensuring consistent parameters during deposition processes and can control the thickness of layers being placed. The issue of uniformity is not limited to numeric objectives; it illustrates a commitment to creating surfaces which are not only defined by their technical specifications but also have a glossy to silky finish that is virtually indistinguishable from one segment to the next. Work done by Smith et al. [10] on thin-film deposition categorically underlines the above assertion by calling for standardization of material deposition

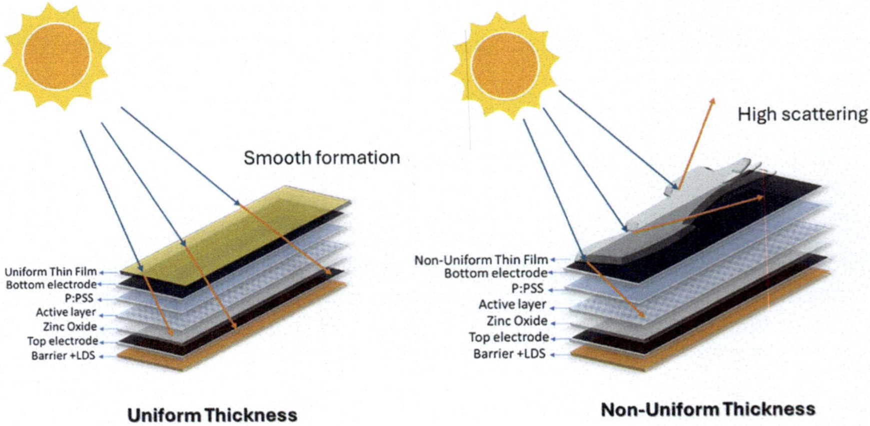

Fig. 2 The impact of uniform film thickness on performance

techniques for enhanced performance and stability of end products. The uniformity is one of the key factors that defines the quality and applicability of thin films to boost up their functioning in diverse industries [11].

2.1.2 Precision

The degree of precision of thickness control increases as one travels into the literal realms of the optical, where less than a few atomic layers are crucial. This accuracy is not a mere job of guesswork, but an imperative called for by certain requirements and acceptance in the market. It is essential to have a thin layer and maintain substantial control in thickness when working in fields like optics where coatings of thin films are used in the modification of light parameters [12]. The thickness of each layer influences its refractive index, which in turn affects the required optical effects when deviated from the established value [13]. Table 1 shows this relationship as a chart and illustrates how variation in precision shows the extent of impact on functionality [14, 15]. The table shows the effects of different levels of precision of the thickness of the film of the part on the performance of the device, where the high accuracy ensures perfect and stable functioning of the device while lowering its accuracy will lead to a reduction in the efficiency and soft reliability of the device.

2.1.3 Adhesion

Entering the domain of adhesion, durability, and performance of the material are closely linked to the significance of appropriate thickness. The adhesion of the film becomes a crucial factor in determining the device's efficacy in biomedical applications, for example, when biocompatible coatings interact with biological tissues.

Table 1 Consequences of precision levels in film thickness control.

Precision level	Consequences	References
High precision	– Enhanced electrical performance in electronic devices	[16]
	– Optimal light transmission in optical coatings	[13]
	– Improved adhesion and durability in thin films	[17]
	– Minimal variability in optical characteristics	[18]
Medium precision	– Satisfactory electrical performance in electronics	[19]
	– Adequate light transmission in optical coatings	[20]
	– Substantial adhesion and durability in thin films	[17]
	– Moderate variability in optical characteristics	[21]
Low precision	– Compromised electrical performance in electronics	[22]
	– Reduced light transmission in optical coatings	[23]
	– Risk of adhesion issues and limited durability	[17]
	– Significant variability in optical characteristics	[18]

Figure 3 illustrates the connection between adhesion strength and film thickness, highlighting the fine balance needed to attain ideal performance [17]. In contrast to other materials like glass, stainless steel, polymers, and metals, silicon exhibits the maximum strength when coated with diverse substrates, giving them a thin thickness and ease of usage.

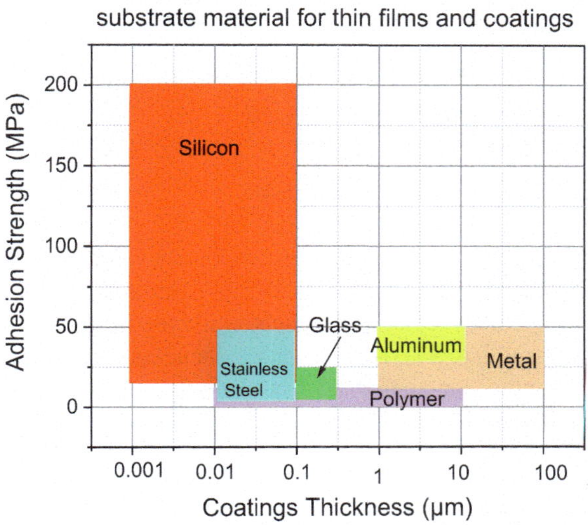

Fig. 3 The relationship between film thickness and adhesion strength

Table 2 Impact of film thickness on optical properties

Film thickness	Optical property impact
Increased	**Transparency**: Decreases in thicker films
	Reflectivity: Enhances reflectivity in mirrors
	Color: May lead to interference effects and vivid colors
	Light Absorption: Can increase, depending on material
Decreased	**Transparency**: Increases in thinner films
	Reflectivity: Decreases, as observed in anti-reflective coatings
	Color: May alter interference effects and color perception
	Light Absorption: May decrease, but material-specific
Variable	**Color**: Engineered for specific colors in variable-thickness films
	Light Absorption: Tailored for applications like solar cells

2.1.4 Optical Properties

The impact of film thickness on optical properties extends beyond the realm of technical issues and encompasses the visual and aesthetic qualities of materials. Transparency, reflectivity, and color are all determined by thickness, which affects the finished product's appearance as well as its usability [24]. Table 2 illustrates how several optical features are impacted by film thickness in a comprehensive way, emphasizing how these aspects are interdependent [21].

2.2 Important Features Used to Control Thickness: Techniques Unveiled

2.2.1 Deposition Techniques

The art of depositing thin films includes a variety of methods, each with its own benefits and uses. Because Physical Vapor Deposition (PVD) can create high-purity films, it is useful in sectors where material integrity is crucial. Contrarily, Chemical Vapor Deposition (CVD) is excellent at producing conformal coatings on complex surfaces. Dynamic spin coating works especially well for obtaining consistent thickness on a variety of substrate geometries [25] (Table 3).

2.2.2 Monitoring Systems

In the intricate realm of thin film and coating deposition, real-time monitoring and feedback systems stand forth as vigilant guardians, entrusted with the sacred duty of safeguarding the precise film thickness levels. They are not simply used for

Table 3 Comparative overview of Physical Vapor Deposition (PVD), Chemical Vapor Deposition (CVD), and spin coating techniques with respect to thickness

Criteria	Physical Vapor Deposition (PVD)	Chemical Vapor Deposition (CVD)	Spin coating	References
Principle	Physical processes such as evaporation or sputtering	Chemical reactions to deposit thin films	Rotating substrate while applying liquid precursor	[26–28]
Thickness control	Precise control through monitoring deposition rate	Controlled by adjusting precursor gas flow and reaction conditions	Limited control, influenced by spin speed, viscosity, and precursor concentration	[29–31]
Deposition rate	Generally slower	Can achieve higher deposition rates	Faster than PVD but typically slower than CVD	[32–34]
Uniformity	Good uniformity for small areas	Better uniformity over larger areas	Provides good uniformity for small to moderate-sized substrates	[31, 35, 36]
Substrate compatibility	Compatible with a variety of substrates	May require high temperatures, limiting substrate choices	Suitable for a wide range of substrates	[31, 37, 38]
Cost and complexity	Generally lower cost and simpler equipment compared to CVD	Higher cost and more complex equipment due to chemical reactions	Relatively low cost and simple equipment	[34, 39, 40]
Applications	Electronics, optics, wear-resistant coatings	Semiconductor manufacturing, coatings	Thin organic films, photoresist in microfabrication	[41–43]

measuring; they become the bearers of unique precision, and the ritual of measurement transforms into the pervasive consistency that weaves through the fascinating journey of film development [44]. Thin film and coating deposition can be compared to painting where the base is a thin material and the final artwork is created with layers on top of each other. Here in this creative endeavor, these monitoring and feedback systems, equipped with various advanced measuring tools, become active entities. They assume the roles of active managers that not only monitor the thickness measurement but also the core of films creating process—that thin layer of glue that unites the layers with precision and accuracy to the manufacturers' plans. Furnished with such devices as thickness monitors, optical spectrometries, QCM, and ellipsometry, these guardians perform the intricate and continuous tango of acquiring information [45, 46]. It is not only for the numbers they deal but also for the detailed accuracy needed to give form to each layer of the thin film or coating just as a painter does on a canvas [47, 48].

2.2.3 Substrate Preparation

The achievement of the best film thickness type starts with the initial steps in the preparation of the substrate. These industries include semiconductor manufacturing where the substrate surface has to contain fine circuitry; the surface has to be treated effectively to ensure that it adheres and is of an even thickness. Figure 4 depicts the process of substrate preparation, which is essential in any technique and highlights the complexity of the procedure [49]. This is a figure depicting planarization steps for thin-film deposition with five required basic process steps. First, the right substrate material has to be chosen to match the wanted film properties. Subsequently, the substrate is subjected to a specific cleaning process, which may employ solvents or ultrasonic washing baths, etc., to remove the unwanted debris. Depending on the type of sparkler, it will require washing in warm or cold water; after washing the substrate, there should not be any moisture left on it. Further treatments are carried out if necessary on the surface in order to alter the surface characteristics of the substrate. Lastly, the structure is created on the prepared substrate by the film deposition process with the help of sputtering, CVD or PVD, and the uniform and high-quality film is developed on the substrate (Fig. 4).

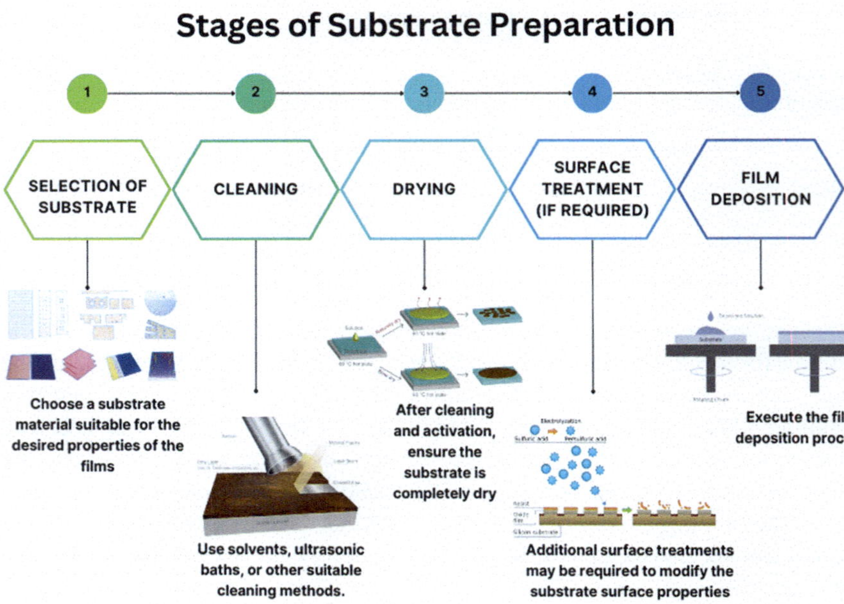

Fig. 4 Steps involved in substrate preparation for film deposition

2.2.4 Masking and Patterning

Selective deposition belongs to the sphere of deposition techniques, which underscores the interconnection between precision and control; pity masking and patterning had to be mentioned. These techniques make engineers have control of the thickness of films, especially in certain areas and they can come up with complicated patterns and special additional layers. The strategic use of different masking and patterning techniques is described in detail in Fig. 5 below, where all the approaches are presented with a view of creating a more convenient layout [50]. From the established image, it is evident that the process of thin-film processing is a stepwise one. First, blade coating is used for applying a thin layer of coating material onto a substrate. The blade helps in fixing an even distribution of the coating material and comes with features that regulate the thickness and uniformity of a deposited film, which is very critical towards the intended specialty of optical and electrical rejuvenization. After the deposition process, methods that make use of lasers are applied in the manufacturing process. Laser phase transformation is the first technique wherein a 532 nm laser with a 30 ns pulse is used to cause a phase transformation in the material and improve its properties while not affecting the substrate. The second technique of the thin-film processing, laser ablation, is based on the usage of a 355 nm laser which has a 10 ps pulse and serves to remove only the required sections of the thin film to form the necessary patterns for further structuring of the devices. Lastly, needle patterning is carried out with a tip width of 40 μm with the needle to develop thin strings and patterns for the fine wiring of the electrical paths. Each of them is important to control the accurate property changes of the thin film for the application the thin film in photovoltaic and electronic devices.

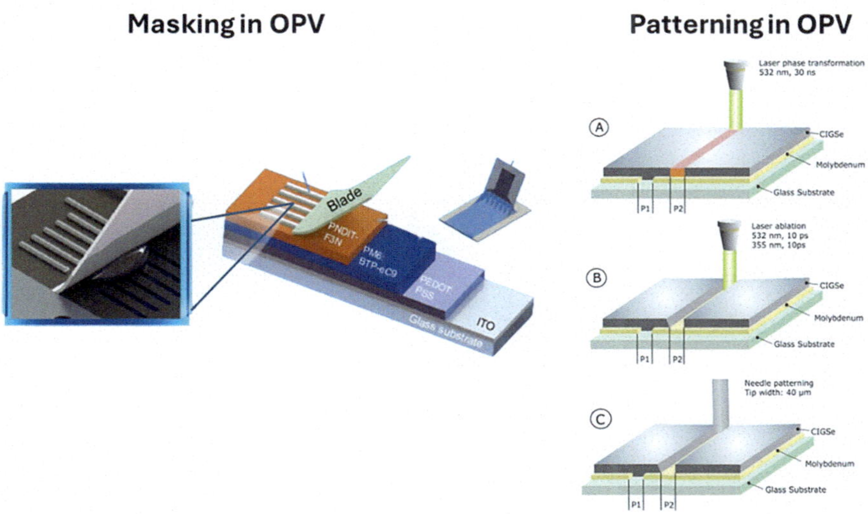

Fig. 5 Steps involved in masking and patterning for selective film deposition

3 Techniques for Thickness Measurement

In a multitude of industries where films are applied or incorporated in certain products, thickness measurement is critical when it comes to checking the quality or in improving the process. Thus, strategic implementation of the techniques for thickness measurement is of paramount importance for achieving accuracy. Several widely employed methods include.

3.1 Ellipsometry

The ellipsometry is a non-invasive method used to determine the variation in polarization state of the light reflected off a thin layer sample. By examining these changes, ellipsometry can effectively quantify the film thickness as well as the refractive index (as shown in Table 6). This method is applicable to both the transparent as well as the opaque films and is adopted in the formulation of semiconductors and thin-film industry [51]. Ellipsometry is a technique of analyzing the changes in light polarization when the light interacts with a surface of the analyzed sample (shown in Fig. 6). At the sample, the given light is close to a purely polarized beam, and its polarization plane changes according to the sample's manifestations, such as refractive index and thickness, and then passes through an analyzer containing a polarizer and detector. By measuring values including tan Ψ and Δ the characteristics of the thin films such as the refractive index or thickness and structural parameters are obtained. This makes ellipsometry very important in areas such as material science and semiconductor industries due to its accuracy in the measurement of optical properties (Table 4).

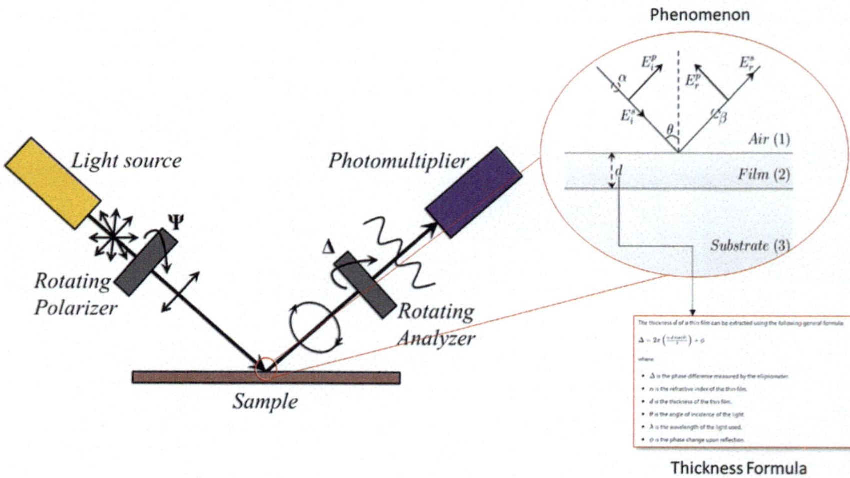

Fig. 6 Schematic diagram of working principle ellipsometry and their thickness formula

Table 4 Ellipsometry measurements

Film type	Thickness range	Application	References
Transparent films	1 nm–1 μm	Semiconductor wafer fabrication	[52]
Opaque films	10 nm–10 μm	Thin-film research	[53]
Organic coatings	5 nm–100 nm	Optical coating optimization	[54]

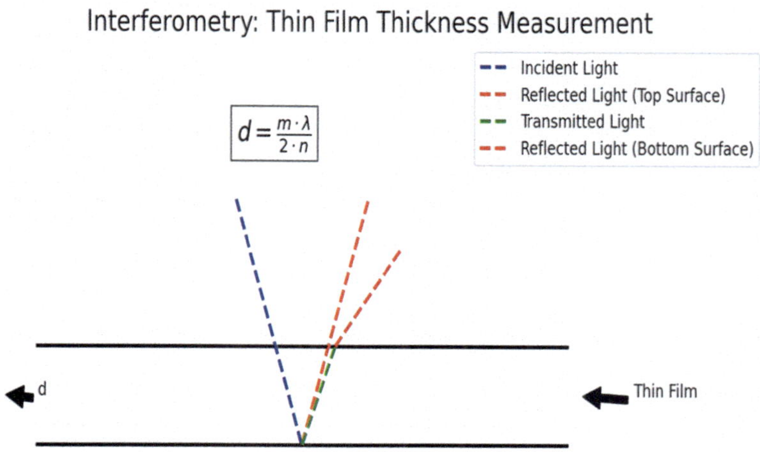

Fig. 7 Fundamental working of interferometry applied in the determination of the thickness of a thin layer

3.2 *Interferometry*

Interferometric techniques rely on patterns of interference to determine film thickness features and give off high precision and resolution. Optical and white-light interferometries are known to be used in measurements. Interferometry is based on the principle of combining two or more light waves to compare among the samples differences of merely distance or changes in materiality (as shown in Fig. 7). The technique is based on the matrix of the interference fringes that light waves form and thus, allows one to measure thin-film thickness, surface roughness, and refractive index in order to characterize thin films and coatings. These techniques are used extensively where the accuracy of feature sizes and specifications of components is measured in nanometers such as in microelectronics and optics industries [55].

3.3 *X-Ray Reflectometry (XRR)*

XRR employs the Bragg's law that relates with the phase shift of the reflected X-ray wave to quantify the thickness, density and roughness of the thin film in amperes (refer to Table 5). This technique is most effective for the thin films that are in the substrate and any attempts in the material science, semiconductor, as well as thin-film studies (schematic of the working principle is shown in Fig. 8) [56]. X-ray Reflectometry (XRR) deals with illuminating the thin film with X-rays and then measuring the intensities of the reflected rays at different master angles. From the obtained reflectivity curve, the information on the film thickness, density, and roughness can be derived from XRR analysis. This non-destructive microtechnique finds application in multi-layer systems, thin films on different substrates, etc.

Table 5 X-ray reflectometry applications

Thin-film type	Substrate material	Application	References
Metal oxide films	Silicon substrates	Semiconductor device characterization	[57–59]
Polymer thin films	Glass substrates	Material science research	[60, 61]
Nanocomposite films	Flexible substrates	Thin-film characterization for flexible electronics	[62, 63]

Table 6 Applications of variable thickness in optoelectronics

Film type	Application	References
Thin-film solar cells	Enhanced light absorption	[68, 69]
LED displays	Improved light emission	[57]
Display optics	Controlled light transmission	[67, 70]

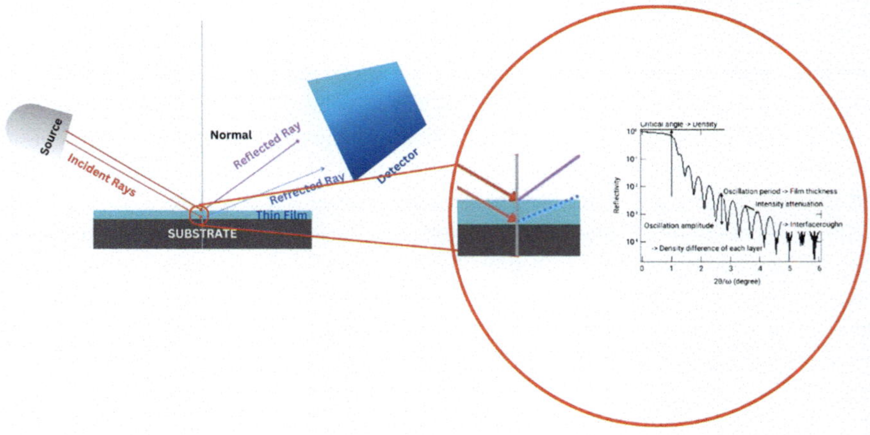

Fig. 8 X-ray reflectometry setup for thin-film analysis

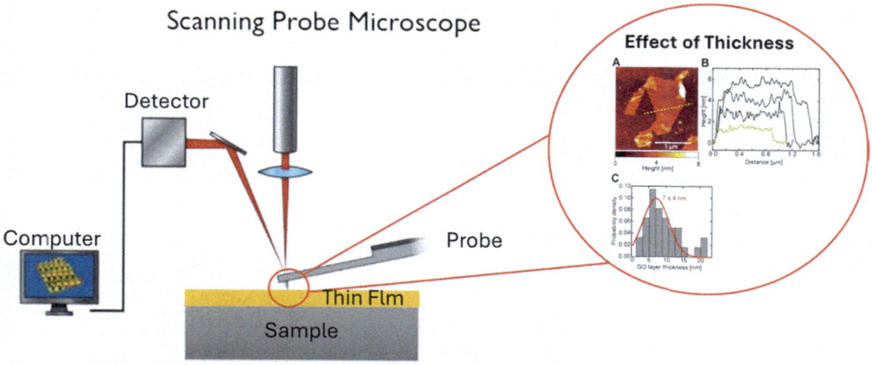

Fig. 9 Illustrates the indirect relation of film thickness on surface property

3.4 *Scanning Probe Microscopy (SPM)*

Among other microscopies, surface microscopy consists of such types as scanning probe microscopies [Atomic Force Microscopy—AFM and Scanning Tunneling Microscopy—STM] giving the information of surface topography (schematic diagram is shown in Fig. 9). These techniques though not direct thickness measurement techniques can be used to analyze the features on the surface of the coats which helps one to indirectly gauge the thickness of the film and any changes in it [64]. Atomic force microscopy gives information at the near-surface level mainly by recording forces between a sharp probe and a given sample. This technique comprises AFM and STM, two types that provide atomic or molecular level resolution in the sample. It is applied in analysis of the surface morphology, mechanical and electrical properties of thin films and nanostructures.

4 Applications of Variable Thickness

Flexible films that can be made with a variable thickness and possess special features serve as invaluable materials in various spheres. Because these films may have their thickness customized, they offer special functions that make them invaluable in a variety of fields.

4.1 *Optoelectronics*

Variable-thickness films will appear as critical elements in the rapidly evolving field of optoelectronics and are likely to create new enabling technologies for a range of device applications including, inter alia, solar cells, LEDs, and displays. The notable accuracy in controlling the thickness of the film is an ideal foundation that enhances performance since it directly manipulates the light absorption, emission, as well as transmission of the films' properties. One of the most important applications of variable-thickness films is in the development of solar cells whereby the aim is to convert light into electricity. These films are primarily very thin indeed, and the thickness of these films is evidently very central to the absorption spectrum of the material. This can be achieved through better selection and control of the film thickness that determines the scope of light absorption from that coming into the solar cell, and hence increases conversion efficiency. The enhancement of film thickness using this degree of accuracy facilitates the creation of new solar cells that can capture more light, which in turn can increase the overall energy intake of solar devices [65].

Thin film illustrates its various applications in LED, where variable thickness enhances the management of light output (shown in Table 6). These films, used at semiconductor material integrated within the structure, allow controlling the emission wavelength. Consequently, engineers are able to manipulate colors with impressive accuracy and control the thickness of the thin films, solving many problems in the use of LEDs in display technologies and complex lighting applications. Such control is crucial to guarantee high quality of displayed colors, which is critical for consumer electronics, signs, and automotive lighting applications. Organizations that are into the manufacturing of display screens in smart phones, TV sets, and any other electrical gadgets should grasp the importance of employing variable-thickness films in their production [66, 67].

The incorporation of optoelectronic application of variable-thickness films shows how material science and engineering has sealed the fates of high-end apparatus. While researchers investigate the independence of the optical thickness of the film and the mechanical thickness of the film, there is an increasing opportunity to study advanced aspects of energy conversion efficiency in multispectral and advanced lighting-led display panels. Thus, the control of thickness for the optoelectronic films becomes another important issue in the development of more efficient and multidimensional devices promising with the progress of high technologies.

4.2 Biomedical Engineering

"Flexible films" have gained much attention in the field of biomedical engineering and known to be actively involved in the progress of numerous applications, for example, drug delivery systems, tissue engineering scaffolds, biosensors, and so on (shown in Fig. 10). Due to their ability in specifically controlling the thickness of the film, these coatings have found their application in the complex network of biomedical technology having an enabling capacity for controlling some of the vital biological processes with an extreme measure of accuracy. Using variable-thickness material in drug delivery systems improved the approach to delivering therapeutic agents. To avoid such complications, it is possible to deliberately design the thickness

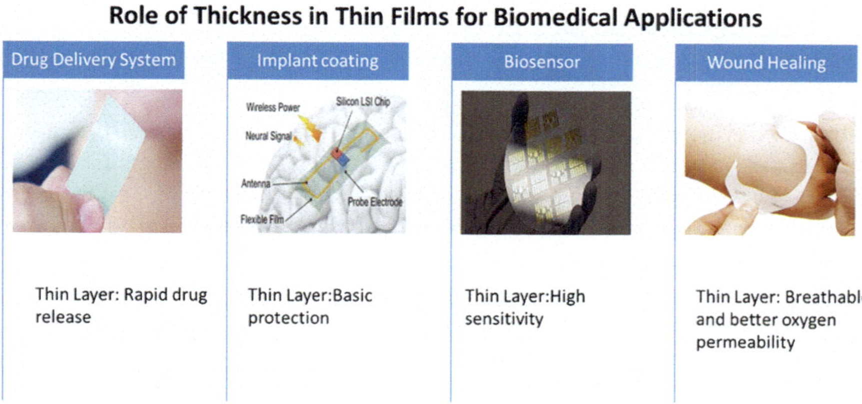

Fig. 10 Pictorial representation of thin film in biomedical application with role of thickness

of such films and thus closely control the release profile of drugs. It allows for the long-acting, well-proportioned, predictable, and constant performance of drugs in the body to provide the best treatment, reduce toxic effects, and improve patient compliance. Whether it involves controlled release of a drug over a definite time period or the administration of a random amount of dosage, variable-thickness films symbolize opportunities to approach each patient individually and achieve successful outcomes. In the realm of tissue engineering, these films form quite crucial structural subcomponents of scaffolds intended to facilitate and direct cells' differentiation. Through controlling the thickness of the film being used in the experiment, then the scientists can control the physical and biochemical signals that are offered to the cells such as adhesion, proliferation as well as differentiation. Such control plays a crucial role in designing biomimetic scaffold that recreates the natural tissue environment to promote repair and regeneration of the diseased or injured organs [71, 72].

In the field of biosensors, variable-thickness films indeed provide a complex and elaborate structure for sensing and detecting the biological molecules. Such thickness preliminarily enables the sensor to possess a documented sensitivity towards distinct biomarkers or analytes in the least amount of time possible. This capability is useful in diagnostics of illnesses at early stage, the progression and growth of diseases among other important activities. The roles listed above must have been undertaken by biomedical engineering students, thus demonstrating the importance and versatility of variable-thickness films in different fields due to their interdisciplinary nature. There remains the question of how to apply these coatings—these precisely constructed layers or films—regarding even the most complicated problems in the healthcare sector to advance research, development and application of drugs, regenerative medicine, and diagnostic technologies. With the development of the technology, the scope of the applications of variable-thickness films in biomedical field is also set to grow marking further advancement in the area towards enhanced welfare of patients [73].

4.3 Coatings and Surface Engineering

In automobiles, aerospace, manufacturing industries, and among others, the use of coating of varying thickness has remained compulsory in improving surface properties as well as enhancing the performance of important components (shown in Table 7). These coatings which may be in form of thin films are some of the most important in enhancing the service longevity, performance, and general efficiency of various machinery and equipment [74].

The thin-film coatings, applied at optimum thickness to individual layers or stacks of layers, act as a protective barrier from the environmental conditions seen in the harsh industries. These coatings address frictional forces such as those the automotive engine undergoes, as well as harsh conditions such as high heat and pressure like aerospace engines or turbines by setting up a protective shield thereby reducing part wearage and increasing durability of important components. Furthermore, coatings that are correctly applied for the right threatened thickness promise a superior corrosion protection. Thin films of coatings are used in practical applications, these are effective in marine environment and chemical processing industries, thin-film coatings offer protection to the underlying metal part preventing rusting and guaranteeing durable structures.

Besides, in the aspect of efficiency and high-performing operation, it is equally important that coatings greatly contribute to the minimization of friction and increase of lubrication. Through careful regulation of thickness and creating different forming compositions for these coatings, it becomes possible to reduce losses from friction in mechanical systems and, therefore, increase energy efficiency and extend the service life of machinery. Surface engineering is also the field that grows with new developments in materials science and nanotechnology that create new opportunities for improvement of coating systems. These could range from micron and sub-micron surface coatings with incredibly strong bonding characteristics to component designs involving strake layers, each of which is optimized to provide certain performance characteristics. Recent advances in coating techniques with variable-thickness and thin films have become key tools in the optimized surface engineering of products. By making use of these enhanced surface coatings, the manufacturers are then able to deliver products that offer superior performance, durability, and efficiency under current usage conditions [11, 75] (Fig. 11 and Table 7).

4.4 Semiconductor Manufacturing

The other fundamental factor emerging in the complex web of semiconductor production is the issue of thin-film deposition, which is critical for the development of complex tech formations. This precision is crucial in manufacturing microchips, thin-film transistors, and MEMS devices used in communication gadgets and other

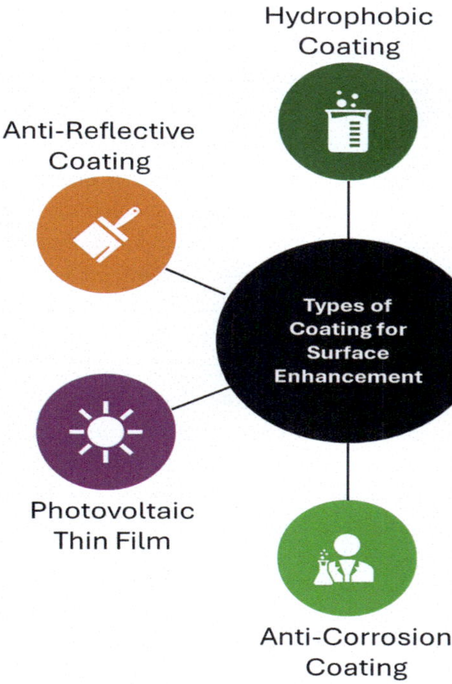

Fig. 11 Show the types of coating for surface enhancement

Table 7 Surface engineering applications of variable-thickness coatings

Coating type	Application	References
Wear-resistant coatings	Automotive components	[76, 77]
Corrosion protection	Aerospace structures	[78]
Lubrication coatings	Industrial machinery	[79, 80]

appliances which require the best electrical and mechanical properties. Semiconductor chips that form the basis of other electronic equipment require the use of films with varying-thickness measurements. The method of fabrication includes sequential deposition and chemical etching of thin layers to create a desired topology of structures. The possibility to accurately control the thickness of the film is used to achieve the needed level of electrical conductivity, insulation, and interconnection which allows to accommodate millions, or even billions of transistors in a single chip [81].

Thin-film transistors (TFTs) that are used in creating liquid crystal displays (LCDs) and in other applications illustrate the significance of the thickness regulation. Film thickness can be controlled to lengths that allow engineers to pre-program the property of TFTs that affect the basic performance including carrier mobility and on/off ratios for display technologies. This level of precision is essential for thin-film transistors in the manufacturing of high-quality screens and active matrix backlights typically found in electronic devices. Film thickness control is keenly even more important, particularly to MEMS devices which consist of both mechanical as well as electrical components in miniature. Some of these minuscule elements are typically employed in sensors, actuators, and microphones and the operation of these devices depends on thin-film deposition and patterning. The maintainability of the specific thickness of the motion picture guarantees the mechanical solidity and electrical competency of the MEMS gadgets, and thus enhancing their usability in different places. When it comes to the incredible maze of semiconductor manufacturing, one comes realizable the usual adage, the secret to control the thickness of the film offers not only technical proficiency but also an element of art. Balancing rigor, in other words interdisciplinary, becomes scale to improving and progressing the science and engineering of materials. The sections that follow will extend the discussion of the recent developments of occurrence control techniques, discussing tendencies, trends, and probable future perspectives of this particular area within the constantly developing sphere of microelectronics engineering. This exploration will make anew and discover other possibilities and innovations in the coming generation of electronic devices and technology [33, 82].

5 Conclusion

Film thickness actually does not represent quantity; it is an unspoken command that determines the fate of the materials and products. It is especially noteworthy to see the essentiality of thickness control because thinning is an important parameter which determines both the uniformity and functionality of an extreme number of materials used in various fields and industries. Whether carrying out a coating to enable the part to perform cohesively or to guarantee adherence for durability or meet strict tolerances, it requires a careful control of the quantity of the coating material being applied in terms of thickness.

The properties of optical, transparency, reflectivity, and color which are influenced by the thickness of the top film are thus a sign of importance in industries that require aesthetic appeal and functionality. The fact that these properties can be controlled with thicknesses means that solutions can be found, new applications can be developed, or existing applications enhanced, especially in areas such as electronics and displays. Having an array of methods of determining film thickness helps in adding more elite methods into the manufacturing process. Techniques like ellipsometry and interferometry are linear, while the X-ray reflectometry and the scanning probe microscopy are nonlinear methods and are suitable for different applications and materials. These techniques are very crucial in quality assurance that checks and implements the thickness of the films produced, to meet the prescribed and acceptable thickness.

As a result, variable-thickness films are introduced as a versatile application in various industries which provides the specific characteristics and performance. In optoelectronics, they also enhance the capture, release, and manipulation of light energy in solar power cells, LEDs, etc. Biomedical engineering expands uses of variable-thickness films including in drug delivery systems for controlled release, cell adhesion, and sensing in tissues engineering scaffolds as well as in biosensors. In automotive and aerospace fields, coatings/surface engineering uses variable-thickness films to improve wear-resistant, corrosive-protective, and lubricative properties. As in the case of fabless semiconductor manufacturing, thickness of films used in manufacturing is a key parameter in integrating complex circuits and micro-electromechanical systems.

When closing the curtain on this film thickness search, two observations can be made: Film thickness is one of the key drivers for tackling quality and functionality issues as well as the cooking up of innovative solutions. The transition from realizing the factors that define the thickness control to the continued dimension including the employment of sophisticated methods of deposition, different tracks of monitoring systems, preparing the substrate, and the masking and patterning all provides an enlightening view of the elaborate interplay of science and technology needed to perfect this critical facet of material engineering.

Thus, it is critical to emphasize that film thickness is not an isolated characteristic but rather a part and parcel of a more extensive storyline—the storyline of material development and technological progress. Manufacturers who understand this factor have the possibilities of bringing their products to another level of quality and performance with the available technologies like measurement techniques, variable-thickness films, etc. Through appreciating the precise details of controlling film thickness, industries can provide the best examples through which future innovations can be built, allowing created materials and products to reflect the desire for precision, performance, and a clear propose.

Reference

1. Gomes, S., Fernandes, J., Anacleto, P., Mendes, P., Gultepe, E., Gracias, D.: Ultra-small energy harvesting microsystem for biomedical applications (2014). https://doi.org/10.1109/EuMC.2014.6986520
2. Zhang, W., et al.: Experimental study on the thickness-dependent hardness of SiO_2 thin films using nanoindentation. Coatings **11**(1), 1–12 (2021). https://doi.org/10.3390/coatings11010023
3. Karki, S., Kim, H., Na, S.-J., Shin, D., Jo, K., Lee, J.: Thin films as an emerging platform for drug delivery. Asian J. Pharm. Sci. **11**(5), 559–574 (2016). https://doi.org/10.1016/j.ajps.2016.05.004
4. Maruddin, F., Malaka, R., Baba, S., Amqam, H., Taufik, M., Sabil, S.: Brightness, elongation and thickness of edible film with caseinate sodium using a type of plasticizer. IOP Conf. Ser. Earth Environ. Sci. **492**(1) (2020). https://doi.org/10.1088/1755-1315/492/1/012043
5. Lee, C., et al.: Ultimate permeation across atomically thin porous graphene. Science (80-) **344**(6181), 289–292 (2014). https://doi.org/10.1126/science.1249097
6. Channa, I.A., Distler, A., Zaiser, M., Brabec, C.J., Egelhaaf, H.: Thin film encapsulation of organic solar cells by direct deposition of polysilazanes from solution. Adv. Energy Mater. **1900598**, 1900598 (2019). https://doi.org/10.1002/aenm.201900598

7. Lipovšek, B., et al.: Detailed optical modelling and light-management of thin-film organic solar cells with consideration of small-area effects. Opt. Express **25**(4), A176 (2017). https://doi.org/10.1364/oe.25.00a176
8. Eustace, I.J.: Some factors affecting oxygen transmission rates of plastic films for vacuum packaging of meat. Int. J. Food Sci. Technol. **16**(1), 73–80 (1981). https://doi.org/10.1111/j.1365-2621.1981.tb00998.x
9. Li, L.H., Xing, T., Chen, Y., Jones, R.: Boron nitride nanosheets for metal protection. Adv. Mater. Interfaces **1**(8), 1–21 (2014). https://doi.org/10.1002/admi.201300132
10. Smith, A.D., et al.: Applying a combination of laboratory X-Ray diffraction and digital image correlation for recording uniaxial stress-strain curves in thin surface layers. Int. J. Mech. Sci. **183**, 105731 (2020). https://doi.org/10.1016/j.ijmecsci.2020.105731
11. Wang, T., Shao, T., Zhao, Q., Shang, H.: Effect of magnetic field on the thickness uniformity of thin film deposited on inner surface. Surf. Coatings Technol. **429**, 127974 (2021). https://doi.org/10.1016/j.surfcoat.2021.127974
12. Solodovnyk, A., et al.: Optimization of solution-processed luminescent down-shifting layers for photovoltaics by customizing organic dye based thick films. Energy Technol. **4**, 385–392 (2016). https://doi.org/10.1002/ente.201500404
13. Solodovnyk, A.: Design and in-depth optical studies of advanced material concepts for luminescent down-shifting coatings for photovoltaics design und tiefgehende optische Studien von hochentwickelten Materialienkonzepten für lumineszierende Konversionsbeschichtungen für (2016)
14. Sudarsanan, B.Y.K.: Significant Precision in Crystal Structural Details : Holly Springs Hydroxyapatite, pp. 1534–1543 (1969). https://doi.org/10.1107/S0567740869004298
15. Sun, Y., Zhao, H., Wang, Z.: Influence of surface unevenness on thickness measurement precision of thin film. **26**, 112–114 (2005)
16. Kwon, J., Takeda, Y., Shiwaku, R., Tokito, S., Cho, K., Jung, S.: Three-dimensional monolithic integration in flexible printed organic transistors. Nat. Commun. **10**(1), 54 (2019). https://doi.org/10.1038/s41467-018-07904-5
17. Myshkin, N., Kovalev, A.: Adhesion and surface forces in polymer tribology—a review. Friction **6**(2), 143–155 (2018). https://doi.org/10.1007/s40544-018-0203-0
18. Alrajhi, A.H., Ahmed, N.M., Al Shafouri, M., Almessiere, M.A., Ahmed Mohammed Al-Ghamdi, A.: Green synthesis of zinc oxide nanoparticles using salvia officials extract. Mater. Sci. Semicond. Process. **125**, 105641 (2021). https://doi.org/10.1016/j.mssp.2020.105641
19. Sun, P., Wang, K., Zhu, H.: Recent developments in graphene-based membranes: structure, mass-transport mechanism and potential applications. Adv. Mater. **28**(12), 2287–2310 (2016). https://doi.org/10.1002/adma.201502595
20. Guglielmi, M., Martucci, A.: Sol-gel nanocomposites. In: Handbook of Sol-Gel Science and Technology: Processing, Characterization and Applications, pp. 3041–3063 (2018). https://doi.org/10.1007/978-3-319-32101-1_100
21. Alaneme, K.K., Adu, O.P., Oke, S.R., Falodun, O.E., Olubambi, P.A.: Densification characteristics, microstructure and wear behaviour of spark plasma sintering processed titanium-niobium pentoxide (Ti–Nb_2O_5) based composites. J. King Saud Univ. Eng. Sci. **34**(4), 282–289 (2022). https://doi.org/10.1016/j.jksues.2020.10.005
22. Sun, W., Liu, J., Chu, H., Dong, B.: Pretreatment and membrane hydrophilic modification to reduce membrane fouling. Membranes **3**(3), 226–241 (2013). https://doi.org/10.3390/membranes3030226
23. Channa, I.A., et al.: Solution processed oxygen and moisture barrier based on glass flakes for encapsulation of organic (opto-) electronic devices. Flex. Print. Electron. **6**(2), 25006 (2021). https://doi.org/10.1088/2058-8585/ac0716
24. Channa, I.A., Ashfaq, J., Siddiqui, M.A., Chandio, A.D., Shar, M.A., Alhazaa, A.: Multi-Shaded Edible Films Based on Gelatin and Starch for the Packaging Applications (2022)
25. Wasa, K., Kitabatake, M., Adachi, H.: CHAPTER 2 thin film materials technology. Thin Film Mater. Technol. 17–69 (2004) [Online]. Available: http://www.sciencedirect.com/science/article/pii/B9780815514831500034

26. Oleinik, I.I., et al.: Diamond CVD growth mechanisms and reaction rates from first-principles. MRS Proc. **616** (2011). https://doi.org/10.1557/PROC-616-123
27. Polok-Rubiniec, M., Dobrzański, L.A., Adamiak, M.: Comparison of the PVD coatings. Arch. Mater. Sci. Eng. **38**(2), 118–125 (2009)
28. Sabzehmeidani, M.M., Gafari, S., Jamali, S., Kazemzad, M.: Concepts, fabrication and applications of MOF thin films in optoelectronics: a review. Appl. Mater. Today **38**, 102153 (2024). https://doi.org/10.1016/j.apmt.2024.102153
29. Liang, F., et al.: A review of thin film electrolytes fabricated by physical vapor deposition for solid oxide fuel cells. Int. J. Hydrogen Energy **47**(87), 36926–36952 (2022). https://doi.org/10.1016/j.ijhydene.2022.08.237
30. Carlsson, J.O., Martin, P.M.: Chemical vapor deposition. In: Handbook of Deposition Technologies for Films and Coatings, pp. 314–363 (2010). https://doi.org/10.1016/B978-0-8155-2031-3.00007-7
31. Liu, Q., Yan, Y., Meng, L., Zhang, Z., Zhou, P.: Influence of airflow disturbance on the uniformity of spin coating film thickness on large area rectangular substrates. Coatings **12**, 1253 (2022). https://doi.org/10.3390/coatings12091253
32. Guo, X., et al.: Studies of pecvd and ozone CVD deposition rate, uniformity and step coverage. MRS Proc. **363** (2011). https://doi.org/10.1557/PROC-363-113
33. Gutpa, J., Shaik, H., Naveen Kumar, K., Sattar, S.A.: PVD techniques proffering avenues for fabrication of porous tungsten oxide (WO3) thin films: a review. Mater. Sci. Semicond. Process. **143**, 106534 (2022). https://doi.org/10.1016/j.mssp.2022.106534
34. Benrezgua, E., et al.: Synthesis and properties of copper doped zinc oxide thin films by sol-gel, spin coating and dipping: a characterization review. J. Mol. Struct. **1267**, 133639 (2022). https://doi.org/10.1016/j.molstruc.2022.133639
35. Ballestas, K., Zapata, J., Ramirez, D.: Quantification of coverage, uniformity and residues for CVD monolayer graphene transfer process based on image analysis. Appl. Surf. Sci. **638**, 158074 (2023). https://doi.org/10.1016/j.apsusc.2023.158074
36. Zhong, Z., Wang, J.: Uniformity and characterisation of PVD aluminium films. Surf. Eng. **21**, 119–124 (2005). https://doi.org/10.1179/174329405X40939
37. Osokin, V., Kurapov, Y., Stel'makh, Y., Shpak, P.: Practice aspects of deposition of nanostructured composite NaCl–Fe films by EB-PVD on a rotating substrate. J. Nano- Electron. Phys. **13**, 5011 (2021). https://doi.org/10.21272/jnep.13(5).05011
38. Konstantiniuk, F., Schiester, M., Tkadletz, M., Czettl, C., Schalk, N.: Annealing activated substrate element diffusion and its influence on the microstructure and mechanical properties of CVD TiN/TiCN coatings. Surf. Coatings Technol. **488**, 131079 (2024). https://doi.org/10.1016/j.surfcoat.2024.131079
39. Hung, T.-H., Liu, P.-J., Wang, C.-Y., Chung, T., Chen, K.-N.: A low-cost passivation for low temperature Cu–Cu bonding using PVD-deposited Cu3N. IEEE J. Electron Dev. Soc. 1 (2023). https://doi.org/10.1109/JEDS.2023.3308368
40. Sivanatham, P., Essakky, M.S.S., Ghosh, S., Mehndiratta, A., Kar, S.: Cost-effectiveness of implementing risk-based cardiovascular disease (CVD) management using updated WHO CVD risk prediction charts in India. PLoS One **18**, e0285542 (2023). https://doi.org/10.1371/journal.pone.0285542
41. Sagalovich, A., Sagalovich, V., Popov, V.: Application of CVD Coatings on the Inner Surfaces of Cooled GTD Blades (2024). https://doi.org/10.20944/preprints202404.1343.v1
42. Vorobyova, M., Biffoli, F., Giurlani, W., Martinuzzi, S., Linser, M., Caneschi, A.: PVD for decorative applications: a review. Materials (Basel) **16** (2023). https://doi.org/10.3390/ma16144919
43. Mulder, K., Lee, S., Chen, W.: A triangular model of fractal growth with application to adsorptive spin-coating of polymers. PLoS ONE **19**, e0298916 (2024). https://doi.org/10.1371/journal.pone.0298916
44. Seul, H.J., Kim, H.G., Park, M.Y., Jeong, J.K.: A solution-processed silicon oxide gate dielectric prepared at a low temperature via ultraviolet irradiation for metal oxide transistors. J. Mater. Chem. C **4**(44), 10486–10493 (2016). https://doi.org/10.1039/c6tc03725a

45. Chen, J.T., et al.: Tuning nanostructure of graphene oxide/polyelectrolyte LbL assemblies by controlling pH of GO suspension to fabricate transparent and super gas barrier films. Nanoscale **5**(19), 9081–9088 (2013). https://doi.org/10.1039/c3nr02845c
46. Alsaad, A.M., Ahmad, A.A., Al Dairy, A.R., Al-anbar, A.S., Al-Bataineh, Q.M.: Spectroscopic characterization of optical and thermal properties of (PMMA-PVA) hybrid thin films doped with SiO_2 nanoparticles. Results Phys. **19**, 103463 (2020). https://doi.org/10.1016/j.rinp.2020.103463
47. Rukchon, C., Nopwinyuwong, A., Trevanich, S., Jinkarn, T., Suppakul, P.: Development of a food spoilage indicator for monitoring freshness of skinless chicken breast. Talanta **130**, 547–554 (2014). https://doi.org/10.1016/j.talanta.2014.07.048
48. Dodero, A., Escher, A., Bertucci, S., Castellano, M., Lova, P.: Intelligent packaging for real-time monitoring of food-quality: Current and future developments. Appl. Sci. 11(8) (2021). https://doi.org/10.3390/app11083532
49. Li, Z., et al.: Review of thin-film resistor sensors: exploring materials, classification, and preparation techniques. Chem. Eng. J. **477**, 147029 (2023). https://doi.org/10.1016/j.cej.2023.147029
50. Dub, S., Ramsey, B.J., Harrison, D.: Low-cost patterning of thin film. Proc. Inst. Mech. Eng. Part B-journal Eng. Manuf. Proc. Inst. Mech. Eng. B-J Eng. Ma. **220**, 1555–1558 (2006). https://doi.org/10.1243/09544054JEM523
51. Cox, L.J.: Ellipsometry and polarized light. Opt. Acta Int. J. Opt. **25**(3), 270–271 (1978). https://doi.org/10.1080/713819761
52. Güler, E.N., Distler, A., Basu, R., Brabec, C.J.: Ultraflexible organic solar cells organic solar cells
53. Smijs, T.G., Pavel, S.: Titanium dioxide and zinc oxide nanoparticles in sunscreens: focus on their safety and effectiveness. Nanotechnol. Sci. Appl. **4**, 95–112 (2011). https://doi.org/10.2147/NSA.S19419
54. Uddin, A., Upama, M.B., Yi, H., Duan, L.: Encapsulation of organic and perovskite solar cells: a review. Coatings **9**(2) (2019). https://doi.org/10.3390/coatings9020065
55. Chatterjee, S., Bhardwaj, R.: A short review on optical interferometry techniques for characterization of a thin liquid film on a solid surface. Sādhanā **48** (2023). https://doi.org/10.1007/s12046-023-02091-6
56. Koppel, L., Parobek, L.: Thin-film metrology by rapid x-ray reflectometry. AIP Conf. Proc. **449** (1998). https://doi.org/10.1063/1.56834
57. Nzereogu, P.U., Omah, A.D., Ezema, F.I., Iwuoha, E.I., Nwanya, A.C.: Anode materials for lithium-ion batteries: a review. Appl. Surf. Sci. Adv. **9**, 100233 (2022). https://doi.org/10.1016/j.apsadv.2022.100233
58. Sick, N., Krätzig, O., Eshetu, G.G., Figgemeier, E.: A review of the publication and patent landscape of anode materials for lithium ion batteries. J. Energy Storage **43**, 103231 (2021). https://doi.org/10.1016/j.est.2021.103231
59. Ye, M., Biesold, G.M., Zhang, M., Wang, W., Bai, T., Lin, Z.: Multifunctional quantum dot materials for perovskite solar cells: charge transport, efficiency and stability. Nano Today **40**, 101286 (2021). https://doi.org/10.1016/j.nantod.2021.101286
60. Channa, I.A., Ashfaq, J., Gilani, S.J., Shah, A.A., Chandio, A.D., Nasser, M.: UV Blocking and Oxygen Barrier Coatings Based on Polyvinyl Alcohol and Zinc Oxide Nanoparticles for Packaging Applications (2022)
61. Channa, I.A., et al.: Sustainable and eco-friendly packaging films based on poly (Vinyl Alcohol) and glass flakes. Membranes 12(7) (2022). https://doi.org/10.3390/membranes12070701
62. Tebaldi, M.L., Belardi, R.M., Montoro, S.R.: Chapter 8—Polymers with nano-encapsulated functional polymers: encapsulated phase change materials. In: Thomas, S., Shanks, R., S.B.T.-D., A. of N.P.B., Chandrasekharakurup, N.S. (eds.) Micro and Nano Technologies, pp. 155–169. William Andrew Publishing, Boston (2016). https://doi.org/10.1016/B978-0-323-39408-6.00019-4
63. Laxmiprasanna, N., et al.: A review on nano composite polymer electrolytes for high-performance batteries. Mater. Today Proc. (2022). https://doi.org/10.1016/j.matpr.2022.07.291

64. Browning, N.D., Chisholm, M.F., Pennycook, S.J.: Atomic-resolution chemical analysis using a scanning transmission electron microscope. Nature **366**(6451), 143–146 (1993). https://doi.org/10.1038/366143a0
65. Gopal Krishna, B., Tiwari, S.: Chapter 19—Bioinspired solar cells: contribution of biology to light harvesting systems. In: Fraga, M.A., Amos, D., Sonmezoglu, S., V.B.T.-S.M.S. for Subramaniam, S.E.T. (eds.). Solar Cell Engineering, pp. 593–632. Elsevier (2021). https://doi.org/10.1016/B978-0-12-821592-0.00006-6
66. Greene, J.E.: Review Article: Tracing the recorded history of thin-film sputter deposition: From the 1800s to 2017. J. Vac. Sci. Technol. Vac. Surf. Film **35**(5) (2017). https://doi.org/10.1116/1.4998940
67. Huang, Y., Hsiang, E.-L., Deng, M.-Y., Wu, S.-T.: Mini-LED, Micro-LED and OLED displays: present status and future perspectives. Light Sci. Appl. **9**(1), 105 (2020). https://doi.org/10.1038/s41377-020-0341-9
68. Makhdoom, M.A., Sgobba, V., Channa, I.A., Ghewins, N.: Producing oxide free silicon nanocrystals—a novel and benign approach. Optik (Stuttg) **246**, 167789 (2021). https://doi.org/10.1016/j.ijleo.2021.167789
69. Channa, I.A., Distler, A., Brabec, C.J., Egelhaaf, H.-J.: Solution-Coated Barriers for Organic Electronics. LTD (2021). https://doi.org/10.1016/b978-0-12-818890-3.00009-6
70. Zeumault, A., Mendez, J., Brewer, J.: Innovations in thin-film electronics for the new generation of displays. J. Soc. Inf. Disp. **32**, 121–135 (2024). https://doi.org/10.1002/jsid.1274
71. Prestat, M., Thierry, D.: Corrosion of titanium under simulated inflammation conditions: clinical context and in vitro investigations. Acta Biomater. **136**, 72–87 (2021). https://doi.org/10.1016/j.actbio.2021.10.002
72. Thanka Rajan, S., Subramanian, B., Arockiarajan, A.: A comprehensive review on biocompatible thin films for biomedical application. Ceram. Int. **48**(4), 4377–4400 (2022). https://doi.org/10.1016/j.ceramint.2021.10.243
73. Kong, I.: Chapter 7—polymers with nano-encapsulated functional polymers. In: Thomas, S., Shanks, R., S.B.T.-D., A. of N. P. B., Chandrasekharakurup, N.S. (eds.) Micro and Nano Technologies. William Andrew Publishing, Boston, pp. 125–154 (2016). https://doi.org/10.1016/B978-0-323-39408-6.00007-8
74. Rahaman, M.S., Thérien-Aubin, H., Ben-Sasson, M., Ober, C.K., Nielsen, M., Elimelech, M.: Control of biofouling on reverse osmosis polyamide membranes modified with biocidal nanoparticles and antifouling polymer brushes. J. Mater. Chem. B **2**(12), 1724–1732 (2014). https://doi.org/10.1039/c3tb21681k
75. Ng, L.Y., Mohammad, A.W., Ng, C.Y.: A review on nanofiltration membrane fabrication and modification using polyelectrolytes: Effective ways to develop membrane selective barriers and rejection capability. Adv. Colloid Interface Sci. **197–198**, 85–107 (2013). https://doi.org/10.1016/j.cis.2013.04.004
76. Yadav, S., Kumar, A., Paramesh, T., Sunita, K.: A review on enhancement of wear resistance properties of titanium alloy using nano-composite coating. IOP Conf. Ser. Mater. Sci. Eng. **455**(1) (2018). https://doi.org/10.1088/1757-899X/455/1/012120
77. Liu, X., Chu, P.K., Ding, C.: Surface modification of titanium, titanium alloys, and related materials for biomedical applications. Mater. Sci. Eng. R. Rep. **47**(3–4), 49–121 (2004). https://doi.org/10.1016/j.mser.2004.11.001
78. Höhlein, S., König-Haagen, A., Brüggemann, D.: Macro-encapsulation of inorganic Phase-Change Materials (PCM) in metal capsules. Mater. (Basel, Switzerland) **11**(9) (2018). https://doi.org/10.3390/ma11091752
79. Li Sip, Y.Y., et al.: Slippery lubricant-infused silica nanoparticulate film processing for anti-biofouling applications. J. Appl. Biomater. Funct. Mater. **21**, 22808000231184690 (2023). https://doi.org/10.1177/22808000231184688
80. Pernice, G.: The evolution and application development of high-performance, thin-film. Dry-Lubricant Coatings (2015). https://doi.org/10.4043/26307-MS
81. Salvatore, G.A., et al.: Wafer-scale design of lightweight and transparent electronics that wraps around hairs. Nat. Commun. **5**(1), 2982 (2014). https://doi.org/10.1038/ncomms3982

82. Park, J.S., Maeng, W.-J., Kim, H.-S., Park, J.-S.: Review of recent developments in amorphous oxide semiconductor thin-film transistor devices. Thin Solid Films **520**(6), 1679–1693 (2012). https://doi.org/10.1016/j.tsf.2011.07.018

Microstructure and Composition

Muhammad Ali Shar and Abdulaziz Alhazaa

Abstract The chapter discusses various facets of thin films and its composition with relation to microstructure and how such factors determine its features and characteristics. Thin films which are used in applications such as solar cells, medicinal implants, and protective and semiconductor devices have distinct characteristics due to the inherent structural features as well as since they are nanoscale in thickness. There are characteristics such as mechanical strength, electrical and thermal conductivity, optical transparency, and chemical stability among others which play a very large role in their usage and these are dictated by the composition of the material. The former techniques like XPS, EDS, XRD, and SEM/TEM behave well in the determination of the composition of the material but the techniques such as XRD, SEM, TEM, and AFM are used for the analysis of microstructure of the particular material. This chapter explores several ways in which compositional changes can increase many desirable and favorable properties, such as increasing the optical transmission/absorption by modifying the material's composition; strengthening the tensile strength through alloying; modifying the material's electrical conductivity by doping. By comprehending, modifying, and synchronizing such variables, thin films can be optimized for specific applications and pave the innovations in electronics, optoelectronics, energy storage, and much more and thereby bridging the scientific advancement of material science and technology.

List of Abbreviations

SEM Scanning Electron Microscopy
TEM Transmission Electron Microscopy
AFM Atomic Force Microscopy
XPS X-ray Photoelectron Spectroscopy
EDS Energy-Dispersive X-ray Spectroscopy
AES Auger Electron Spectroscopy

M. A. Shar (✉) · A. Alhazaa
King Abdullah Institute for Nanotechnology, King Saud University, Riyadh 11451, Saudi Arabia
e-mail: mashar@ksu.edu.sa

A. D. Chandio and I. A. Channa (eds.), *Thin Films and Coatings*, Engineering Materials,
https://doi.org/10.1007/978-981-96-0005-2_4

SIMS	Secondary Ion Mass Spectrometry
GD-OES	Glow Discharge Optical Emission Spectroscopy
RBS	Rutherford Backscattering Spectrometry
CTE	Coefficient of Thermal Expansion
ALD	Atomic Layer Deposition
CVD	Chemical Vapor Deposition
PVD	Physical Vapor Deposition
LEDs	Light-Emitting Diodes
OLED	Organic Light-Emitting Diode
TFT	Thin-Film Transistor

1 Introduction

Thin films occupying a rather wide spectrum in the branch of material science are seen as nature's miracle within a drop of honey in terms of their features and opportunities. Thin films are severely applicable in many fields such as semiconductor devices, solar cells, protective coatings, and biomedical implants. This also explains why thin films are significant playing a big role in today's advanced technology. Given that the applications of thin films are varied, their microstructure and composition should be investigated properly in order to understand their potential.

This means that the thin films' special properties that many a time arise from their dimensions that are in nanoscales and their distinct structures make up the first point on the thin-film compass. Thin films therefore exhibit properties that are different from bulk materials due to surface phenomena, and effects resulting from small dimensions. Microstructure and composition also influence the functional characteristics of thin films, for instance, the tunable band of a semiconductor thin film or the enhanced electrical conductivity of the metallic thin film. When we begin with this journey through the world of thin films, it becomes even more evident that the approach to transform the effective potential of these items is through appreciation of the part being played by composition and the operation of microstructure [1].

The ability to pressure and detail the microstructure is the touchstone of thin-film science. Windows of microstructural characterization techniques that can be used to observe the internal structure and the crystalline phase distribution of thin films are the grain boundaries and the flaws. Thin-film characterization technique is characterized by an X-ray diffraction (XRD) which can offer crystallographic information. Information on the grain size, shape, and distribution can be obtained from scanning electron microscopy (SEM) and transmission electron microscopy (TEM), which provides elucidated morphological characteristics of thin films. The information about surface features and roughness is taken by Atomic Force Microscopy (AFM) that is fundamental for understanding thin-film formation. Thus, by synchronizing different characterization techniques, the researchers are likely to obtain more information on thin films' microstructure and bring about original design approaches. Actually, this is the culmination of the process of searching for the answer [2].

However, composition is a major factor that influences the thin-film characteristics across the four characterization techniques. Thickness of the thin films mainly influences the behavior and is dictated by elements that make up materials and have mechanical, chemical, optical, and electrical characteristics. The last correlation demonstrated that doping concentration and elemental composition are an excellent tool for bandgap engineering and regulating conductivity and carrier mobility in thin films of semiconductors. In the same manner, alloying elements are used to regulate the catalytic properties, mechanical hardness, and resistance to corrosion as used in thin films of metals. The ability to change output of a material slightly to enhance magnetic features or alter its phases of reflectance of light makes it highly useful for developing more innovative uses for the material. It is essential to understand how such relationships between composition and properties in order to create thin films that will be equipped with useful properties ensuring the desired and accurate performance in specific application.

Thin films are needed for a number of applications because thin-film structures themselves are reputed for their inherent benefits inherent to thin-film structures and for the ability of thin-film structures for versatile usage. It is important to note that, thin films are used in many fields cut across electronics, energy, optics, biomedicines, signal, and doping flexibility. This indicates that like in the case of bulk deposition processes, thin films can be tailored to exactly correlate these requirements depending on the precursor materials, process parameters, and the deposition techniques used. There is nothing like designing and creating thin films or it's simply revolutionary and paves the way for new possibilities and progress on new territories: From the compatibility of the body tissue to implant that would be coated with the appropriate layers to enhance the effectiveness of the solar cells by designing the appropriate layers of the absorber. Thus, in terms of usefulness and versatility, actual potential in thin films is yet to be seen, which is because this setting only serves as a playground for constant material science development.

More detail and emphasis on the thin films and their contrite as well as the relationships between all the factors governing the performance and characteristics of the thin films are described in the next chapters. This is a sort of brief survey of the major focal areas from the fundamentals to the innovations that can prove really enlightening at every turn of the journey and which will provide the status not only of the historic but also the current and the future thin-film science and technology. Well, let us now get inside the structure and take a brief look at what can be seen as this exciting trip starts now. This is where creative and imaginative spirits are fostered, and it goes without saying that in this area, there is truly little that cannot be accomplished.

2 Characterization Technique for Microstructure

The structure of thin layers is characterized by the methods that were briefly described above, and each of them allows having a different view on some aspects of the atomic structure, orientation, topography, and defects that constitute the layer. Among the methods that are most frequently employed are:

2.1 Scanning Electron Microscopy (Sem)

Principle

SEM works by projecting a beam of high-energy electrons onto a specimen and the incident beam brings about a variety of interactions that result in emission of secondary electrons, backscattered electrons, and characteristic X-rays. These emissions are received by the detectors and are then used to enhance images and also to determine the composition. Electrons are focused and directed in a systematic and orderly manner across the face of the sample in a raster fashion for fine detail and data about the surface contours and chemistry. These signals are then made to correspond with the movement of the beam to make images of the objects with clear structural detail (shown in Fig. 1). SEM is highly utilized for such a purpose since it offers enhanced topographical imaging in addition to the elemental mapping of various specimens.

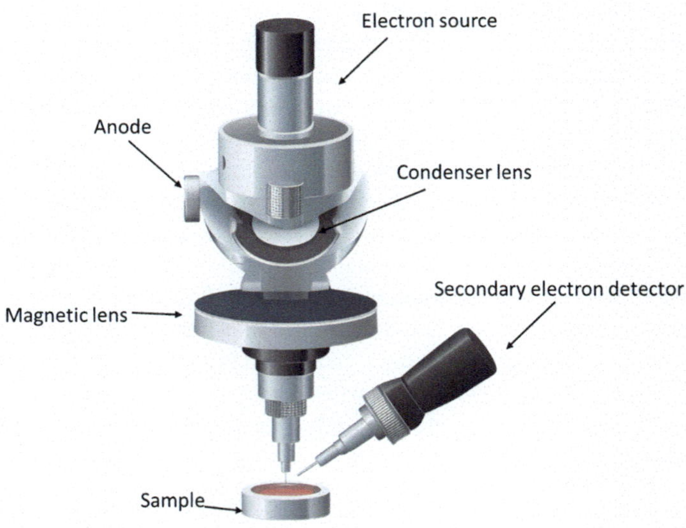

Fig. 1 **a** SEM ray diagram with SEM equipment

Procedure

The first and most important process in SEM investigation of thin films is sample preparation, and in this arrangement, a thin film sample is deposited on an adequate substrate. After that, the sample is placed on the SEM device and then mounted with a thin layer of conductive material on a conductive holder for further analysis. A rather intricate equipment used in SEM is an electron cannon, electromagnetic lenses, and a detector to pick signals that are emitted.

Imaging uses the electron beam collected from the sample surface and, in most cases, it is conducted with a resolution of a few nanometers in size. As the beam scans by means of some type of raster across the surface of the sample, it interacts with the latter's atoms. A SEM detection system manages to capture multiple signals from these interaction events, such as backscattered electrons, secondary electrons, and characteristic X-rays.

In this context, SEM is useful for studying thin film microstructure and gathers more information on the surface of such materials at the nanoscale. SEM offers researchers a phenomenally better view in terms of the morphology, defects, and interconnectivity of the thin-film materials by providing both the topographical and chemical images [3].

Here is a case study, in which ZnS thin films doped with varying concentrations of Fe were exposed to RF magnetron sputtering on quartz glass substrates, and they were then annealed at 600 °C in sulfur vapor by Xinpeng et al. The undoped film has relatively small grains, as indicated by the SEM data in Fig. 2, which indicates low crystallization with uniformly sized ZnS grains. The grains enlarge and the particles align and disperse over the surface as more Ag is applied. As the Ag content rises, tiny pores begin to form and expand [4].

2.2 *Transmission Electron Microscopy (Tem)*

Principle

Transmission Electron Microscopy (TEM) is a technique that involves guiding an electron beam through a very minute sample; the technique provides significantly detailed images that are near atomic scale. The functioning of the TEM is based on two factors which are the reaction of the atoms present in the sample under investigation to the incoming electron stream. A source fires electrons through the matter and the electrons form an image on a phosphorescent screen or other detection measures as the electrons pass through and refract.

Procedure

Before TEM thin film analysis, the actual thin film to be investigated is first used to prepare a placement of the sample on a thin TEM grid, commonly made of gold or copper. Afterwards, thin foil preparation is carried out by doing ion milling or FIB

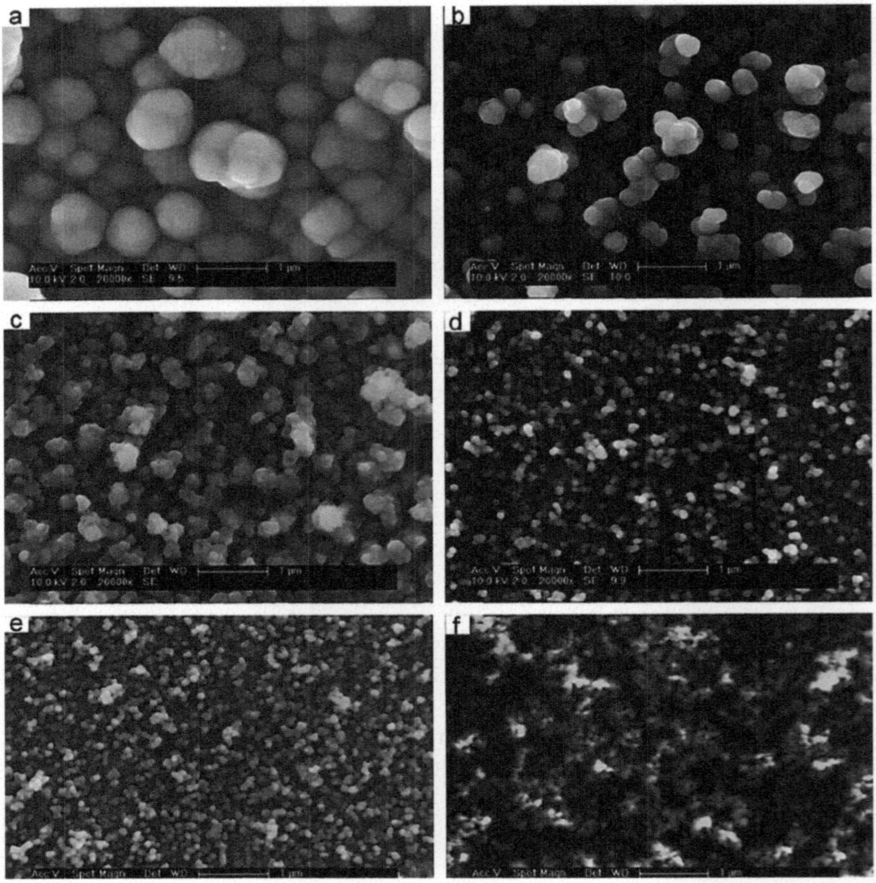

Fig. 2 SEM images of ZnS films doped with different content of Fe

milling to make the thin-film sample electron transparent. Lastly, after preparation, the TEM sample is crushed and located in device and is willing for use.

The thin-film sample is exposed to a beam of electrons and these begin to interact with the various atoms in the sample. Another device, a fluorescent screen or detector, then intercepts this transmitted electron and, by the intensity of the electrons, creates an image. Occasionally, enabling microstructure such as atomic plan, grain boundaries as well as defects within a crystal lattice depending on the beam energy and the image forming could be created using TEM (shown in Fig. 3).

In conclusion, it has been established that TEM can be applied aiding in the investigation of thin films to unravel the atomic arrangement and profile of the films. So, the high-resolution images and the quantitative data provided by TEM make it possible to study the morphology, defects, and crystal structures of the thin-film materials in much detail [2].

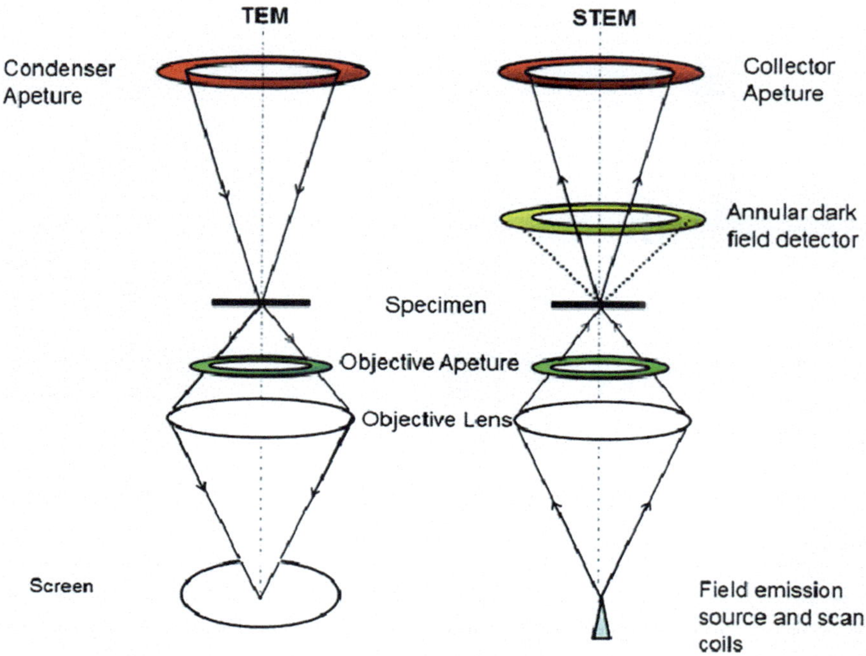

Fig. 3 TEM versus SEM ray diagram CC BY 4.0

The real-life example of TEM in which V. B. Ranmale investigated the effects of doping bismuth ferrite ($BiFeO_3$) with nickel (Ni) using the spray pyrolysis process to create thin films. Strong structural integrity is indicated by the TEM investigation in Fig. 2e–h, which reveals that the lattice fringe spacing in nickel-doped $BiFeO_3$ is around 0.284 nm. The pictures demonstrate minute atomic alterations brought about by nickel doping, verifying nickel's existence and impact on the crystal lattice [5] (Fig. 4).

2.3 Optical Microscopy

Principle

Optical microscopy refers to imaging techniques which use visible light to make a sample larger and reveal its structure. The process whereby light affects a sample in optical microscopy by illuminating it and either transmitting through or reflecting off as a lens forms an image is known as the optical microscopy theory. Reflecting on the variety of both industrial and scientific processes, optical microscopes are indispensable tools, for they can provide a wealth of knowledge regarding the shape, surface properties, and distribution density of materials.

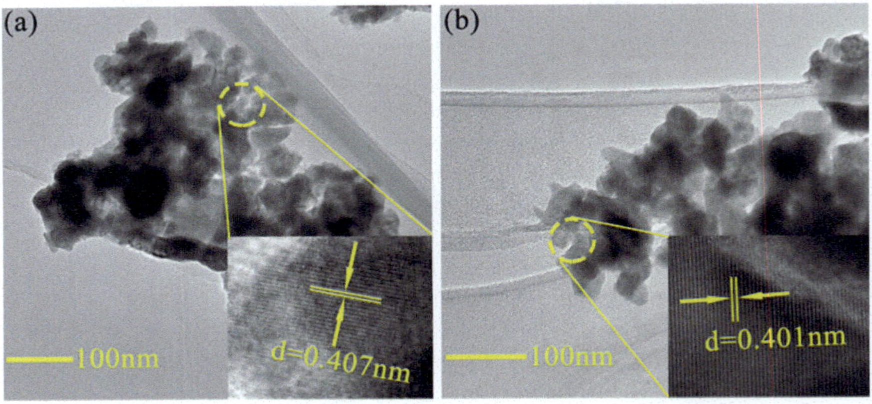

Fig. 4 TEM images of as-synthesized thin films of $BiFeO_3$ with different magnifications CC BY 4.0

Procedure

First, the sample needs to be prepared by making a thin film coating method to the right substrate for further optical microscopy of a thin film. Following that the prepared sample is placed on the stage in an optical microscope and it is left there so that they make observations. An optical microscope comprises a light source through which light passes through the condenser lens and the objective lens; eyepiece or camera for image viewing.

The objective lens times downlight from the source on the sample surface having passed through the condenser lens during the formation of an image. The sample shown here interacts with light and either reflected off the surface of the film sample or can be transmitted through the thin layer. The objective lens collects light that either passes through or is reflected from the sample and then brings this light through to form an image of the sample onto the eyepiece or camera.

In addition to resolving power, various modes of optical microscopes can be employed, such as polarized light microscopy, dark-field microscopy, and bright-field microscopy among others due to the fact that they all provide certain characteristics of the microstructure of the specimen under analysis. Dark-field microscopy by only leading the scattered light from the material is of high contrast, while bright-field microscopy is applied with the high-contrast samples with homogenous densities. Based on the polarization of light that passes through the sample, thin crystal films and other birefringent polarizing microscopes can be observed.

Overall, optical microscopy is a useful method for determining the microstructure of thin films and their surface features. Optical microscopes facilitate easy and precise analysis of the morphology, flaws, and interactions of thin-film materials by offering high-resolution pictures and contrast enhancement capabilities [6].

2.4 Atomic Force Microscopy

Principle

Atomic force microscopy, which falls under the scanning probe microscopy, is a powerful tool for surface characterization and imaging of materials with low surface roughness in the nanoscale. The theoretical base of AFM is in contact between a sharp tip and the surface of a sample. During the journey of the tip across the surface, a number of forces come into play, which include chemical binding forces, transformer van der Waal forces, and electrostatic forces. AFM utilizes a very sharp tip positioned at the end of a flexible cantilever to record surface topographies at the atomic level by detecting the bending of the cantilever in response to the surface of the sample.

Procedure

The first step in AFM microscopic examination of the thin film is the sample preparation where the thin film is placed on a suitable substrate. Then the sample is placed on the AFM stage and shifted to the required position of further analysis. The AFM equipment includes a sharp tip mounted on a cantilever to scan the sample surface Integral Globe, 2008. By using laser beam and photodetector equipment and allowing cantilever to vibrate and aiming at observing how it affects the tip and sample, the tip–sample interaction is determined.

The AFM tip is positioned very close to the sample during imaging, in most cases, it is within several nanometers. The structure bends owing to the forces exerted from the surface of the sample on the tip or vice versa while it scans the surface in a matrix manner. The method used to ensure topographic imaging of the sample surface is the measurement of variations in the position of the laser beam on the photodetector in order to detect this change in deflection.

The information about mechanical, electrical, or even magnetic characteristics of the sample can be obtained with the help of AFM in addition to topography (shown in Fig. 5). For example, mechanical properties such as stiffness of thin layers and adhesion can be quantified by AFM which employs tools like force spectroscopy or contact mode pictures. Similarly, magnetic force microscopy (MFM) and conductive AFM can give information about magnetic and electrical parameters of thin films, respectively [3].

To understand more about AFM, let's discuss the example, Daniela Predoi created thin films of hydroxyapatite (20CrHAp) doped with chromium and coated them on silicon substrate using the spin coating technique. According to AFM measurements, the surface of the CrHAp coating is remarkably consistent and smooth with very little variation in surface adhesion and porosity. The antifungal investigations show that the 20CrHAp thin films successfully stop Candida albicans from forming fungal biofilms. Less fungal cells stuck to the covering over time; clusters formed initially, but after 48 and 72 h, they disintegrated. These results demonstrate the coating's potent antifungal qualities and capacity to inhibit the growth of biofilms [7].

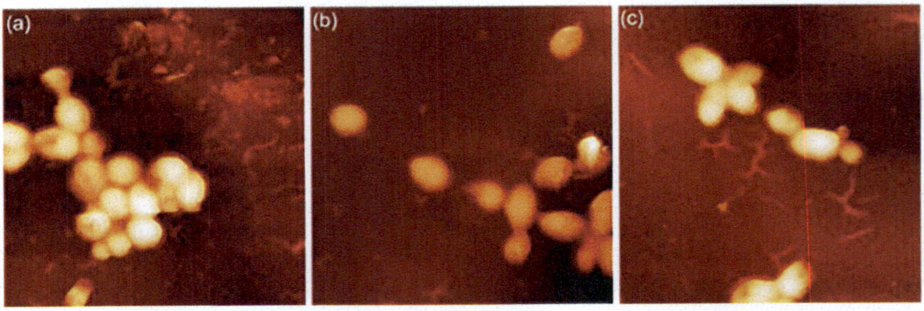

Fig. 5 AFM topography of 20CrHAp thin films incubated with *Candida albicans* ATCC 10,231 fungal cells for 24 (**a**), 48 (**b**) and 72 h (**c**). CC BY 4.0

3 Characterization Technique for Composition

A variety of methods are used to ascertain the composition of thin films, each with special advantages and capabilities for examining the material's chemical composition and elemental components. Among the methods that are frequently employed are:

3.1 X-Ray Photoelectron Spectroscopy (XPS)

Principle

The electron surface analysis is also known as ESCA or X-ray Electron Spectroscopy for Chemical Analysis and XPS or Photoelectron Spectroscopy where it is used for determination of oxidation state of substances, their chemical structure, and the element of which the substance is made. XPS depends strongly on the probe's photoelectric effect; here the X-ray is used to ionize or stimulate the core electrons of the sample material. However, when a photon is capable of imparting sufficient energy to a sample and expelling an electron from an atom from the inner shell such as the K or L shell, it is defined as a core-level photoelectron. The energy and the number of these emitted photoelectrons allow for counting or measuring in order for one to determine the nature of the sample, that is, the elements that may exist in the sample and the state they are in.

Procedure

Thin-film sampling for the XPS process means that preparation of the thin film is done on the right substrate to enable characterization. Subsequently, the ready sample is transferred into vacuum chamber of the XPS apparatus to irradiate it with a monochromatic X-ray beam. These atoms contain core electrons in this sample and, when struck by X-rays, the energy imparted to these atoms is enough to knock out what are known as photoelectrons from this sample surface.

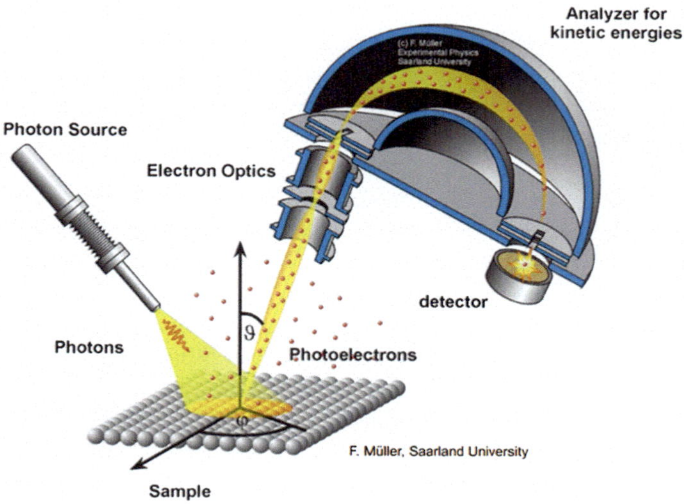

Fig. 6 **a** X-ray photoelectron spectroscopy ray diagram CC by 4.0

Thin-film characterization for XPS analysis entails positioning the thin film in the right substrate where it rightly belongs. The prepared sample is then inserted into the XPS device having a monochromatized X-ray source that serves to irradiate the material. They are uniquely suited to removing enough energy within core electrons for certain atomic constituents of the sample and evicting these photoelectrons from the surface (shown in Fig. 6).

In addition, XPS could provide information about the chemical environment as well as the type of oxidation state of the elements which define the thin film. Given that different chemical species in the sample have different oxidation states and chemical bonding geometries connected to the photoelectrons, it is possible to distinguish between many chemical species within the samples [8].

3.2 Energy-Dispersive X-Ray Spectroscopy

Principle

Energy-dispersive X-ray spectroscopy is an analyzing tool that operates through an interaction of an electron beam used in the generation of particular X-rays originating from the sample. This makes it possible to quantify the amount of elements present in all the materials found on the location with a high level of certainty. EDS uses both a sample's energy and its emitted X-ray fluorescence when incident electrons knock out the target material's atomic shells and the emitted X-ray uniquely identifies the elements in the sample.

Procedure

For preparing the thin film for the EDS analysis, further actions should be fulfilled: issues relating to the deposition of the thin film onto the appropriate substrate. Afterward, the sample is mounted on the SEM stage which has an EDS detector connected to it.

Once the sample is prepared, it is mounted on the stage of the SEMs which has an EDS detector. Electrons are finely focused and are allowed to bombarding a sample, some of the atoms in the sample and the electrons interact, while the rest of the atoms emit X-rays at certain specific energy levels.

The energy and intensity of the emitted X-rays are measured and monitored with the help of EDS detector and, in the detector, the energy and intensity of emitted X-rays, also called the intensity of the X-ray spectra, are gotten. This technique gives all the elemental constituents of the thin film whether directly or indirectly through determining the disperse spectrum of this line of energies that various elements in a given sample radiate at. The intensity of the outgoing X-ray photons as well as the type of signal collected can also be used to determine the identity and concentration of the particular element in the sample.

EDS can provide information where different elements are located inside the thin film, and whether there are other phases or particles there. EDS has the capacity to demonstrate the signal of the elements on the surface in order to express homogeneity and distribution of the thin film material.

Since properties of EDS provide elemental information in a quantitative manner about the presence and distribution of elements with high sensitively and resolution in space, it is a more suitable technique of analysis for thin films in general [9].

In the above case study, SEM analysis of ZnS films with Ag was discussed. Its EDS analysis of that sample The Ag-doped ZnS thin film's EDS spectrum reveals the presence of B, Zn, S, originating from the quartz substrate. Figure 7 shows that the doped film has a larger B content than the goal, which is probably because Zn, S, and B sputter at different rates. In comparison to the undoped film, the doped films' S/(Zn + B) ratios are closer to 1, indicating improved composition [4].

3.3 Auger Electron Spectroscopy (AES)

Principle

AES is also a type of surface analytical method that allows identifying the elemental and chemical composition of the sample, as well as their bonding state. The AES idea derives from the Auger effect, which means that an atom moves from two electronic energy states and gives out an Auger electron. In AES, an intense and narrow probe beam of electrons is employed to knock out core level electrons from the surface atoms of the sample. When subsequent electrons from the higher-energy energy levels drop down into the vacancies in the core levels, other race electrons, known as Auger electrons, are emitted. The energy of these Auger electrons can be employed

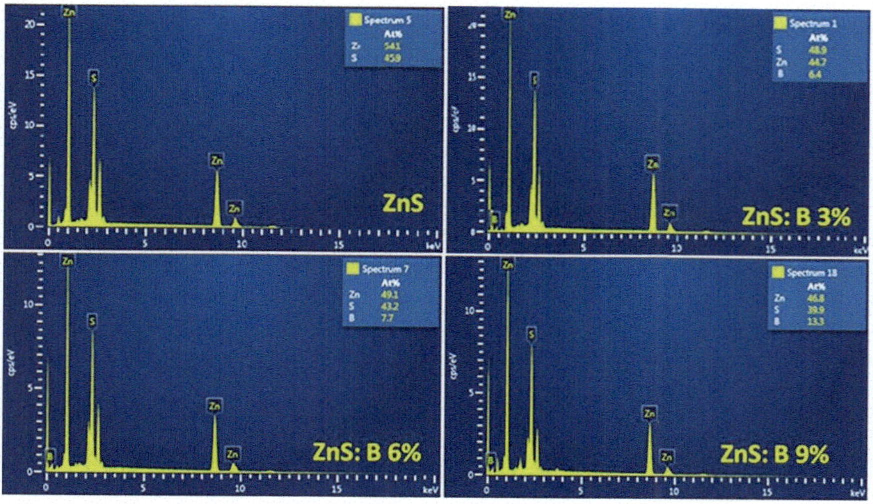

Fig. 7 EDS analysis of ZnS films doped with B film CC BY-NC-ND

to spectroscopically differentiate the various elements present in the sample and to delineate chemical status of the atoms present in the sample.

Procedure

The sample of thin film is first deposited on the appropriate substrate for the purpose of analyzing the thin films through the help of Auger Electron Spectroscopy technique. The prepared sample is then transferred to the vacuum chamber of the AES instrument and irradiated by an electron source referred to as electron canon. The energy of the electrons, thus, is capable of penetrating through the surface layers of the sample and knocking out the core-level electrons from atoms of the material.

The Auger electrons are emitted and are analyzed for kinetic energy and intensity through an electron analyzer device. Since each element in the sample generates the characteristic Auger electron peaks with specific energy signature, it become possible both to detect and to determine concentration of the elements incorporated in the thin film. The energy of the detected Auger electrons is analyzed, and the results are compared to a reference information base, which allows for identifying the elemental composition and chemical state of the analyzed material.

AES can be again useful in yielding information concerning the surface chemistry of thin films, the type of chemical bond involved in thin films and the existence of pollutants or contaminants. Through analysis of Auger electron spectra, it may be possible to find out the content of the definite chemical kinds in the accepted sample and their identification. Taking all this into consideration, AES give comprehensive detail of the chemical composition and surface morphology of a material and is a time-saving method used in analyzing the chemical properties and the elements of formation of very thin layers—Thin films.

3.4 Secondary Ion Mass Spectroscopy

Principle

Specific chemical elements present at the surface and the subsurface of the material can also be determined by the use of Surface Sensitive Ion Mass Spectrometry (SIMS). In as much as SIMS, the general concept is such that distinct sample surface atoms are sputter-blasted by a basic primary ion beam, thereby discharging secondary ions from the surface of the sample. SIMS uses and primary ion beam which is made up of monoatomic or polyatomic ions, and which is directed at the sample during the SIMS experiment. A mass spectrometer evaluates the secondary ions produced after the primary ions cause other ions off the sample surface.

Procedure

Thin-film SIMS requires sample preparation and the sample is prepared for thin-film SIMS analysis by depositing the thin film onto an appropriate substrate. Following that, the prepared sample is inserted into the SIMS instrument and exposed to the primary ion beam as well as placed and minutes are allowed for stabilization under ultra-high vacuum conditions. By the bombardment of main ions, atoms are knocked off from the surface of the sample and the secondary ions are produced.

A mass spectrometer that differentiates the ions on base of the mass to charge ratio or m/z is used to move the released secondaries from the sample surface and simultaneously analyze them. The isotopic distribution and the elemental abundance of the sample can be identified from the respective mass spectrum of the detected ions. Since a thin film sample is made up of several chemical elements, researchers are able to determine the elemental composition of the thin film material using the intensity of certain ions and link them to specific elements (Fig. 8).

Furthermore, it can provide depth profile using Secondary Ion Mass Spectrometry (SIMS). In this method, thin layers of material are gradually sputtered from the exposed surface with the help of the primary ion beam being scanned across the sample. Thus, due to the identification of the strength of certain ions depending on

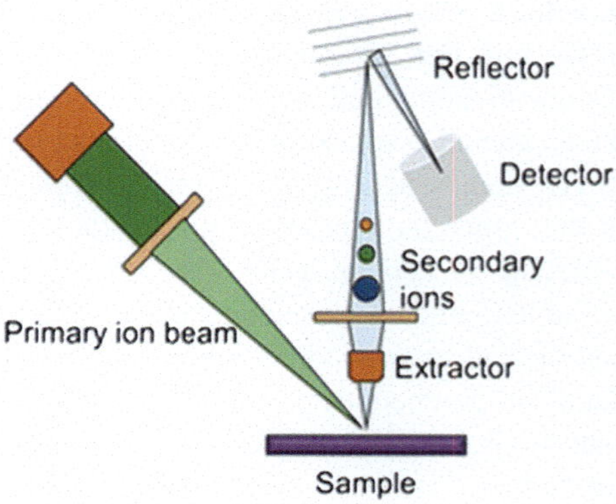

Fig. 8 SIMS ray diagram CC by 4.0

the depth, it is possible to gain a comprehension regarding the distribution of the elements in the thin film and thicknesses of the layers [10].

3.5 Glow Discharge Optical Emission Spectroscopy (GD-OES)

Principle

GD-OES stands for glow discharge optical emission spectroscopy, an analytical technique used for determining a material's elemental content and composition probed. GD-OES is a method where a low-pressure glow discharge plasma is produced promoting excitation of atoms in the sample and leading to emission of characteristic optical wavelengths. The low-pressure inert gas such as argon and helium is in the vacuum chamber with the sample in GD-OES. In the case of the sample, it is subjected to either DC or RF voltage in order to ionize the gas which forms a glow discharge plasma. An optical spectrometer measures and identifies the optical emissions spectra of the sample's excitable atoms due to their interaction with plasma.

Procedure

Sample preparation in thin film GD-OES is carried out by placing the thin film on a suitable substrate onto which the thin film has to be deposited. After that, the prepared sample is introduced into the vacuum chamber of the GD-OES instrument with the help of glow discharge plasma in a low pressure. The atoms in the sample are ionized and then emit specific optical wavelengths because of the plasma.

An optical spectrometer is an instrument that is used in collecting the optical wavelengths emitted out and categorizing it depending on its energy as well as the intensity of the rays. By final spectrum obtained, the basic elemental matrix and its relative mass concentration of the sample are determined. As the intensities of specific spectral lines are recorded and compared to the specific elements, scholars are able to determine the thin film material's elements.

Depth profiling is another area in which GD-OES is employed; in this process, the sample is bombarded with a focused ion beam causing layers of material to be eroded. The intensity of such particular lines with respect to the depth assists scientists and researchers in discovering more about the layer thicknesses of the thin film as well as the elements present. Taking all the factors into consideration, GD-OES is one of the most efficient ways of analyzing a thin film's elemental content and its distribution in depth [11].

3.6 Rutherford Backscattering Spectrometry (RBS)

Principle

Rutherford Backscattering Spectrometry (RBS) can be described as an analytical tool to determine the content of elements and the depth scale of materials. The principle of RBS mainly depends on the elastic scattering of high-energy ions which interacts with sample atom nuclei. In RBS, a high-energy ion beam, which may comprises of hydrogen or helium ions, is directed towards the sample surface. Ions get into the sample and experience only scattering with the nuclei of the given substance's atoms. The heavier nucleus in the sample also bounces some of the incident ions at high angles which become backscattered ions which are then measured and analyzed to determine the elemental makeup as well as the depth distribution of the substance.

Procedure

However, to perform the RBS analysis of thin films, the thin film has to be initially deposited on a suitable substrate. Subsequently, the ready sample is positioned in the RBS instrument's vacuum chamber and gets exposed to the high-energy ion beam bombardment. The ions in turn penetrate the sample and participate in the scattering process with the atoms' nuclei in the substance.

Employ, a particle detector to locate those ions in the sample to measure the energy and angle of incidence of the backscattered ions. The nature and density of the sample's elements may be determined by analyzing the ions' energy spectrum and the angle to which they are scattered. The other method for profiling of the depth distribution of the thin film material is to determine the intensity of the ions back scattered from the sample while rastering the ion beam across the surface of the sample.

The elemental quantitative analysis of thin films and their thickness levels are therefore described by RBS.

4 Effect of Composition on Properties

One must note that the type of composition significantly affects thin-film characteristics and therefore the range of its possible applications. They define the composition of the thin-film material as the specific chemical parts that the thin film is made of and the proportions in which the parts are used. Some of the film characteristics that can be greatly affected by the composition include mechanical, optical, electrical, chemical as well as physical. Recognizing these effects is instrumental in defining the thin-film materials to suit particular application requirements. We examine how composition affects different attributes below:

4.1 *Mechanical Properties*

Tensile Strength and Ductility

Compositional effects are also observed to affect tensile strength and ductility of thin films but to a lesser extent than the effects of thickness on thin films. That is to say, alloying elements or additions can improve ductility because they either suppress the dislocation motion or strengthen the grain boundary cohesion. On the same note, depending on the kind of material system and the processing condition, some of the compositions lead to higher Ts at the expense of ductility.

4.2 *Electrical Properties*

Electrical Conductivity

It is alternatively seen that the electrical conductivity of thin films is more or less dependent on its composition. In case of increasing or decreasing the valency of dopant, the carrier concentration or mobility inside the film changes, and this affects the conductivity properly. One can control the electrical conductivity and therefore it becomes possible for thin-film application in optoelectronic and electronic devices like semiconductors, conductive coatings, and sensors.

Semiconducting Properties

In semiconductor thin films, namely the physical features like bandgap and carrier mobility, are highly dependent on its composition. Thin films can be modified by engineers in terms of semiconducting properties suitable for photovoltaics or LEDs, or for transistors, by forming an alloy or doping the existing material. Thus, the composition optimization enhances the device's performance and power efficiency via fine control of electrical transitions and charging events [12].

4.3 *Optical Properties*

Transparency and Absorption

The material composition is one of the important factors that influence thin-film optical transmittance and absorptance. It often involves changes to the energy band levels and electronic transitions that are directly related to the film's ability to absorb or pass light. In this way, to achieve the corresponding degrees of transparency or absorption, the composition can be tuned and is crucial for such applications as optical coatings, photovoltaics, and display technologies.

Refraction Index and Dispersion

Closely related to the refractive index and dispersion, two properties which are extremely important for application of thin films in lenses, mirrors, and waveguides, are the compositional fluctuations. Due to the ability of engineers to manipulate composition, the refractive index can be adjusted to correlate with a certain optical scenario for instance the wavelength selective filter or the anti-reflective coatings. The possibility to control the dispersion also allows to manufacture optical components having certain spectral and dispersion parameters [13].

4.4 Chemical Properties

Chemical Stability

Chemical nature plays a significant role in thin films' ability to be stable in many an environ circumstances. In extreme environments, thin-film coatings may be longer-lasting and more reliable where certain ingredients or chemicals, elements, or compound characteristics are incorporated to enhance the thin-film coating's chemical stability, corrosion protection, or oxidation durability. Thin-film engineers can choose right compositions to stop deterioration mechanisms and thin-film performance can be guaranteed to function for the intended use such as in the electronics industry for thin-film devices and coatings, protective coatings, and automobile industry among others.

Reactivity

Another characteristic of thin films is reactivity to gases and liquids or to another type of material and it is also affected by composition. Chemical functions, for instance, those affecting catalysis, sensing, and surface alteration may be enabled by surface functionalization or modification through compositional changes supportive of particular chemical bonding or reactions. The reactivity aspect of thin-film material design implies that there is a need to understand how composition influences the interaction of materials with other substances or the environment for the desired functionalities and chemical characteristics in a specific application [14].

4.5 Structural Properties

Crystal Structure

It is important to note that the structure of a material and its ability to phase shift depends on the thin-film content. All-age adjustments have resembled the capacity to promote certain crystallographic phases or degrade distortions like dislocations, vacancies, or the grain boundaries. Thin films' properties are generally optimized for various performance and reliability concerns used in applications in protective coatings, MEMS devices, and microelectronics through control of their content.

Texture and Grain Size

Depending on the film's formation process, its mechanical, electrical, and optical characteristics can be influenced significantly by the compositional features of the texture and grain size. Grain growth and its orientation can be controlled to levels by controlling composition and this can affect properties such as strength, conductivity as well the light scattering. Understandably identifying how composition relates to microstructure plays a significant role in improving thin film performance and tailoring materials for a specific purpose-clientele which includes but is not limited to high-performance optical films, wear-resistant coatings etc. [15].

4.6 *Thermal Properties*

Thermal Conductivity

Deposition of thin films affects their composition, and thus their thermal conductivity which is critical for thermal applications. Phonon scattering mechanisms reveal that with the help of alloying and doping the process of heat conduction in semiconductors can be either enhanced or reduced. This enables the engineers to control the thermal properties of the thin film to give the best thermal performance in heat exchangers, thermal barrier coatings, electronics, etc.

Coefficient of Thermal Expansion (CTE)

Depending on the specific elements within the alloy, CTE can be adjusted resulting in modifications in the thin film's dimensional stability and potential use as a substrate material. Changes in composition enable engineers to design ways of reducing thermal stress and eliminating the possibility of delamination or crack from multi-layer structures by ensuring the thin-film coefficients approximate that of the substrate. If thin-film coatings are to stand up to the heat cycle environments of automotive, aerospace, and electronics uses, CTE must be controlled [16].

4.7 *Magnetic Properties*

Magnetization and Magnetic Ordering

Some properties of thin films including the magnetic ordering, coercivity, and magnetization are usually dictated by the composition of the material. In data storage or as a nanosensor or in spintronics, the magnetic behavior can be tuned either by alloying or by using magnetic impurities. With a proper choice of thin films, the engineers

Table 1 Comparison of properties on different composition

Composition	Electrical conductivity	Optical transparency	Mechanical strength	Chemical stability
Silicon	High	Moderate	Excellent	Stable
Gallium Nitride	Moderate	High	Good	Reactive
Indium gallium nitride	Enhanced	Variable	Excellent	Stable
Silicon carbide (Nanocomposite)	Tunable	Moderate	High	Stable

can develop new high-performance magnetic storage media, magnetic sensors, and spintronic devices [17] (Table 1).

The characteristics of four materials are contrasted in the table: silicon, silicon carbide (nanocomposite), indium gallium nitride, and gallium nitride. High electrical conductivity, good mechanical strength, moderate optical transparency, and stable chemical characteristics are all present in silicon. Despite being chemically reactive, gallium nitride has a high optical transparency and good mechanical strength despite its moderate electrical conductivity. With its improved electrical conductivity, changeable optical transparency, superior mechanical strength, and stable chemical characteristics, indium gallium nitride is a highly useful material. The versatile properties of silicon carbide (nanocomposite) include tunable electrical conductivity, sustained chemical stability, high mechanical strength, and moderate optical transparency.

5 Tailoring Thin Films for Specific Applications

The qualities, structure, and composition of thin films can be changed in order to fit the requirements of a specific application or use case. In order to achieve the required functionality, this procedure is often carried out in conjunction with the selection of materials, manufacturing technologies, and treatments of the finished product. I'll go into great depth about each process of customizing thin films for certain uses below:

5.1 Define Application Requirements

The first process that needs to be selected in thin-film manufacturing for a particular application is to determine the needs of the application. This requires understanding the characteristics of the processes and operations, the context which governs it, the legal and organizational requirements, and the specifications up to which it is desired to perform. For instance, when designing thin-film coatings for protection against corrosion in marine-related circumstances, mechanical stress, variation in temperatures, and interaction with salt water have to be taken into account. By defining the particular requirements for application which are examined in an explicit manner, it is possible for researchers and engineers to set the goals as well as to design the standards for the succeeding stages of thin-film production.

5.2 *Material Selection*

Criticality of selecting the right material for thin-film production: Depending on the special qualities and functions required for the specific application, it is most important to choose the appropriate material. One must adopt several important factors; thermal properties, mechanical properties, chemical properties, optical properties, and electrical properties of potential materials. For example, to apply in thin-film electrode preparation of energy storage gadgets, materials with high conductivity and electrochemical stability such as carbon-based nanomaterials or transition metal oxides are selected. The selection of the material underpinning the thin film production influences the technological properties of the final thin film.

5.3 *Film Deposition Techniques*

Thus, the deposition processes should be proper for creating thin films with the required characteristics such as qualities. Resist thickness, uniformity, compatibility, and microstructure as well as the film thickness and uniformity can be accurately tailored with processes like atomic layer deposition (ALD), chemical vapor deposition (CVD), and physical vapor deposition (PVD). For example, it is possible to use the sputtering or sol–gel processes to deposit layers for the final optical parts and at the same time to obtain the necessary smoothness of the working surface and optical transparency. In relation to this, thin-film deposition parameters have to be optimized to ensure that the thin films developed as per specific applications meet the desired quality and utility standards.

5.4 *Structural and Morphological Control*

Thin-film performance can be optimized for certain applications by modifying its morphological and structural characteristics. This entails managing variables such as surface roughness, film morphology, orientation, crystal structure, and grain size. Thin-film surfaces and microstructures are manipulated by methods such as annealing, ion implantation, and surface functionalization. To encourage cell adherence and tissue integration, for example, surface roughness and topographical features can be manipulated during the production of thin film coatings for biomedical implants. Control over structure and morphology is essential to improving the performance and dependability of customized thin films.

5.5 Composition Engineering

Thus, it can be concluded that the performance of thin film can be tailored based on the requirements of a particular application through the control of morphological and structural properties. This is because, during manufacturing, parameters have to be controlled to give specific characteristics like surface roughness, film morphology, orientation, crystal structure, and grain size. By the use of processes such as annealing, ion implantation, and surface functionalization, thin-film surfaces and microstructures are used. In this regard, approaches such as cell adherence and tissue integration may be enhanced through control of properties such as the surface roughness within the thin-film coatings applied to biomedical implants. It is here that the ability to control structural and morphological properties is a critical element to enhance and guarantee the secured performance of the bespoke thin films.

5.6 Surface Modification and Functionalization

Changes on the surface or functional groups may be introduced to enhance the interaction with thin-film environment or to facilitate specific uses. This involves the restriction of surface energy, wettability, adhesion, biocompatibility, or bioactivity through the use of coated layer, nanostructures/layered monolayer, or chemical treatment. For instance, functional molecules or biomolecules are immobilized on the surface with the purpose of certain molecular identification and quantification in the preparation of thin film coatings for biosensing. The fabrication of thin films is partly due to the surface functionalization and modification for different applications like drug, biosensing, sensor, and catalyst applications.

5.7 Characterization and Testing

One has to characterize and test new materials to analyze their characteristics and functionality in relevant situations when it comes to evaluating characteristics and functionality of the designed customized thin films. Depending on whether the properties to be measured are structural/morphological, chemical, electrical/optical, or mechanical, the characterization techniques employed are spectroscopy, microscopy, X-ray diffraction, electron microscopy, atomic force microscopy, and mechanical testing. Essential testing such as reliability, durability, and performance standards confirms that thin films tailored on specific application are suitable. Thus, testing and characterization are valuable activities for finding out the best strategies and methods to enhance the thin film fabrication process.

5.8 Optimization and Iteration

This means that progressive and systematic enhancement in the design as well as fabrication of thin films to suit clients' specifications is safeguarded through empirical feedback from testing and characterization of the final product (illustrated in Fig. 9). The execution of prescribed performance objectives and addressing of issues or constraints encountered in development are due to the ability to adjust the various processing parameters, material components, and surface modulations. The production methodology of thin films and the material used are further developed to increase the thin-film functionality and affordability as well as repeatability for various uses in target applications. New thinking and improvements in thin-film technology are due to the linear iterative design cycle which is a part of very thin-film formation.

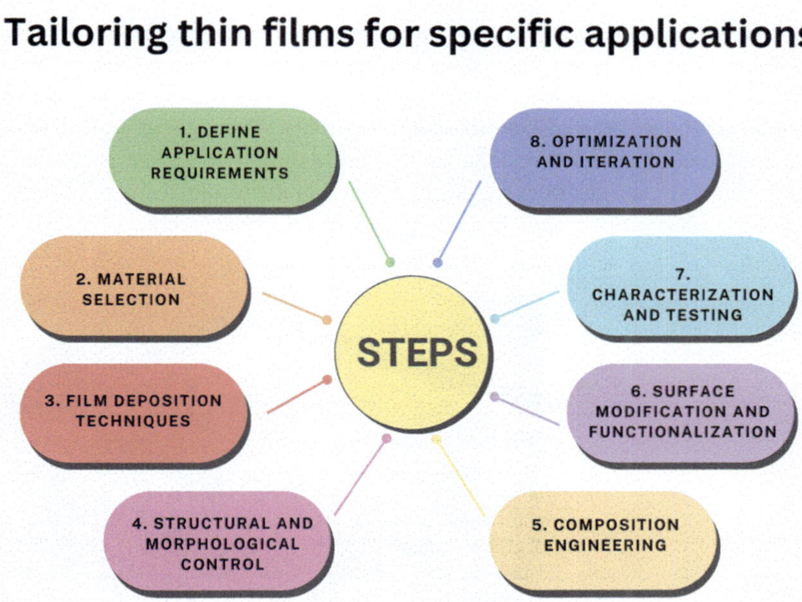

Fig. 9 Tailoring thin films for specific applications steps

6 Tailoring of Thin Films for Different Applications

Deposition of thin films and coatings can be easily altered to fit various industries and provides new and higher performance features. Here, we explore various applications where tailored thin films and coatings play a crucial role:

6.1 *Electronics and Semiconductor Devices*

Thin films have especially played a crucial role in the area of electronics whereby thin films are used as parts of some of these devices as it will be illustrated below. Industry scientists can grow semiconductor materials with precise electrical properties, which means bandgap engineering and the pinnacle of carrier mobility are possible through changes in the thin film's composition and structure. Further, the dielectrics are applied in a selective manner to preserve and shield circuits and components to minimize the vulnerability of the circuit from signals, and boost the reliability of the equipment [18] (Fig. 10).

Through interfacial engineering, thin-film semiconductor devices such as OFETs, OSCs, PVSCs, and OLEDs have made major advancements. Traditional interfacial materials, however, have hindered future advancements, especially in the areas of sensitivity, flexibility, and stability. These problems have been solved by self-assembled monolayers (SAMs), which provide advantages such surface passivation, stability, diversity, transparency, and sensitivity. Through improved molecular packing and chemical functionalization, SAMs improve device performance. Proper formation on substrates is ensured by manufacturing techniques like as solution-based deposition, vapor deposition, and microcontact printing [19] (Fig. 11).

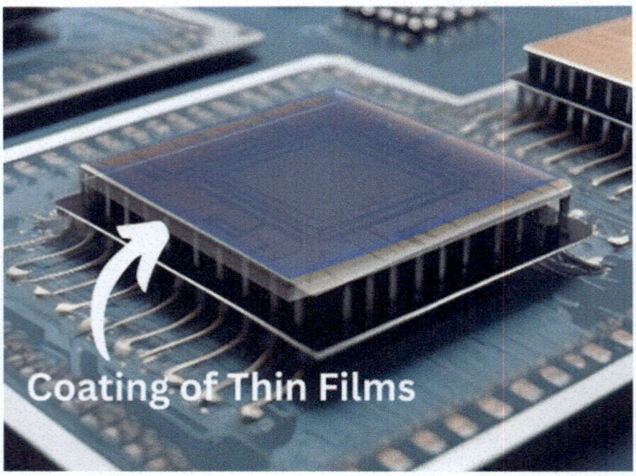

Fig. 10 Thin film coating on Semi-conductor device

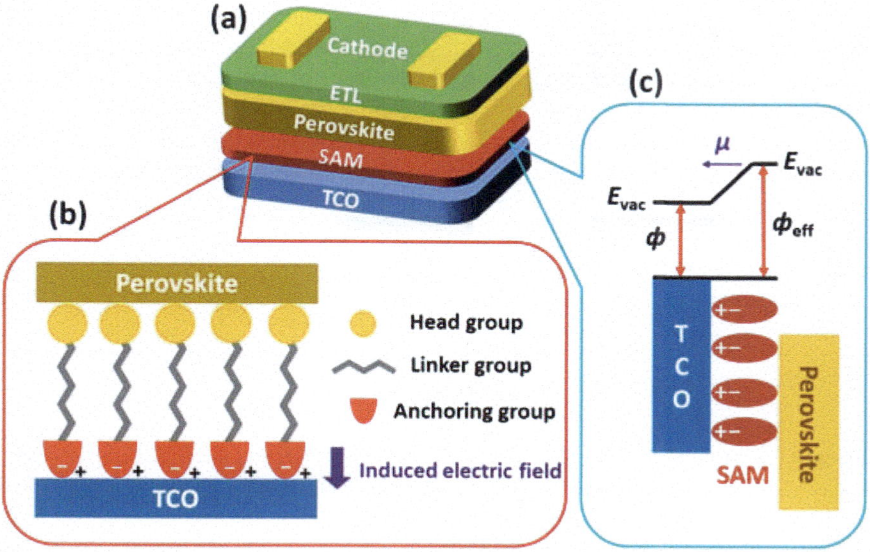

Fig. 11 Structure of perovskite solar cells CC by 4.0

6.2 Optoelectronics and Photonics

Thin films and coatings are essentially found in numerous devices used in optoelectronics and photonics such as light emitting diodes, solar cells, photodetectors, and optical and other under coatings. Light absorption, emission, transmission, and reflection can be regulated, when engineers consider the optical parameters of thin movies, which consequently stimulates inventions in photonics. In addition, coatings are applied professionally to improve spectrum selectivity of optical surfaces, and light trapping, and antireflective characteristics, which leads to the improvement of the effectiveness, and efficiency of photonic systems [20] (Fig. 12).

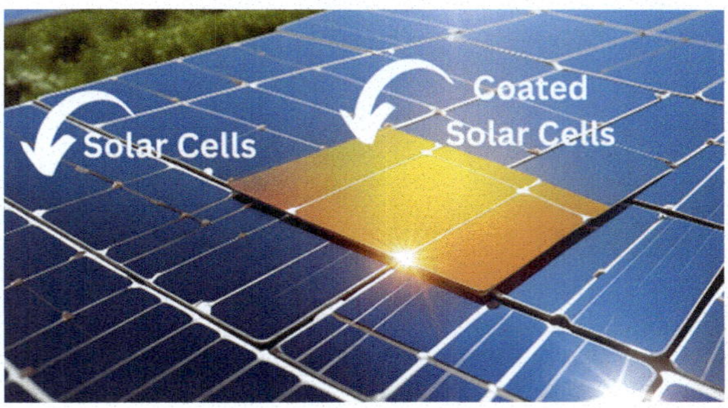

Fig. 12 Silicon nitrate coating on solar cells

In latest research about MOF (metal–organic frameworks) thin films, whose tunable adsorption and emission properties make them suitable for use in optoelectronic applications such as photodetectors, LEDs, sensors, batteries, supercapacitors, and solar cells, combine the characteristics of solid materials with molecular building blocks. These films are useful for gadgets because they have the ability to transmit charges and conduct electricity. Layer-by-layer, solvothermal, hydrothermal, and different coating procedures are examples of fabrication techniques. Optoelectronics hold great promise, but refining their composition and ensuring steady performance remain formidable difficulties [21].

6.3 Displays and Touchscreens

Currently, a number of flat panel displays and touch screens such as liquid crystal display (LCD), organic light emitting diode (OLED) display and touch panel are based on thin film transistor (TFT). Thin films tailored for the purpose such as Graphene or Indium tin oxide allows for touch screens and high image densities. In addition, the coatings that provide anti-glare, anti-scratch or anti-fingerprint characteristics enhance the reliability and readability of the surfaces of the display, enhancing the users' experience.

6.4 Energy Generation and Storage

Supercapacitors, fuel cell, batteries and the solar cells all require thin films and are all applied in energy conversion and energy storage fields. Energy researchers can achieve better stability and efficiency, and enhance the capacity of the thin films used in energy harvesting, conversion and storage though controlling the compositional and structural features. Moreover, the functional coatings with catalytic, ion conductive, or corrosion properties improve performance and life expectancy of energy devices in adverse environments [22].

6.5 Biomedical Devices and Implants

Drug delivery systems, implantable devices, biosensors, and the scaffolds to be used in tissue engineering are fields that take advantages of thin films and coatings. Thin films contribute to the development of medical technology through the biofunctionalization, biocompatibility and basic foundation of temporal release of quite a number of medicinal agents achieved through the fine manipulation of the surface properties. Moreover, the anti-microbial, anti-fouling or bio-active coatings behave as essential shields to avoid infections, nuts and bolts healing and improvement of the working of the biomedical instruments and tools [23].

6.6 *Automotive and Aerospace Applications*

Auto-mobiles and aircrafts make use of thin films and coatings for the purpose of increasing wear, protection against corrosion, heat dissipation and also the sleekness of the body. Nanocoatings that are tailored to act as hydrophobic layers, low friction layers or self-healing layers enhance the hardness, appearance and functionality of engine parts, automobile paints and aircraft paints. Furthermore, thin films which possess thermal barrier or heat resistant ability are significant for exhausts systems, aerospace structures and engine part because they can sustain high temperatures and thermal cycling [24].

6.7 *Protective and Functional Coatings*

Closely, enveloped and incorporated and protective barrier layers shield surfaces against comparative wear, abrase, chemicals, and corrosion (shown in Fig. 13). Conversion coatings act as a barrier or are antirust, provide corrosion inhibitors, or serve to be sacrificial to the metal, ceramic or composite material being coated, or to obtain a specifically designed color. Moreover, functional coatings, possessing the ability to lubricate, convey electricity, or insulate, imbue new functions into surfaces in the auto, consumer, and industrial goods markets, increasing products' utility and longevity [25].

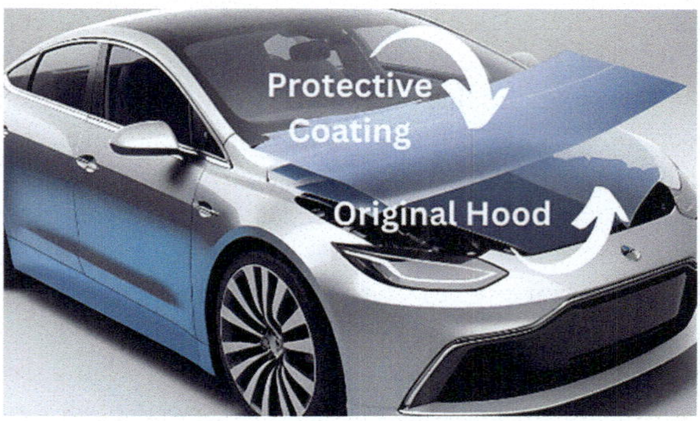

Fig. 13 Protective ceramic coating on car hood

6.8 Environmental and Energy-Efficient Coatings

Thin films and coatings evolve to become competent allies for combating environment decay and energy extravagance, curtailing energy consumption and pollutant and material emission. Special coatings for managing the reflection and transmission of light and heat increase the efficiency of windows, household appliances and buildings. Also, such functional coatings as photocatalytic, self-cleaning or pollutant-degrading ones are crucial for pilotless struggle against air and water pollution, enhancement of IAQ and development of the sustainable urban environment. Intentional engineering of thin films and coatings have a modest yet pertinent role towards the protection to the global environment by offering solutions for sustainability.

7 Conclusion

This chapter is devoted to thin film customization for specific applications and is also concluded with describing the complexities of changing the properties of thin film coatings and materials to meet certain requirements of different fields. This chapter has shown how scientists and engineers can modify the material content and arrangement of thin films' properties to enhance their application range, improve their efficiency, and diversify applications across numerous areas.

Each aspect of thin film tailoring concerning the basics of defining application requirements and selection of appropriate materials; fine processes of film deposition, structure regulation, and surface management enhances thin films for a specific use. Introducing thin films in applications entails an accurate control in the thickness and uniformity together with the morphology properties of the films through the fabrication techniques such as physical vapor deposition, chemical vapor deposition and atomic layer deposition. This is a prospect to solve problems of various industries individually, such as electronics, optoelectronics, energy, biomedical, automotive, aerospace, etc.

Therefore, the chapter underlines the significance of thin films to many significant technologies like semiconductors, optoelectronics, displays, solar cells, automotive coatings, biomedical implants, and environmental coatings. Systematically changing the optical, electrical, mechanical, and chemical properties of thin films, through advanced science, authors enhance creativity while addressing some of the major problems in the world's environmental responsibility, power consumption, diseases, transportation, and more.

The chapter also lays importance on characterization and testing methods to validate dependability and performance of designed thin films. Spectroscopy, microscopy, X-ray diffraction, and mechanical testing for example as well as other methods of analysis that assist in the optimization process and offer certification on effectiveness of solutions that are tailored. These are very valuable techniques, providing information regarding structure–property relations in thin films.

The chapter in essence presents thin film technologies as versatile, modifiable solutions as well as the potential future developments in solving complex engineering challenges and advancing the fields of material science, nanotechnology, and others related. It is an endless path for creating new solutions for specific applications because the enhancement of the thin-film technology is a continuous process and has tremendous potential showing that the future is made of perfect thin films for new and advanced technologies, which will even shape our environment.

References

1. Pshyk, A., et al.: High-entropy transition metal nitride thin films alloyed with Al: microstructure, phase composition and mechanical properties. Mater. Des. **219**, 110798 (2022)
2. Castano, I., Evans, A.M., Reis, R.D., Dravid, V.P., Gianneschi, N.C., Dichtel, W.R.: Mapping grains, boundaries, and defects in 2D covalent organic framework thin films. Chem. Mater. **33**(4), 1341–1352 (2021)
3. Güzelçimen, F., et al.: The effect of thickness on surface structure of rf sputtered TiO_2 thin films by XPS, SEM/EDS, AFM and SAM. Vacuum **182**, 109766 (2020)
4. Qu, X., Zhang, R., Zhang, P., Cao, X., Yu, R., Wang, B.: Effect of Ag incorporation on the microstructure and properties of ZnS thin films. Opt. Mater. **148**, 114879 (2024)
5. Ranmale, V., Kadam, L., Shinde, T.: Exploring the influence of Ni doping thin films for supercapacitor application. Energy Storage **6**(4), e637 (2024)
6. Al-Asbahi, B.A., Qaid, S.M., Hezam, M., Bedja, I., Ghaithan, H.M., Aldwayyan, A.S.: Effect of deposition method on the structural and optical properties of $CH_3NH_3PbI_3$ perovskite thin films. Opt. Mater. **103**, 109836 (2020)
7. Predoi, D., et al.: Synthesis, characterization, and antifungal properties of chrome-doped hydroxyapatite thin films. Mater. Chem. Phys. 129690 (2024)
8. Krishna, D.N.G., Philip, J.: Review on surface-characterization applications of X-ray photo-electron spectroscopy (XPS): recent developments and challenges. Appl. Surf. Sci. Adv. **12**, 100332 (2022)
9. Manickam, K., Muthusamy, V., Manickam, S., Senthil, T., Periyasamy, G., Shanmugam, S.: Effect of annealing temperature on structural, morphological and optical properties of nanocrystalline TiO_2 thin films synthesized by sol–gel dip coating method. Mater. Today: Proc. **23**, 68–72 (2020)
10. Greczynski, G., Haasch, R.T., Hellgren, N., Lewin, E., Hultman, L.: X-ray photoelectron spectroscopy of thin films. Nat. Rev. Methods Primers **3**(1), 40 (2023)
11. Vesel, A., Zaplotnik, R., Primc, G., Mozetic, M.: Fundamentals of thin film depth profiling by glow discharge optical emission spectroscopy. J. Vac. Sci. Technol. A **41**(4) (2023)
12. Salunkhe, P., AV, M.A., Kekuda, D.: Investigation on tailoring physical properties of Nickel Oxide thin films grown by dc magnetron sputtering. Mater. Res. Express **7**(1), 016427 (2020)
13. Ezike, S.C.: Effect of concentration variation on optical and structural properties of TiO_2 thin films. J. Modern Mater. **7**(1), 1–6 (2020)
14. Kardile, H., Somvanshi, S.B., Chavan, A.R., Pandit, A., Jadhav, K.: Effect of Cd2+ doping on structural, morphological, optical, magnetic and wettability properties of nickel ferrite thin films. Optik **207**, 164462 (2020)
15. Song, X., et al.: Intrinsic nanoscale structure of thin film composite polyamide membranes: connectivity, defects, and structure–property correlation. Environ. Sci. Technol. **54**(6), 3559–3569 (2020)
16. Nehate, S., Saikumar, A., Prakash, A., Sundaram, K.: A review of boron carbon nitride thin films and progress in nanomaterials. Mater. Today Adv. **8**, 100106 (2020)

17. Waseem, S., Anjum, S., Zeeshan, T., Shah, I.: Effects of Cr-doping on structural, morphological, magnetic, optical and electrical properties of Ti0. 9Fe0. 1-xO_2 thin films. Optik **212**, 164630 (2020)
18. Harish, V., et al.: Cutting-edge advances in tailoring size, shape, and functionality of nanoparticles and nanostructures: a review. J. Taiwan Inst. Chem. Eng. **149**, 105010 (2023)
19. Li, M., Liu, M., Qi, F., Lin, F.R., Jen, A.K.-Y.: Self-assembled monolayers for interfacial engineering in solution-processed thin-film electronic devices: design, fabrication, and applications. Chem. Rev. **124**(5), 2138–2204 (2024)
20. Pathak, D., et al.: New tailored organic semiconductors thin films for optoelectronic applications. Euro. Phys. J. Appl. Phys. **95**(1), 10201 (2021)
21. Sabzehmeidani, M.M., Gafari, S., Kazemzad, M.: Concepts, fabrication and applications of MOF thin films in optoelectronics: a review. Appl. Mater. Today **38**, 102153 (2024)
22. Fan, L.-Z., He, H., Nan, C.-W.: Tailoring inorganic–polymer composites for the mass production of solid-state batteries. Nat. Rev. Mater. **6**(11), 1003–1019 (2021)
23. Donahue, M.J., et al.: Tailoring PEDOT properties for applications in bioelectronics. Mater. Sci. Eng. R. Rep. **140**, 100546 (2020)
24. Cordill, M.J., Kreiml, P., Mitterer, C.: Materials engineering for flexible metallic thin film applications. Materials **15**(3), 926 (2022)
25. Zhu, X., et al.: Toward tailoring nanofiltration performance of thin-film composite membranes: novel insights into the role of poly (vinyl alcohol) coating positions. J. Membr. Sci. **614**, 118526 (2020)

Engineering Applications in Electronics

Hareem Zubairi

Abstract The current and main concern "Engineering Applications in Electronics" is dedicated on the applicability of thin films in different high-end applications. In this chapter, the following applications of the thin films are discussed further; it is observed that the thin films have many more applications in other semiconductor devices, display technologies, sensors, and detectors. Thin films are important in manufacturing aeroplane and other semiconductor devices since they allow for a better miniaturization and upgrading of circuits in the devices to be made. In the case of display technologies, thin films play a role in better screens like OLEDs and LCDs to enhance the resolution and energy utilization. The chapter also touches on the applications of thin films specifically in sensors and detectors due to its ability to detect even slight changes in the environment together with the biological markers. Through the review of these applications in this chapter, the emphasis is placed on the core concept of thin films which enhances the technological growth of electronic devices, promotes innovation and optimizes performance in numerous gadgets. The information gathered presents the information enumerating the significance of thin films in today's circuit boards and display how they are absolutely crucial to engineering.

1 Introduction

In the realm of uses thin films have become essential elements due to the increasing need for creative solutions in electronics. This section delves into the range of applications enabled by thin films focusing on their significance in shaping electrical devices. The advancement of adaptable devices has largely depended on the utilization of methods for depositing advanced metallic thin films despite the presence of complex techniques for producing flexible thin-film systems. Thin films play a role in a multitude of applications spanning from everyday gadgets to cutting-edge aerospace technologies [1, 2]. Their inherent mechanical, thermal, electrical, and

H. Zubairi (✉)
University in Manchester, Manchester, England
e-mail: hareem.zubairi@postgrad.manchester.ac.uk

A. D. Chandio and I. A. Channa (eds.), *Thin Films and Coatings*, Engineering Materials,
https://doi.org/10.1007/978-981-96-0005-2_5

magnetic characteristics that make them highly adaptable are mainly attributed to their thickness, and diameter, e.g., one cannot measure the dielectric and piezoelectric properties without area, and thickness [3]. Moreover, when the thin-film coatings are subjected and/or applied to the hard base, like aerospace gear as well as bendable displays and sensors, where multiple layers are applied, it makes the metal-like hard surfaces bendable [1]. This section will explore the uses of thin coatings that drive innovation in the realm of electronics. Such applications include sensors, displays, smart wearables and automotive electronics. The widespread integration of electronics into items and surfaces has led to transformations in sectors like healthcare, consumer tech and transportation [2]. This integration has unlocked levels of convenience and functionality, for users.

In the ever-evolving landscape of engineering applications, thin films play a pivotal role, particularly in the applications and innovations of electronics where demands for innovation, flexibility, and efficiency are ever-increasing. This chapter serves as a thorough explanation of the intricate world of engineering applications in electronics, shedding light on the profound impact of thin-film technology in shaping modern electronic devices. Additionally covered are the obstacles related to transferring layers onto thinner as well as much more versatile polymer films. In addition to that, the thin films' relevance in specifying the electro-mechanical integrity of using them in versatile systems.

1.1 Overview of Thin Films in Engineering Applications

Thin films, which vary in thickness from nanometers to micrometers have ended up being crucial components in a considerable range of design applications. Their unique properties, which include electric conductivity as well as mechanical adaptability make them flexible elements for a variety of systems and devices [4]. Within the world of electronic devices, thin films operate as the fundamental parts for a substantial range of applications incorporating display technologies, sensing units, along semiconductor devices [1].

1.2 Importance of Thin Films in Electronics

Thin films are an essential component in electronic devices, operating as vital components in screen modern technologies, semiconductor devices, sensing units and detectors. The capability of these elements to present tailored attributes consisting of optical transparency, electric conductivity, and mechanical flexibility, provides them important for achieving planned procedures in digital systems [1, 2, 5–9]. Their use of microchips to power smartphones to generate flexible displays adorning wearable devices, and thin films is at the forefront of the innovation in the electronic industry.

1.3 Semiconductor Devices: Thin Films at the Heart of Electronics

Everything from communication systems to computer technologies is powered by semiconductor devices, which work as the structure for modern electronic devices [10, 11]. Thin films are important in the fabrication of semiconductors as they help with the integration as well as miniaturization of digital parts onto silicon substrates. Incorporated circuits (ICs) [9], thin-film transistors (TFTs) [9, 12] as well as photovoltaic (PV) cells are a couple of applications of semiconductor devices in which thin films play an essential function in ensuring ideal functionality, performance, and efficiency.

1.4 Display Technologies: Enabling Visual Communication

Current years have observed a revolution in screen modern technologies, which has been driven by advancements in thin-film innovation. Thin-film transistors (TFTs) are utilized in everything from organic light-emitting diodes (OLEDs) to liquid crystal display screens (LCDs), promoting energy-efficient procedures as well as high-resolution images in modern display panels [8, 13, 14]. Thin-film displays have been executed in a wide range of devices such as car display screens, TVs, and mobile phones [15–17], therefore essentially revolutionizing the way we interact with visual information.

1.5 Sensors and Detectors: Thin Films for Sensing Solutions

Environmental tracking, as well as biomedical diagnostics, are just two of the several areas that depend on sensing units as well as detectors. The outstanding attributes of thin films are made use of by thin-film sensing units and are determined very carefully by detecting and measuring chemical, biological, and physical properties. Therefore, many types of thin film-based sensing devices are fabricated in areas such as medical care, environmental tracking, and industrial process control [2, 7, 18, 19]. For example, for biocompatible coatings, Titanium Nitride (TiN), Hydroxyapatite (HAP), and Silicon Dioxide have been used, for drug delivery, Polylactic acid has been used commonly, and for diagnostic sensors, gold and graphene are the most suitable materials [20–22]. These devices may consist of gas sensors, biosensors, pressure-sensing units as well as temperature sensors.

1.6 Challenges and Future Directions

Although there are several benefits to utilizing thin-film modern technology in electronic devices, there are still particular challenges that remain to be addressed. These may include the optimization of device efficiency, reliability, durability, and fabrication of the flexible substrate. Nonetheless, with continual research study and development on overcoming these barriers and launching new progression opportunities coupled with advancement, the future of thin-film modern technology in electronic devices is appealing.

The following chapter will delve into the various applications of thin films in electronic devices, like semiconductor devices, display modern technologies, sensing units as well as detectors. By comprehensive examination of the real world with its existing propensities, our objective is to provide readers with a thorough understanding of the role of thin films in shaping the future of electronics engineering.

2 Thin Films in Semiconductor Devices

2.1 Role of Thin Films in Semiconductor Fabrication

Several technologies created in the second half of the twentieth century rely extensively on microelectronics, in which microprocessors have been playing a major role in wearable electronics, flexible devices, medical devices, automotive applications, energy harvesting devices like solar cells, and smart textiles [20, 21]. Consider a typical car as a representation: microprocessors control the equipment involved in processing materials and producing parts. Microprocessor-equipped machines are used during assembling, for instance, for automotive applications, transparent conductive oxides and silicon carbides have been used, for energy harvesting devices, the solar cells are made up of Cadmium Telluride materials, for smart devices like cards, polyimide and silicon are used, and for smart textile, conductive polymers are the most essential material [20]. Quality is assessed in data collection processes utilizing microelectronic devices at different points [23]. Microprocessor-based instrumentation continuously monitors and alters the engine in real time while it is operating. While traveling, we may use maps to accurately determine our position and direct us to our intended destination. Computers also monitor inventory and delivery dates. Microelectronic equipment facilitates connections between customers, dealers, and credit institutions. An example receiving significant attention now is the "information highway," which depends on cost-effective, high-performance desktop computers owned by users in their homes and enterprises [24]. Meeting these economic and technological standards needs advanced software and precise engineering of materials, device structures, and machinery designed for them.

These engineering and technical marvels are built upon a base of fascinating but inadequately understood fundamental issues in solid-state and interface physics and chemistry. The importance of the chemistry involved is shown by the acronyms assigned to procedures used in producing electronic materials and devices. For instance, chemical vapor deposition (CVD) and metalloorganic chemical vapor deposition (MOCVD) are two frequently used methods in microelectronics production that include intricate chemical reactions taking place at or close to emerging surfaces [8, 13, 18, 25, 26]. Recently, academic institutions have started researching the chemistry of these important materials (like CdTe, graphene, MXene, Indium Gallium Zinc Oxide, etc.) and processes, although research chemists have traditionally been involved in the industry.

To generate the optimized composition, researchers have been focusing on these materials since the last century. The precise fabrication of various materials, especially thin films, and structures with very small dimensions, is a major challenge in today's industry, even though silicon is widely used and well-established in microelectronics technology. Optoelectronics and photonics are notable among non-silicon electrical materials. These disciplines pertain to materials exhibiting optical activity, which involves either emitting light or sensing it. Compound semiconductors made from elements in columns 13 and 15 (III–V) or columns 12 and 16 (II–VI) are highly sought after in this scenario, particularly for applications such as light-emitting diodes and solid-state lasers [23, 24, 27]. The III–V group includes GaAs, whereas the II–VI group includes ZnSe. Unlike elemental semiconductors such as silicon, which do not pose challenges in terms of stoichiometric control, compounds like these may be highly challenging to manipulate. Interface stoichiometry will provide unique control challenges in comparison to bulk stoichiometry, albeit being similarly important. Nevertheless, there are evident advantages as compared to semiconductors composed of separate parts. Bandgap engineering refers to modifying the emission wavelength by altering the stoichiometry (composition mole fractions) of the components [28, 29]. Lighting and display applications, including light-emitting diodes, may benefit from materials with controllable color, accomplished by precise regulation of the stoichiometry of mixtures of Al, Ga, and In, with N, P, As, and Sb [25, 30–32]. These modern alternatives are up to 10 times more efficient in converting electrical energy into light compared to classic light bulbs that employ filaments. LED light bulbs have a longer lifespan compared to incandescent lights due to the absence of filaments [33–35]. Chemists continuously work to achieve the required stoichiometric control.

Modern silicon-based microelectronics use a diverse range of materials in addition to silicon and its oxides, and issues related to these materials are attracting significant attention. To enhance speed and frequency responsiveness, the wire connecting the active transistors in Ultra Large-Scale Integration (ULSI) must expand in length and thickness along with the active components. Lowering length scales below 0.25 μm may significantly impact charge transfer speed and device performance by improving connection and packaging materials [8, 9, 12, 17, 36]. A complex technical issue in surface chemistry is the precise deposition and control of the thickness of barrier layers, such as transition metal nitrides, at the interfaces of insulators, conductors, and semiconductors. This is a crucial element in integrated circuit fabrication.

Therefore, semiconductor devices form the foundation of modern electronics, driving technological advancements in computing, communication, and beyond. Thin films play a pivotal role in semiconductor fabrication, enabling the miniaturization and integration of electronic components onto silicon substrates. By depositing thin layers of materials with tailored properties, such as conductivity and bandgap, onto semiconductor wafers, manufacturers can engineer devices with enhanced performance and functionality.

2.2 *Overview of Thin-Film Deposition Techniques in the Semiconductor Industry*

Thin-film deposition techniques are indispensable processes within the semiconductor manufacturing industry as they empower manufacturers to precisely control the material deposition onto semiconductor substrates. The techniques are critical in the fabrication of thin-film layers, which form the fundamental structure of electronic components within semiconductor devices. The following are some of the most prevalent thin-film deposition techniques utilized in the semiconductor industry:

(a) ***Physical Vapor Deposition (PVD)*** is a widely employed technique utilized in the fabrication of semiconductor substrates with thin coatings. A thin film is produced through the physical discharge of a substance from a solid source, followed by its condensation onto a substrate, illustrated in Fig. 5.1a. PVD techniques encompass various methods, including sputtering, which employs energetic ions to dislodge atoms from a target material in preparation for deposition onto the substrate, and evaporation, which raises the material to its melting point [37–43].

(b) ***Chemical Vapor Deposition (CVD)***: Particularly for substances such as silicon dioxide and silicon nitride, chemical vapor deposition (CVD) is a conventional technique for producing thin films. Chemical Vapor Deposition (CVD) involves the introduction of gaseous precursors into a compartment containing

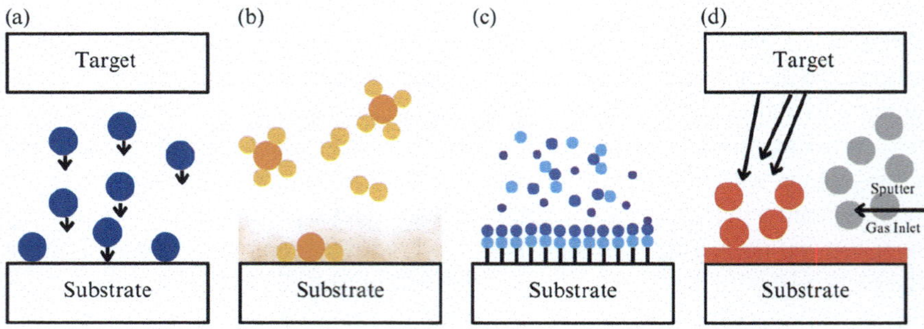

Fig. 5.1 **a** PVD **b** CVD **c** ALD **d** Sputtering

the substrate. Within this chamber, the precursors undergo chemical reactions that result in the formation of a solid, thin deposit on the surface of the substrate, as shown in Fig. 5.1b. CVD can be categorized into distinct methods such as low-pressure chemical vapor deposition (LPCVD) and plasma-enhanced chemical vapor deposition (PECVD), each of which offers distinct advantages to deposition rate, film quality, and uniformity [44–46].

(c) ***Atomic Layer Deposition (ALD)***: ALD is an exceptionally precise technique utilized in the fabrication of thin films, enabling meticulous regulation of both layer thickness and composition at the atomic level. The process of atomic layer deposition (ALD) entails the development of thin coatings through the application of alternating pulses of gaseous precursors to the substrate. This results in the creation of a monolayer of material on the surface of the substrate, Fig. 5.1c. Atomic Layer Deposition (ALD) is ideally suited for applications requiring flawless and consistent thin films, as it permits precise control over thickness and exceptional conformity in the production of such films [47, 48].

(d) ***Sputtering*** is a versatile technique for thin-film deposition in which atoms are displaced from a target material by energetic ions; the displaced atoms then settle on a substrate to form the thin film, as shown in Fig. 5.1d. Sputtering encompasses various modes, including DC sputtering, RF sputtering, and magnetron sputtering. Each of these modes offers unique advantages to the rate of film deposition, film quality, and the ability to manipulate film characteristics. In the fabrication of semiconductors, sputtering is frequently used to deposit thin coatings of metals, transparent conductive oxides (TCOs), and other materials required for device production [49–51].

Thin-film deposition techniques afford manufacturers an extensive array of options when it comes to fabricating thin films, all the while ensuring meticulous regulation of thickness, composition, and properties. By selecting the optimal deposition method for a given application, manufacturers can ensure the production of high-performance semiconductor devices that possess optimal functionality and durability.

2.3 *Material Selection for Thin-Film Semiconductor Devices*

The selection of materials for thin-film semiconductor devices is a critical aspect of the fabrication process, influencing device performance, reliability, and manufacturability. Materials such as silicon, gallium arsenide, and indium phosphide are commonly used in semiconductor fabrication due to their superior electrical and optical properties [52–54]. Additionally, the integration of novel materials, such as organic semiconductors and two-dimensional materials, holds promise for unlocking new functionalities and applications in thin-film devices.

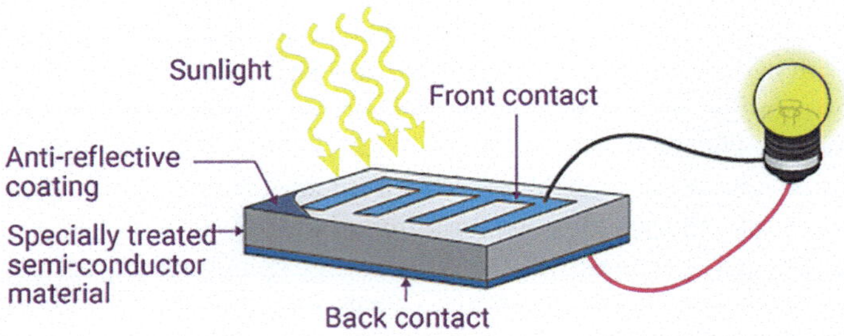

Fig. 5.2 Generation of electricity by photovoltaic cell

2.4 Applications of Thin Films in Semiconductor Devices

Thin films find a myriad of applications in semiconductor devices, ranging from microchips and memory storage devices to photovoltaic cells and sensors. Thin-film transistors (TFTs) serve as the building blocks of integrated circuits (ICs), enabling the amplification and switching of electrical signals in electronic devices [55, 56]. This is possible due to the precise control over material properties that thin films offer, such as conductivity and bandgap. Materials commonly used in thin films include silicon for its excellent semiconductor properties, silicon dioxide for insulation, and high-k dielectrics like hafnium oxide to reduce leakage current. Additionally, materials like indium gallium zinc oxide (IGZO) are used for their superior electron mobility in display technologies, while organic semiconductors are explored for flexible electronics. Furthermore, photovoltaic cells utilize thin-film layers to convert sunlight into electricity, offering a renewable energy solution for power generation as shown in Fig. 5.2 [6, 8, 57]. Moreover, thin-film sensors enable the detection and measurement of physical, chemical, and biological phenomena with high sensitivity and precision, facilitating advancements in healthcare, environmental monitoring, and industrial automation.

2.5 Challenges and Future Trends in Thin-Film Semiconductor Devices

Despite the significant advancements in thin-film semiconductor devices, several challenges remain to be addressed. These include the development of reliable high-k dielectric materials, the scalability of manufacturing processes, and the optimization of device performance and reliability. Furthermore, emerging trends such as flexible and stretchable electronics present new opportunities and challenges for thin-film technology, requiring innovative solutions to meet the evolving demands of the semiconductor industry.

In the next chapter, we will shift our focus to display technologies, exploring the role of thin films in enabling vibrant and immersive visual experiences across a diverse range of electronic devices.

3 Display Technologies

3.1 Introduction to Display Technologies

Despite having been in existence for over five decades, liquid crystal displays (LCDs) were regarded as specialized products on account of their inadequate performance, which included minuscule viewing angles, sluggish response times, and the absence of enormous screens. With the advent of active matrix (AM) liquid crystal displays (LCDs), specifically those driven by amorphous silicon (a-Si H) thin-film transistors (TFTs), the landscape has been drastically altered. From 1989 to 2012, the worldwide revenue generated by TFT-LCD grew from approximately $1 billion to nearly $110 billion [58, 59]. The principal manufacturers expanded their business activities from Japan to neighboring countries in the Asia–Pacific region, such as China, Taiwan, South Korea, and Taiwan [60]. A 9% increase in the shipment of LCD television units was anticipated between 2011 and 2012, for a cumulative quantity of 225 million units. During the initial phase of mass production, premium items such as computer monitors, games, and instruments utilized two TFT-LCDs [36]. As manufacturing technology progressed, TFT-LCDs were subsequently integrated into consumer electronics such as televisions and mobile phones. Concurrently, scientists investigated novel TFTs to resolve inherent issues associated with device fabrication or performance. Furthermore, TFTs have been examined for applications other than pixel driving in LCDs. Presently is a favorable time to analyze the development and progression of TFT technology, as well as to speculate on possible future developments [36, 58, 59]. Figure 5.3 illustrates generic and oxide-based TFT landmarks schematically.

Therefore, display technologies encompass a wide array of technologies and techniques used to present visual information to users. From the traditional cathode-ray tube (CRT) displays to the latest organic light-emitting diode (OLED) screens, display technologies have evolved significantly over the years, driving innovations in various industries such as consumer electronics, automotive, healthcare, and more. The primary goal of display technologies is to provide users with clear, vibrant, and immersive visual experiences across different devices and applications.

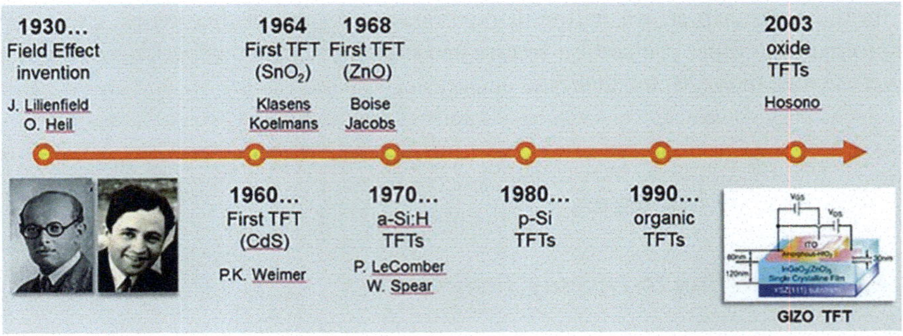

Fig. 5.3 Key milestones reached with TFTs

3.2 *Thin-Film Technology in Display Manufacturing*

Thin-film technology plays a crucial role in modern display manufacturing, enabling the creation of thin, lightweight, and energy-efficient displays. At the core of many display panels lie thin-film transistors (TFTs) [59], which serve as the active switching elements responsible for controlling individual pixels. These transistors are typically fabricated using semiconductor materials deposited in thin layers onto glass or flexible substrates, forming intricate arrays that govern the behavior of millions of pixels across the display surface [60].

3.3 *Thin-Film Transistors (TFTs) in Display Panels*

Thin-film transistors (TFTs) are semiconductor devices that regulate the flow of electrical current within display panels, enabling precise control over the brightness, color, and contrast of individual pixels. Common semiconductor materials used in TFT fabrication include amorphous silicon (a-Si), polycrystalline silicon (poly-Si), and metal oxide semiconductors such as indium gallium zinc oxide (IGZO) and amorphous indium gallium zinc oxide (a-IGZO) [59]. These materials offer high electron mobility and stability, making them well-suited for applications requiring fast response times and high-resolution imagery.

3.4 *Types of Displays Utilizing Thin-Film Technology*

Thin-film technology enables the creation of various types of displays, each offering unique advantages and applications [36, 58–61]:

Liquid Crystal Displays (LCDs)

- LCDs utilize thin-film transistor arrays to control the alignment of liquid crystal molecules, selectively modulating the passage of light through each pixel.
- TFT-LCDs are widely used in devices such as smartphones, tablets, and televisions due to their high resolution, wide viewing angles, and relatively low manufacturing costs.

Organic Light Emitting Diodes (OLEDs)

- OLEDs consist of thin organic layers sandwiched between electrodes, where the emission of light occurs upon the application of an electric current.
- Thin-film transistor technology enables precise control over individual OLED pixels, resulting in displays with vibrant colors, high contrast ratios, and flexible form factors.
- OLED displays are increasingly prevalent in smartphones, smartwatches, and televisions, offering superior image quality and energy efficiency compared to traditional LCDs.

MicroLEDs

- MicroLEDs represent an emerging display technology that utilizes arrays of microscopic light-emitting diodes (LEDs) to create high-resolution displays with excellent brightness and color accuracy.
- Thin-film transistor arrays enable the precise control and addressing of individual microLED pixels, allowing for seamless integration into various display applications, including large-scale video walls, augmented reality (AR) headsets, and automotive displays.

3.5 Applications of Thin-Film Displays

Thin-film displays find applications in a diverse range of electronic devices and industries, including consumer electronics, automotive, aerospace, healthcare, and more. They enable immersive visual experiences, enhance user interaction, and drive innovation across various sectors. From smartphones and smartwatches to digital signage and medical imaging systems, thin-film displays play a crucial role in facilitating communication, entertainment, and information dissemination in modern society.

As an example, a TFT display as shown in Fig. 5.4, is a variant of the liquid crystal display (LCD) that controls image generation via thin-film transistors. By modulating the luminance of red, green, and blue sub-pixels with transistors for each pixel on the screen, this method functions. The liquid crystal behavior is comparable to that of a TN display, which regulates light transmission using 90-degree twisted stacks as a waveguide. Red, green, and blue color filters are applied to the interior surface of the top substrate, while ITO electrodes are applied to both the upper and lower glass substrates [61]. A thin-film transistor is present in each sub-pixel and is connected to the data transit line. This enables the display drive system to independently regulate each sub-pixel. Millions of colors can be displayed on the

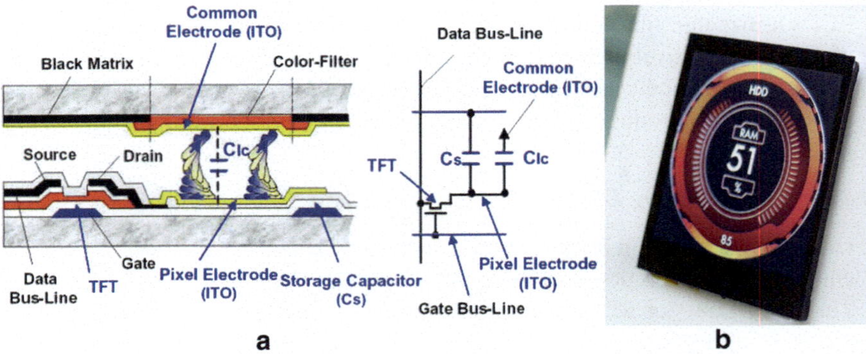

Fig. 5.4 **a** Structure of TFT **b** Thin-film transistor LCDs are used to increase contrast

TFT-LCD through the regulation of light intensity as it traverses each color filter [36, 59]. A storage capacitor is incorporated into each pixel to maintain a constant voltage throughout the cell during pixel updates. Analogous to a TN display, when a signal is transmitted to the pixel, the liquid crystal alters its contorted configuration to align with the electric field and obstruct the path of light [36]. The TFT-LCD exhibits rapid transitions between ON and OFF states, which facilitates swift response times to support high display-update rates that are well-suited for presenting video content or animation with fluidity.

3.6 Challenges and Future Directions in Thin-Film Display Technologies

Despite the significant advancements in thin-film display technologies, several challenges remain to be addressed. These include improving display resolution, increasing energy efficiency, enhancing durability, and reducing manufacturing costs. Furthermore, emerging trends such as flexible and foldable displays present new opportunities and challenges for thin-film technology, requiring innovative solutions to meet the evolving demands of consumers and industries. Continued research and development efforts are essential to overcoming these challenges and unlocking the full potential of thin-film display technologies in the future.

4 Sensors and Detectors

4.1 Introduction to Sensors and Detectors

Sensors and detectors are critical components in a wide range of applications, providing the ability to monitor, measure, and detect various physical, chemical, and biological phenomena. These devices play a crucial role in fields such as healthcare, environmental monitoring, industrial process control, and security. By converting physical signals into measurable electrical or optical signals, sensors and detectors enable real-time data acquisition and analysis, facilitating informed decision-making and enhancing the efficiency and safety of various processes and systems [21, 62].

4.2 Thin-Film Sensors: Working Principles and Design Considerations

Thin-film sensors are a subset of sensors that utilize thin layers of materials to detect and measure specific properties or analytes. These sensors leverage the unique properties of thin films, such as their high surface area-to-volume ratio, sensitivity to changes in environmental conditions, and compatibility with microfabrication techniques [20, 28]. The working principle of thin-film sensors varies depending on the specific sensing mechanism employed, which can include changes in resistance, capacitance, or optical properties in response to the target analyte. Figure 5.5 illustrating the example of thin-film biosensors. Design considerations for thin-film sensors include material selection, sensor geometry, fabrication techniques, and signal processing algorithms, all which impact sensor performance, sensitivity, and reliability.

Recent research highlights the advancements and importance of materials in thin-film sensors, like graphene-based sensors, where the researchers use graphene as

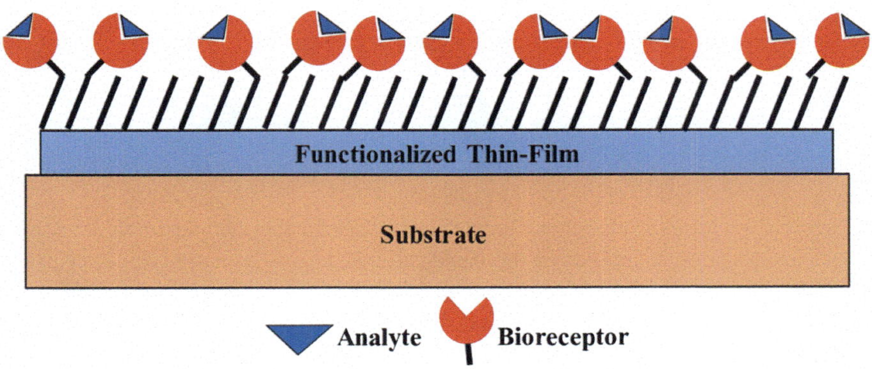

Fig. 5.5 Thin film-based biosensors

a material for detecting volatile organic compounds (VOC) in thin-film sensors. Graphene was chosen because of its high sensitivity and conductivity to detect VOCs at low concentrations, which makes it perfect for environmental monitoring applications. Similarly, another study showed the use of ZnO nanostructured thin films for industrial applications, as they have a high surface area, making them very reactive and sensitive to hazardous gases like ammonia and hydrogen sulfide, making them useful for industrial process control [63].

4.3 *Types of Thin-Film Sensors and Detectors*

Thin-film sensors and detectors encompass a wide range of devices tailored to specific applications and sensing requirements, as shown in Fig. 5.6. Some common types of thin-film sensors and detectors include [20–22, 28, 62, 64]:

(a) ***Gas Sensors***

- Gas sensors detect the presence and concentration of gases in the surrounding environment, making them essential for applications such as air quality monitoring, industrial safety, and emissions control. For example, Metal oxides (e.g., Tin Oxide (SnO2), Zinc Oxide (ZnO)) and conductive polymers (e.g., Polyaniline (PANI), Poly(3,4-ethylenedioxythiophene) (PEDOT)).
- Thin-film gas sensors often utilize metal oxides or conductive polymers as sensing materials, with changes in electrical conductivity or resistance proportional to the gas concentration. For example, Enzymes, antibodies, or other biorecognition elements immobilized on substrates such as gold (Au), silicon dioxide (SiO2), or polymers like polydimethylsiloxane (PDMS).

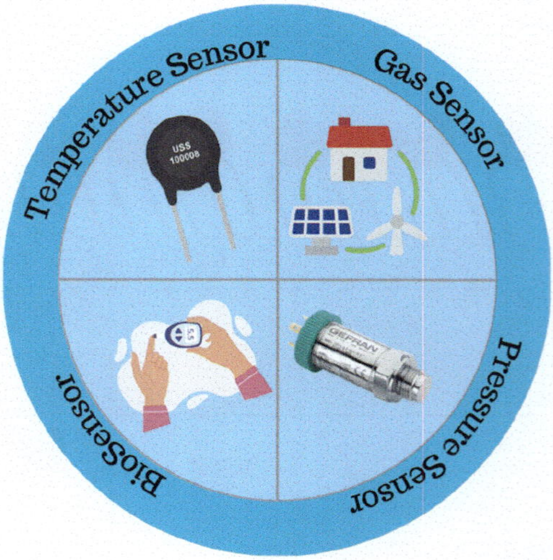

Fig. 5.6 Types of sensors and their applications

(b) ***Biosensors***

- Biosensors are analytical devices that detect biological molecules or biomarkers, enabling applications such as medical diagnostics, food safety testing, and environmental monitoring.
- Thin-film biosensors typically incorporate biorecognition elements, such as enzymes or antibodies, immobilized onto thin-film substrates, with changes in signal intensity or properties indicating the presence or concentration of the target analyte.

(c) ***Pressure Sensors***

- Pressure sensors measure the force or pressure exerted on a sensing element, providing crucial information for applications such as automotive systems, industrial automation, and medical devices. For instance, piezoresistive materials (e.g., Silicon, Polysilicon) or capacitive materials using conductive layers separated by dielectric materials (e.g., Silicon Dioxide (SiO2), Silicon Nitride (Si3N4)).
- Thin-film pressure sensors often utilize piezoresistive or capacitive sensing mechanisms, with changes in electrical resistance or capacitance proportional to the applied pressure.

(d) ***Temperature Sensors***

- Temperature sensors monitor temperature changes, enabling precise control and regulation of thermal processes in various applications, including HVAC systems, automotive engines, and consumer electronics. The materials that are suitable for these applications are Metals with temperature-dependent resistivity (e.g., Platinum (Pt), Nickel (Ni)), semiconductors like silicon (Si) or germanium (Ge), or temperature-sensitive polymers.
- Thin-film temperature sensors can utilize materials with temperature-dependent electrical properties, such as resistance or capacitance, to measure temperature changes accurately.

4.4 Applications of Thin-Film Sensors and Detectors

Thin-film sensors and detectors find applications in a diverse range of industries and fields, including healthcare, environmental monitoring, industrial process control, aerospace, and defense. They enable real-time monitoring of critical parameters, detection of hazardous substances, and measurement of physical and biological phenomena with high sensitivity and accuracy. From glucose sensors for diabetes management to air quality sensors for pollution control, thin-film sensors and detectors play a vital role in improving human health, safety, and quality of life.

4.5 Challenges and Future Prospects in Thin-Film Sensor Technologies

Despite the significant advancements in thin-film sensor technologies, several challenges remain to be addressed. These include enhancing sensor sensitivity, selectivity, and stability, reducing power consumption, improving miniaturization and integration capabilities, and ensuring compatibility with emerging applications such as wearable devices and Internet-of-Things (IoT) platforms [21, 62]. Additionally, the development of sustainable and scalable fabrication processes, as well as the integration of smart materials and advanced signal processing techniques, will be crucial for unlocking the full potential of thin-film sensor technologies in the future.

5 Conclusion

5.1 Recap of the Importance of Thin Films in Electronics

In this chapter, we have explored the multifaceted role of thin films in engineering applications, with a particular focus on their significance in electronics. Thin films serve as versatile building blocks for semiconductor devices, display technologies, sensors, and detectors, enabling the creation of compact, energy-efficient, and high-performance electronic systems across a diverse range of applications and industries.

6 Summary of Key Points Discussed in the Chapter

Throughout this chapter, we have discussed various aspects of thin-film technology, including deposition techniques, material selection, device fabrication, applications, challenges, and future directions. We have examined the role of thin films in semiconductor devices, display technologies, sensors, and detectors, highlighting their importance in driving technological innovation and addressing societal needs.

6.1 Anticipation of Future Developments and Trends in Thin-Film Technology for Electronics

Looking ahead, we anticipate continued advancements in thin-film technology, fueled by ongoing research and development efforts aimed at overcoming existing challenges and exploring new opportunities. Emerging trends such as flexible electronics, wearable devices, and Internet-of-Things (IoT) platforms are expected to drive demand for thin-film technologies, spurring innovations in areas such as materials science, device design, manufacturing processes, and system integration.

In conclusion, thin films will continue to play a pivotal role in shaping the future of electronics, enabling the development of next-generation devices and systems that are smaller, lighter, more energy-efficient, and more capable than ever before. By harnessing the unique properties of thin films and leveraging advances in fabrication techniques and materials engineering, we can unlock new possibilities for innovation and advancement in electronics, driving progress towards a more connected, efficient, and sustainable future.

This concluding statement emphasizes the potential of thin-film technology to propel further innovation and advancement in the field of electronics, ultimately contributing to societal progress and the improvement of human life.

References

1. Acosta, E., Acosta, E.: Thin films/properties and applications. Thin Films (2021). https://doi.org/10.5772/INTECHOPEN.95527
2. Cordill, M.J., Kreiml, P., Mitterer, C.: Materials engineering for flexible metallic thin film applications. Materials **15**(3) (2022). https://doi.org/10.3390/MA15030926
3. Damjanovic, D.: Ferroelectric, dielectric and piezoelectric properties of ferroelectric thin films and ceramics. Rep. Prog. Phys. **61**(11), 1267–1324 (1998). https://doi.org/10.1088/0034-4885/61/9/002
4. Zheng, Y., et al.: Multiferroic thin-film integration onto semiconductor devices. iopscience.iop.org **22**, 17 (2010). https://doi.org/10.1088/0953-8984/22/42/423201
5. Iqbal, A., Naqvi, S.A.R., Sherazi, T.A., Asif, M., Shahzad, S.A.: Thin films as an emerging platform for drug delivery. In: Novel Platforms for Drug Delivery Applications, pp. 459–489 (2023). https://doi.org/10.1016/B978-0-323-91376-8.00006-9
6. Luo, Q.: Applications in photovoltaics. In: Solution Processed Metal Oxide Thin Films for Electronic Applications, pp. 109–140 (2020). https://doi.org/10.1016/B978-0-12-814930-0.00008-6
7. Schubert, M.B., Werner, J.H.: Flexible solar cells for clothing. Mater. Today **9**(6), 42–50 (2006). https://doi.org/10.1016/S1369-7021(06)71542-5
8. Petti, C.J., Hilali, M.M., Prabhu, G.: Thin films in photovoltaics. In: Handbook of Thin Film Deposition: Techniques, Processes, and Technologies: Third Edition, pp. 313–359 (2012). https://doi.org/10.1016/B978-1-4377-7873-1.00010-3
9. Bright, C.I.: Transparent conductive thin films. In: Optical Thin Films and Coatings: From Materials to Applications, pp. 741–788 (2018). https://doi.org/10.1016/B978-0-08-102073-9.00021-7
10. Liang, Z., et al.: All-inorganic flexible embedded thin-film capacitors for dielectric energy storage with high performance. ACS Appl. Mater. Interfaces **11**(5), 5247–5255 (2019). https://doi.org/10.1021/ACSAMI.8B18429/SUPPL_FILE/AM8B18429_SI_001.PDF
11. Greene, J.E.: Review article: tracing the recorded history of thin-film sputter deposition: from the 1800s to 2017. J. Vacuum Sci. Technol. A: Vacuum, Surf., Films **35**(5) (2017). https://doi.org/10.1116/1.4998940
12. Qin, M.J., Dou, S.X.: Superconductors, high Tc. In: Encyclopedia of Condensed Matter Physics, pp. 112–120 (2005). https://doi.org/10.1016/B0-12-369401-9/00706-3
13. Benelmekki, M., Erbe, A.: Nanostructured thin films–background, preparation and relation to the technological revolution of the 21st century. Front. Nanosci. **14**, 1–34 (2019). https://doi.org/10.1016/B978-0-08-102572-7.00001-5

14. Matsushima, T., Kanno, I., Wasa, K.: Thin film MEMS. In: Handbook of Sputter Deposition Technology: Fundamentals and Applications for Functional Thin Films, Nano-Materials and MEMS: Second Edition, pp 559–596 (2012). https://doi.org/10.1016/B978-1-4377-3483-6.00008-5
15. Zhang, X., et al.: Ultra-thin multilayer films for enhanced energy storage performance. Nano Energy **121**, 109271 (2024). https://doi.org/10.1016/J.NANOEN.2024.109271
16. Pan, Z., et al.: Substantially improved energy storage capability of ferroelectric thin films for application in high-temperature capacitors. J. Mater. Chem. A Mater. **9**(14), 9281–9290 (2021). https://doi.org/10.1039/D0TA08335F
17. Kuo, Y.: The electrochemical society interface • Spring 2013 55 thin film transistor technology-past, present, and future
18. Hora, J., Hall, C., Evans, D., Charrault, E.: Inorganic thin film deposition and application on organic polymer substrates. Adv. Eng. Mater. **20**(5) (2018). https://doi.org/10.1002/ADEM.201700868
19. Petrov, I., Barna, P.B., Hultman, L., Greene, J.E.: Microstructural evolution during film growth. J. Vac. Sci. Technol., A: Vac., Surf. Films **21**(5), S117–S128 (2003). https://doi.org/10.1116/1.1601610
20. Tricoli, A., Righettoni, M., Teleki, A., Tricoli, A., Teleki, A., Righettoni, M.: Semiconductor gas sensors: dry synthesis and application. Wiley Online Library **49**(42), 7632–7659 (2010). https://doi.org/10.1002/anie.200903801
21. J.A.-S. & T. A.: Vacuum, undefined surfaces, and films, and undefined 1986, 'Thin film transducers and sensors', pubs.aip.org. https://pubs.aip.org/avs/jva/article-abstract/4/3/610/247864. Last accessed 04 Mar 2024
22. G.S.-S. and A. B. Chemical and undefined 1995, 'Recent developments in semiconducting thin-film gas sensors'. Elsevier. https://www.sciencedirect.com/science/article/pii/092540059401278P. Last accessed 04 Mar 2024
23. Sneh, O., Clark-Phelps, R., Londergan, A., Winkler, J., Seidel, T.E.: Solid films, and undefined 2002, 'Thin film atomic layer deposition equipment for semiconductor processing'. Elsevier. https://www.sciencedirect.com/science/article/pii/S0040609001016789. Last accessed 04 Mar 2024
24. Shur, M.S., Rumyantsev, S.L., Gaska, R.: Semiconductor thin films and thin film devices for electrotextiles. Int. J. High Speed Electron. Syst. **12**(2), 371–390 (2002). https://doi.org/10.1142/S0129156402001320
25. Shur, M., Rumyantsev, S., Gaska, R.: Journal of high, and undefined 2002, 'Semiconductor thin films and thin film devices for electrotextiles'. World Scientific, vol. 12, no. 2, pp. 371–390 (2002). Last accessed 04 Mar 2024. https://www.worldscientific.com/doi/abs/https://doi.org/10.1142/S0129156402001320
26. Sproul, W.D., Christie, D.J., Carter, D.C.: Control of reactive sputtering processes. Thin Solid Films **491**(1–2), 1–17 (2005). https://doi.org/10.1016/J.TSF.2005.05.022
27. Y. Qian, X. Zhang, D. Qi, L. Xie, Chandran, B. K.-Sci. C., and undefined 2016: Thin-film organic semiconductor devices: from flexibility to ultraflexibility. researchgate.net. https://doi.org/10.1007/s40843-016-5091-1
28. Kang, B.S., et al.: Wide bandgap semiconductor nanorod and thin film gas sensors. mdpi.com **6**, 643–666 (2006). https://www.mdpi.com/1424-8220/6/6/643. Last accessed 04 Mar 2024
29. Mocherla, P.S.V., Karthik, C., Ubic, R., Ramachandra Rao, M.S., Sudakar, C.: Tunable bandgap in BiFeO3 nanoparticles: the role of microstrain and oxygen defects. Appl. Phys. Lett. **103**(2) (2013). https://doi.org/10.1063/1.4813539
30. Kazmerski, L.: Polycrystalline and amorphous thin films and devices (2012). https://books.google.com/books?hl=en&lr=&id=40tj3r8rEmcC&oi=fnd&pg=PP1&dq=thin+films+in+semiconductor+devices&ots=FbFkQG3vLM&sig=KBFsa4cS82TKpkD0qWFdtWonRwM. Last accessed 04 Mar 2024
31. Kim, J., Kil, H., Lee, S., Park, J., Omega, J.P.-A., and undefined 2022: Interfacial delamination at multilayer thin films in semiconductor devices. ACS Publications. Last accessed 04 Mar 2024. https://pubs.acs.org/doi/abs/https://doi.org/10.1021/acsomega.2c02122

32. Ko, C., Letters, S.R.-A.P., and undefined 2008: Observation of electric field-assisted phase transition in thin film vanadium oxide in a metal-oxide-semiconductor device geometry. pubs.aip.org. https://pubs.aip.org/aip/apl/article/93/25/252101/118079. Last accessed 04 Mar 2024
33. H. K.-J. of V. S. & T. B and undefined 2003: Atomic layer deposition of metal and nitride thin films: current research efforts and applications for semiconductor device processing. pubs.aip.org **21**, 2231–2261 (2003). https://doi.org/10.1116/1.1622676
34. Greiner, M., Materials, Z.L.-N.A., and undefined 2013: Thin-film metal oxides in organic semiconductor devices: their electronic structures, work functions and interfaces. nature.com. https://www.nature.com/articles/am201329. Last accessed 04 Mar 2024
35. Kim, J.H., Kil, H.J., Lee, S., Park, J., Park, J.W.: Interfacial delamination at multilayer thin films in semiconductor devices. ACS Omega **7**(29), 25219–25228 (2022). https://doi.org/10.1021/ACSOMEGA.2C02122
36. Fortunato, E., Barquinha, P., Materials, R.M.-A., and undefined 2012: Oxide semiconductor thin-film transistors: a review of recent advances. Wiley Online Library, vol. 24, no. 22, pp. 2945–2986 (2012). https://doi.org/10.1002/adma.201103228
37. W.W.-M.M. and Processes and undefined 1989: Physical vapor deposition. Springer. Last accessed 04 Mar 2024. https://link.springer.com/chapter/https://doi.org/10.1007/978-94-009-0917-5_4
38. K. H.-J. of V. S. & T. A. Vacuum and undefined 1986: Chemical vapor deposition and physical vapor deposition coatings: properties, tribological behavior, and applications. pubs.aip.org. https://pubs.aip.org/avs/jva/article-abstract/4/6/2832/101460. Last accessed 04 Mar 2024
39. A.R.-T. materials science of semiconductors and undefined 2008: Physical vapor deposition. Springer. Last accessed 04 Mar 2024. https://link.springer.com/chapter/https://doi.org/10.1007/978-0-387-68650-9_11
40. W.S.-S. and C. Technology and undefined 1996: Physical vapor deposition tool coatings. Elsevier. Last accessed 04 Mar 2024. https://www.sciencedirect.com/science/article/pii/0257897295026169
41. Deng, Y., et al.: Physical vapor deposition technology for coated cutting tools: a review. Elsevier. Last accessed 04 Mar 2024. https://www.sciencedirect.com/science/article/pii/S0272884220311123
42. Shahidi, S., Moazzenchi, B., M.G.-T.E. Physical, and undefined 2015: A review-application of physical vapor deposition (PVD) and related methods in the textile industry. epjap.org. Last accessed 04 Mar 2024. https://www.epjap.org/articles/epjap/abs/2015/09/ap140439/ap140439.html
43. Mattox, D.: Handbook of Physical Vapor Deposition (PVD) Processing. 2010. Last accessed 04 Mar 2024. https://books.google.com/books?hl=en&lr=&id=aGUxoVTYjA8C&oi=fnd&pg=PP1&dq=physical+vapor+deposition&ots=bYp-86eI-r&sig=WV2ryTdJ0ZcQgJfWJ0s-QHjqldc
44. Dapkus, P.D.: Metalorganic chemical vapor deposition. Annu. Rev. Mater. Sci. **12**, 243–269 (1982). https://doi.org/10.1146/ANNUREV.MS.12.080182.001331
45. Jasinski, J.M., Meyerson, B.S., Scott, B.A.: Mechanistic studies of chemical vapor deposition. Annu. Rev. Phys. Chem. **38**(1), 109–140 (1987). https://doi.org/10.1146/ANNUREV.PC.38.100187.000545
46. Zhang, Y., Zhang, L., C.Z.-A. of chemical research, and undefined 2013: Review of chemical vapor deposition of graphene and related applications. ACS Publications, vol. 17, no. 10, p. 13 (2024). https://doi.org/10.1021/ar300203n
47. Oviroh, P., Akbarzadeh, R., Pan, D., and technology of, and undefined 2019: New development of atomic layer deposition: processes, methods and applications. Taylor & Francis. Last accessed 04 Mar 2024. https://www.tandfonline.com/doi/abs/https://doi.org/10.1080/14686996.2019.1599694
48. Oviroh, P.O., Akbarzadeh, R., Pan, D., Coetzee, R.A.M., Jen, T.C.: New development of atomic layer deposition: processes, methods and applications. Sci. Technol. Adv. Mater. **20**(1), 465–496 (2019). https://doi.org/10.1080/14686996.2019.1599694

49. Lehmann, C., P.S. status solidi (b), and undefined 1966: On the mechanism of sputtering. Wiley Online Library, vol. 16, no. 2, p. 607 (1966). https://doi.org/10.1002/pssb.19660160216
50. Aumayr, F., H.W.-T. of the R. S. of, and undefined 2004: Potential sputtering. royalsociety-publishing.org, **362**(1814), 77–102 (2004). https://doi.org/10.1098/rsta.2003.1300
51. Sigmund, P.: Theory of sputtering. I. Sputtering yield of amorphous and polycrystalline targets. Phys. Rev. **184**(2), 383–416 (1969). https://doi.org/10.1103/PHYSREV.184.383
52. Lee, S., Shin, J., Jang, J.: Top interface engineering of flexible oxide thin-film transistors by splitting active layer. Adv. Funct. Mater. **27**(11) (2017). https://doi.org/10.1002/ADFM.201604921
53. Wagner, S., Bauer, S.: Materials for stretchable electronics. MRS Bull. **37**(3), 207–213 (2012). https://doi.org/10.1557/MRS.2012.37
54. Rausch, M., et al.: The sputter performance of an industrial-scale planar Mo-target over its lifetime: target erosion and film properties. Surf. Coat. Technol. **381** (2020). https://doi.org/10.1016/J.SURFCOAT.2019.125174
55. Zhao, F., et al.: Silicon-integrated lead-free $BaTiO_3$-based film capacitors with excellent energy storage performance and highly stable irradiation resistance. J. Mater. Chem. A Mater. **9**(26), 14818–14826 (2021). https://doi.org/10.1039/D1TA03049C
56. Jindal, S., Vasishth, A., Devi, S., Anand, G.: A review on tungsten bronze ferroelectric ceramics as electrically tunable devices. Integr. Ferroelectr. **186**(1), 1–9 (2018). https://doi.org/10.1080/10584587.2017.1369315
57. Riggs, B.C., Elupula, R., Grayson, S.M., Chrisey, D.B.: Photonic curing of aromatic thiol–ene click dielectric capacitors via inkjet printing. J Mater Chem A Mater **2**(41), 17380–17386 (2014). https://doi.org/10.1039/C4TA04030A
58. Yamamoto, Y.: Technological innovation of thin-film transistors: technology development, history, and future. Jpn. J. Appl. Phys. **51**(6) PART 1 (2012). https://doi.org/10.1143/JJAP.51.060001/META
59. Kuo, Y.: Thin film transistor technology-past, present, and future. Electrochem. Soc. Interface **22**(1), 55–61 (2013). https://doi.org/10.1149/2.F06131IF/META
60. Sheng, J., Jeong, H.J., Han, K.L., Hong, T.H., Park, J.S.: Review of recent advances in flexible oxide semiconductor thin-film transistors. J. Inf. Display **18**(4), 159–172 (2017). https://doi.org/10.1080/15980316.2017.1385544
61. R.S.-A. Materials and undefined 2009: Thin-film transistors. Wiley Online Library, vol. 21, no. 20, pp. 2007–2022 (2009). https://doi.org/10.1002/adma.200803211
62. Meixner, H., Gerblinger, J., Lampe, U.-S., and A. B., and undefined 1995: Thin-film gas sensors based on semiconducting metal oxides. Elsevier. Last accessed 04 Mar 2024. https://www.sciencedirect.com/science/article/pii/092540059401266K
63. Song, T.H., Randall, C.A.: Copper cofire X7R dielectrics and multilayer capacitors based on zinc borate fluxed barium titanate ceramic. J. Electroceram. **10**(1), 39–46 (2003). https://doi.org/10.1023/A:1024028024779/METRICS
64. Sang, L., Liao, M., M.S.- Sensors, and undefined 2013: A comprehensive review of semiconductor ultraviolet photodetectors: from thin film to one-dimensional nanostructures. mdpi.com **13**, 10482–10518 (2013). https://doi.org/10.3390/s130810482

Optical and Photonic Applications

Inamullah Maitlo

Abstract The chapter titled, "Optical and Photonic Applications," describes the numerous and vital functions that thin films play in the domains of optics and photonics. In this chapter, various uses of thin films including antireflection coatings, optical filters, mirrors, and photovoltaics are explained. Reflective coatings are necessary mostly antireflective to reduce the amount of light that is reflected back by the optical surface and increase the amount that is transmitted through lenses and other optical parts. Multilayer thin films used in optical filters permit or reject certain wavelengths of light which is vital in imaging as well as telecommunication. Moreover, the chapter explores the application of thin films in the construction of mirrors especially in delivering high reflectivity and at the same time improving their life span. Thin films are used in photovoltaics to produce high-performing solar cells by changing sunlight into electricity in a more efficient and economical way. In this chapter, the authors illustrate detailed descriptions of these applications which unveiled the role of thin film as a central factor in the progression of the optical and photonic elements, stimulating advancement and enhanced performance in numerous devices and systems. The discussed information provides a deep insight into the possibilities of thin-film applications in modern optical and photonic engineering.

List of Abbreviations

LEDs	Light-Emitting Diodes
PMMA	Poly (methyl methacrylate)
OLEDs	Organic Light-Emitting Diodes
AR Coatings	Antireflective Coatings

I. Maitlo (✉)
Department of Metallurgy and Materials Engineering, Dawood University of Engineering and Technology, Karachi, Pakistan
e-mail: Inamullah.maitlo@duet.edu.pk

State Key Laboratory of Chemical Resource Engineering and Beijing Laboratory of Biomedical Materials, Beijing University of Chemical Technology, Beijing 100029, China

A. D. Chandio and I. A. Channa (eds.), *Thin Films and Coatings*, Engineering Materials,
https://doi.org/10.1007/978-981-96-0005-2_6

PET	Polyethylene Terephthalate
PC	Polycarbonate
UV Radiation	Ultraviolet Radiation
IR Radiation	Infrared Radiation
WDM	Wavelength Division Multiplexing
DEMUX	Demultiplexing
PVD	Physical Vapor Deposition
CVD	Chemical Vapor Deposition
AR	Augmented Reality
VR	Virtual Reality
UAVs	Unmanned Aerial Vehicles
PV	Polymer Photovoltaics
OPV	Organic Photovoltaics
BIPV	Building-Integrated Photovoltaics
LEED	Leadership in Energy and Environmental Design
BREEAM	Building Research Establishment Environmental Assessment Methodology
LSPR	Light absorption through the Surface Plasmon Resonance
SDGs	Sustainable Development Goals
WDMs	Wavelength Division Multiplexers

1 Introduction

Applied optics and photonics can be described as a very active and constantly evolving branch of science that deals with advanced materials like polymers [1] and the combination of optic and photonic units. These materials which possess the properties like flexibility, transparency, and tunability are quite crucial and useful in enhancing the performance and characteristics of the optical as well as the photonic devices. Optical and photonic systems and devices [2] have become the driving force of the technological evolution and advancement in the society and ordinary people's lives, inventing tools in communication, health care, energy, and many other applications.

In communication, fiber optical and photonic parts make up the framework of the sophisticated technologies [3]. Broadband Internet through which information can be transmitted easily and efficiently with relation to other regions. A number of new optical materials have emerged from polymeric origin that has provided a strong base for lightweight, flexibility of the optical fibers [4] and thus have brought down the installation cost and wide access development of broadband services particularly in the remotest corners of the developing countries. The availability of communications technology cannot be underemphasized as far as social and economic development is concerned.

In medicine, diagnostic and therapeutic applications of optical and photonic technologies have changed the diagnostic, treatment, and imaging processes [5]. Optical sensors [6] and probes belonging to polymer offered accurate and real-time tests that may occasion early disease detection [7] and improved individual treatment solution. Fourth, material engineering of polymers involved in optical imaging has been enhanced through the development of optical coherence tomography OCT [8] and fluorescence microscopy. They have enhanced patient experiences and health facilities globally hence leading to positive results.

It also raises innovation in matters touching on optical and photonic technologies, renewable energies, environmental conservation as well as enhanced production technologies. Organic polymer photovoltaic cells and light-emitting diodes are effective and efficient in converting solar energy to power for illumination. Polymer-based optical sensors are widely used for environmental applications, and these include detection of pollutants in the atmosphere and water, air and water quality, and the impact of climate change. That is why the use of polymer materials is concentrated in the development of solutions to environmental issues.

In advanced manufacturing, photonic polymer materials are used for manipulation and controlling light used in applications such as lithography, 3D printing, and nanofabrication. These technologies make it possible to produce tiny integrated circuits, medical implants, and materials that interact with optical systems, which expands the number of sectors, including fire electronics, contributing to the advancements contained in research papers. The introduction of polymer materials influences the intensification of the production process, its reduction, and the enlargement of the sphere of production structures and materials.

In general, optical and photonic applications offered by polymer materials define the further development of technology. Society that provides fundamental answers to international issues and sets trends in science, engineering, and medicine among pioneers. Research and development sector keeps on discovering new horizons the role of optical and photonic technologies will increase and they will keep on advancing and advancing the world. The use of polymer materials and optical technology is likely to fulfill social demands so as to promote the development of a number of fields.

Figure 6.1 presents several applications of polymers in optical products such as the optical fibers which transmit light, the sensors which detect any change in the environment, the ferro electronics for data storing and manipulation, solar cells that convert light energy into electrical energy, and the mixture of fiber optics and nonlinear optics which makes it possible to introduce several polymers into optical technology.

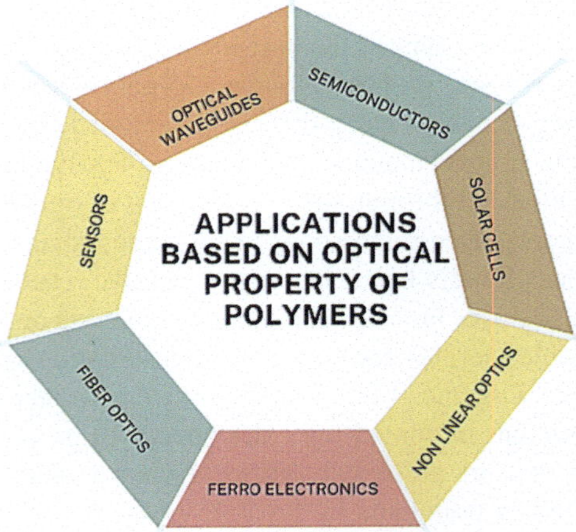

Fig. 6.1 Applications of optical property of polymers

2 Historical Background

One of the most prominent events in the industrial period [1] is the emergence of polymeric material in the optical and photonic fields, which happened in the 1960s. Concerning the identified new feature of the discovered polymer, which was that the polymer was transparent, the other researchers looked at other optical characteristics of the polymer, including transparency, refractive index, and light steering. The first steps were taken towards the creation of optical waveguides that would later be the beginning of the advanced telecommunications technologies. However, historical background of polymer materials goes up to 150 years only. When the first use of plastic for lenses, they were restricted to plastic toys and accessories, minimal optical products due to limitation of the manufacturing process during that time. However, it was not until the 1930s and 1940s the specific acrylics like PMMA [9], for instance gained popularity in such applications as eye and aircraft glazed structure. This was around the 1960s whereby Kodak had been using PMMA and polystyrene products in its economical cameras; this bit was attributed to enhancement of polymer technology. Optical polycarbonate introduced in the later 80 s has been another major advancement. It has been adopted for eyeglasses and safety glasses because of its good optical and like properties. In the 1990s, the promotion in material manufacturing and fabricating technologies expanded polymeric materials and also plastic types. Emerging classes of materials like polyetherimides, polyether sulfones, and cyclic olefins have improved the transmission in the near infra-red region, simultaneously improving thermal stability to make useful for optical and photonics applications. In the 2000s, polymer materials affected the industry in more ways by contributing to the flexible OLED devices [10], polymer PV [11] or photonics and many other devices. These include innovations in energy conservation are in the taxonomy of response to lighting, energy, creation, and technology. The 2010s integrated an improvement to the polymer material into durable, functional, and physically realizable electronic systems, and the improvement of the optical and photonics applications of the material. Metamaterials and plasmonic structures are also emerging based on polymers; photo and manipulation of light waves of even nanodimensions is possible in sensing, imaging, and data processing. Despite this, polymer materials still stand as the only growth opportunity in the realm of optical and photonic flexible, lightweight, and cheaper demanding communications medicine and energy solutions. More updates and more. In the future, more will be done by continued advancement of the R&D polymerous materials will help in making lot of improvements in the global advancement and opportunities for technological and scientific advancement (Fig. 6.2).

Fig. 6.2 Timeline of polymers in applications of optical and photonics

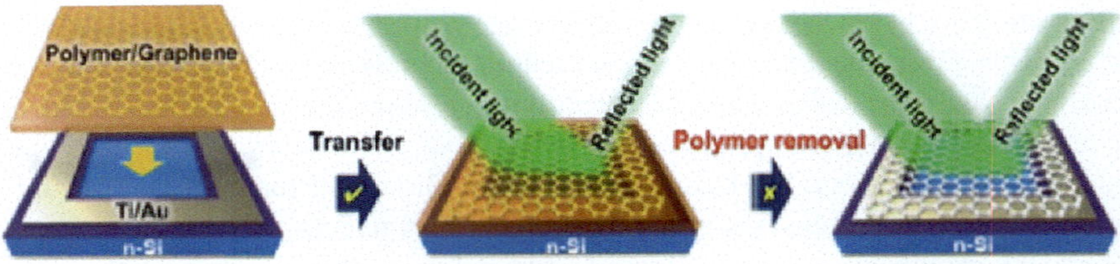

Fig. 6.3 Effect of polymer on antireflection phenomena

3 Antireflective Coatings

3.1 Introduction

Coating, specifically antireflective coatings (AR coatings) are critical for the utilization of polymer material in optical and photonic systems, enhance or optimize output by minimizing negative reflections. These coatings function in the manner that the reflectance of the surface is minimized by measuring and achieving the best possible value of the air–substrate difference so as to allow for instinctively high transmittance of the light and clarity. This optical layer is useful in many applications and provides an extra layer to minimize reflection of the interaction of different media with different thicknesses. Specifically, for the anti-interference treatment, the less air at the interface with flint glass, PET, PMMA, or PC [4] is desirable. In consideration of the fact that reflection presents a major issue on the surface, AR coatings help make crucial contributions to enhancing and optimizing the functionality of systems that incorporate polymer material in their operations; this can be seen from the manner in which AR coatings enhance image quality, boost performance, and minimize instances of glare (Fig. 6.3).

This image shows how the inclusion of polymers in the antireflective coating effectively reduces glare and improves light transmission, resulting in improved optical clarity and efficiency.

4 Applications

Antireflection (AR) coatings are essential coating types of various optical products that aim at minimizing adverse reflections and optimizing the performance of optical accessories. The use of polymers for the preparation of such materials and those with antireflective coatings has been in the focus of great interest in the recent past.

5 Consumer Electronics

The most vital use of polymer systems to counter interference is in the area of electronics, especially in gadgets like smartphones, tablets as well as laptops. These coatings are applied on the surface of the instructions to minimize reflections and glares which enhances the visibility and clarity of the texts when used in well-lit areas. Coatings reduce reflections and thus increase the image contrast and color quality for better visibility for users.

Further, the antireflective polymer coating is applied on the lenses of the digital camera and video camera to avoid reflected light and to increase the clarity of an image. These coatings also enhance the light penetration and make it possible for photographer to take good and clear pictures, especially in poor light area.

6 Optical Lenses

Polymer antireflective coatings in ophthalmology are used on the glasses for good vision and to prevent eye strain. It also makes the user feel that the lens is cleaner and facilitates its functions since haze on the front and back of the lens is eliminated. Moreover, because glare reduction coatings let more light through the lens, it improves the contrast and color quality of the viewer, especially in low-light environment.

In antireflective polymer systems, it is also used in optical product such as binoculars and binoculars for the purpose of improving its quality and resolution images. This technique also assists researchers and astronomers to receive even superior and improved images like microscopical images and objects that are from the sky even as diminishing the science of reflection and enhancing the science of transmittance also at the same time.

7 Solar Panels

The solar panels and photovoltaic cells use polymer antireflective coatings. This is to reduce reflection from the panel surface so that it can trap most of the light it receives greatly enhancing energy efficiency of the panels. The principle of working of the antireflective coating is that it minimizes the reflection hence letting more light into the solar panel for the purposes of trapping solar energy. Also, the layer of polymer that absorbs the impacts may enhance the durability and lifespan of solar panel by shielding the solar panel from conditions such as dust, dirt, and moisture. This coating serves to keep water out and further to reduce collection of any dirt on the surface of the panel and thus, performance and durability are assured.

8 Architectural Glass

AR is also applied in the construction of windows and exterior walls of buildings to exclude glare and reflection and offer a beautiful environment to live human beings. These coatings enhance the visibility indoors and outside the home and also reduce glare and heat while letting in natural light into homes. Moreover, the utilization of the polymer barrier system can enhance the energy transmission at the building and at the same time shield the UV and IR radiation. These coatings make it possible to practice sustainable architecture and construction since energy is saved, as less light and air conditioning is required inside the building.

9 Optical Filters and Coatings

Polymer antireflective coatings apply to filters and mirrors to increase and enhance its performance and performance. These coatings increase their transmission and efficiency by easing various negative impacts from filters and mirror surfaces. Therefore, antireflection coatings prevent or minimize the loss of light in optical systems required in communication, imaging, spectroscopy and sensing among others.

Moreover, the function of the anti-interference polymer coating can be set to specific wavelengths and initial sensitivity. Incidence is a suitable position for altering filters and glasses to meet the required standards. These techniques are also applied in the manufacturing of lasers, cameras, or scientific optical instruments and are also significant to enhance the quality and reliability of the optical products.

Some of the most representative fields of use for the polymeric antireflective coatings are described below, which offer specific solutions in order to minimize reflections where needed. They range from consumer electronics to optical products, solar panels, and glass buildings and meet the need in enhancing visibility, electricity, and workforce. Thus, scientists and engineers are constantly searching for new methods to satisfy the requirements of the modern versatile population using the specific characteristics of polymers.

10 Recent Advancements in Antireflective Coating

Over the past decade, vast improvement has been observed in the polymer-based antireflective coatings due to tremendous progress in the field of material science, nanotechnology, and optical engineering. This coating has the perform batter by enhancing light transmission to lenses by minimizing surface reflectance, optical guides, parallel solar panels and photovoltaic equipment.

11 Novel Polymer Materials for ARCS

The recent activities have principally remained centered on blending novel polymer substances with optical characteristics for ARCS. These polymers have low reflective index, high transparency, and good substrate adhesion; the former offers good light management thus minimizing the effect of the surface [12]. Advances in polymer chemistry and nanotechnology and organic synthesis coupled with advanced molding techniques enable aspect of thickness, optical characteristic, and surface topology of engineered polymer coatings. New polymer-based ARCs offer flexibility, durability, and scalability which are valid for many variants of optical, electronic, and photovoltaic applications.

12 Multilayered Polymer Coatings

The multilayer polymer coatings have shown promising potential to be an effective means of enhancement of the performance characteristics in the whole spectral area. The fabrication of the polymers stopped the researchers from having the layers of polymers with the demonstrative refractive indices for the light wave to go through leaving minimal reflection loss hence they stack them on top of each other. Deposition methods involve layer-by-layer or chemical vapor deposition and other processes are used to deposit a multilayer system of polymers with a specific layering and structure in the materials [13].

Multilayer polymer ARC has the facility for antireflective features; additionally, things such as moisture and UV stability and mechanical damage needed for outdoors and gray cast iron severely threatened work environments also increase ARC's durability.

13 Nanostructured Polymer Surfaces

The electrical resistance properties of polymer systems have been advanced by the nanostructured polymer surfaces. The researchers have succeeded in cutting the reflection of light through light trapping and impedance matching as a result of stamping or etching of nanostructures on the polymer surface [14]. The new generation nanofabrication methods like nanoimprint lithography, electron beam lithography, and block copolymer self-assembly have made it possible to produce nanostructures with size, shape, and aspect ratio in a quantitative manner. Antireflective nanostructures set at the surface of polymer, enhance the coherence of the light passing through, and boost up the efficiency of devices such as solar cells, displays, and optical sensors.

14 Durable and Self-Cleaning Coatings

Novel approaches to polymer-based ARCs have been defined in the recent years with the aim of enabling the systems to be independently foulant and degradation resistant. It involves hydrophobic or superhydrophobic polymers that have low energy and denies water, dust, and other unwanted particle [15] adhesion. Some materials include fluoropolymers, siloxanes, and perfluorinated compounds offer water-repellent, self-cleaning surface, offer stability, durability, and reliability.

AR polymer coatings increase transmittance while cutting down on operations and service life of equipment like solar panels, glass chasers, and air conditioning (Fig. 6.4).

There are recent developments in advancing the polymer-based antireflective coatings in this field, which leads to greater improvement in the properties of the coatings in terms of the optical property, durability and performance. These

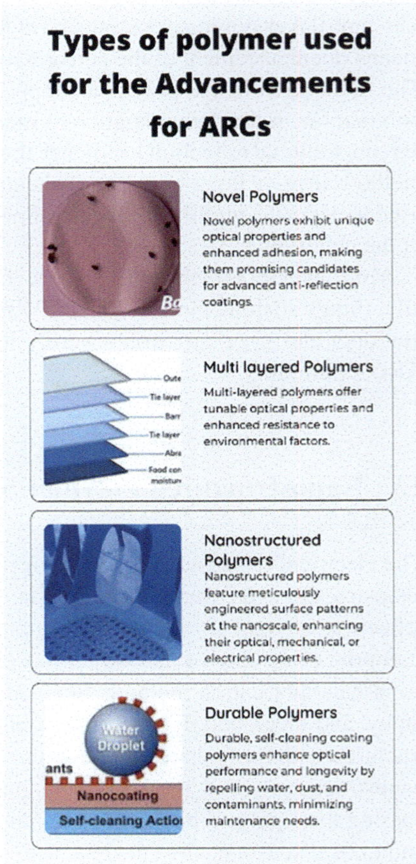

Fig. 6.4 Types of polymers used for antireflective coatings

include new polymer materials, layered coatings, nanostructured surfaces, and self-cleaning technologies are making improvements in ARC and encouraging the utilization in different sectors including lenses, electronics, and solar products among others. Future advancements in the interdisciplinary field of materials and nanotechnology will see a growing prominence of polymer-based ARCs that will significantly enhance the capability, durability, and eco-friendly effects of optical applications in the foreseeable years.

15 Optical Filters

The use of optical filters, especially those based on polymer films is manifested in the various optical systems with optical devices for numerous applications. The evolution of optical filters dates back to the seventeenth century and covers all possible forms, from color filters to nano-thin-layer systems (Fig. 6.5).

The use of optical filters can be dated back to ancient civilizations where elements of art and decorations entailed the use of primitive pigments. Some of the earliest uses were the glass windows of cathedrals in medieval ages that only allowed selective wavelengths to pass through and others to be absorbed causing beautiful effects of what colored light is all about. Optics and the discovery of light and color that was central to the Renaissance was conducted by famous scientists like Isaac Newton and Johannes Kepler. Newton's prism experiment formed the basis for the analysis of light scattering and spectral components and thus made it possible not only for the development of filters with desired spectral characteristics but also for the creation of devices for their study.

The Industrial Revolution was the critical point in the development of other precise reciprocals for several reasons, and it played an important role in the history of optical filters, as it presented production technologies that contributed to the proliferation of optical equipment. Progresses made in the field of thin film during the latter part of the industrial revolution and the two worlds war altered the optics in field. Laue and

Fig. 6.5 Filters (creative common by 3.0)

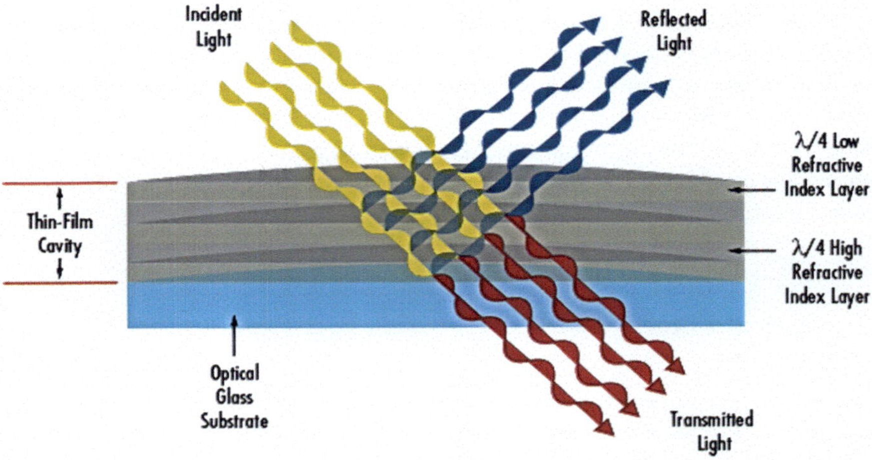

Fig. 6.6 Phenomena of optical filter

Dempster's works were the first that created the basis for the development of thin layers exhibiting optical properties. Thomas Young as well as Augustin-Jean Fresnel for the discovery of the interference effect in thin films that also lifted the possibility of optical filters based on interference effects (Fig. 6.6).

The picture above gives a brief understanding of the optical filters and how it controls which wavelengths to allow to pass through and which ones to reflect. Through the process of reflection and transmission, there lies a great possibility that the filter modifies the spectral distribution of the light thus the color and intensity of the light depends on the design and characteristic of the filter.

Consumer interest today focuses on polymer thin film filters because of the filter's flexibility, effectiveness, and cost. Polymer filters are different from traditional filters in that they possess the following features: flexibility, low weight, and ease of manufacture. These filters will then be manufactured, for instance, to exhibit certain spectral properties like band-pass, long-pass, short-pass and notch filters by precise regulation of the thin film layer's thickness and refractive index.

Furthermore, by depositing polymer thin film optical filters on a substrate of glass, plastic, or semiconductor, WAVE INC brings new capability of polymer thin film optical filters to different types of eye care. The adaptability of the polymer films to the conventional microfabrication processes enables designing miniature and compact optical devices of spectrometers, WDMs or wavelength division multiplexers, optical sensors, and photonic integrated circuits.

Moreover, the enhancement of the optical characteristics of organic polymers has further enhanced the option in using polymer thin film filters. One can modify the refractive index and the absorption coefficient of the organic polymers and the material is compatible with the roll-to-roll technique, which makes the fabrication of these optical filters cheap and large scale.

Optical filters based on polymer thin films have many advantages in various scientific, industrial or even consumer applications Many scientific or industrial or consumer applications where such optical thin film filters peculiarities are valuable. These filters are broadly applied in the industries like telecommunications, spectroscopy, fluorescence microscopy, biomedical imaging, environment monitoring, and technology. For instance, optical filters are valuable in wavelength division multiplexing telecommunications system used in transmitting many optical signals on a single optical fiber.

16 Applications

Polymer-based optical filters and mirrors are used in numerous applications such as telecommunications, information technology, and others with the possibility of efficient regulation of the light passing and reflection/absorption characteristics. Some of their particular characteristics, like flexibility; increased durability, and, in some cases, relatively low price, make the novelties suitable for many uses.

17 Telecommunication

Polymer thin-film filters are widely employed in WDM and DEMUX applications purposes in optical communications today. WDM systems use optical filtering to counteract the multiplexing effect of optical equipment so that a number of data channels can be transmitted over a single fiber simultaneously in the form of parallel and distinct wavelengths. Optical filters including bandpass filters and filters that have thin-film polymer that has specific characteristics must be incorporated in WDM equipment to enable transmission and reception of different signals.

18 Spectroscopy

The polymer thin-film filters are used in the choosing and separation of frequency and or wavelengths in spectrums such as spectroscopy. These allow the researcher to either allow or block particular wavelength of light in an attempt to study sample characteristics. Thin-film-polymer-element filters are actually suitable for fluorescence, Raman, and absorption spectroscopy for function in chemical analysis and characterization of existing goods or in ensuing supervision of the environment.

19 Biomedical Imaging

Thin-film optical filters have a wide range of applications where the biomedical imaging is considered; it is worth that filtering is rather important for enhancement of the images for further comparison. They are used in multiphoton, confocal, and other microscopes to filter of height of fluorophores into biological samples by employing fluorescence staining. Filters created from very thin polymeric layers enable enhancement of the image and signal-to-noise ratio as well as the achievement of high resolutions of cells and different molecular processes.

20 Recent Advancements of Optical Filters

Polymer thin-film filters have garnered much interest mainly because of the perceived application in various sectors like communication, image processing as well as spectral analysis. However, despite the rather important properties exhibited by these elements, many issues have to be addressed to take full advantage of those materials. Also worth considering is that this branch has a rather rosy outlook due to advances in research and technologies.

21 Advanced Materials Development

Further developments in the polymer chemistry and material science will present new polymer structures with optical characteristics, stability, and processability. As close as molecular level modifications of the polymer structure will allow diligent control over optical characteristics, so the performance of fine filters will rise.

22 Nanotechnology Integration

The enhancement of current polymer thin film filters should be done with the help of other nanotechnologies like nanopatterning nanofabrication and plasmon interference. The structures and the nanoscale structures ensure that there is maximum control of the light-related characteristics hence enhancing the performance of the devices.

23 Applications in Emerging Technologies

Polymer thin-film filters are predicted to feature in new-age products like smart gadgets and cars of the future. Since the use of technology cutting edge continues to go up, the use of miniaturized lightweight high-performance optics continues will grow and so will the innovation and investment in this area.

24 Advanced Manufacturing Techniques

Subsequent developments of PVD, CVD, and inkjet printing technologies are expected to achieve better regulation of the film thickness and chemical and physical characteristics of polymer thin-film filters. As a consequence of this situation, manufacturing processes will be scalable and cost-efficient and, hence, the commercialization of this technology (Fig. 6.7).

Polymer thin-film optical filters face challenges such as spectral performance, product stability, manufacturing consistency, integration, and cost-effectiveness thanks to advances in material science, nanotechnology, manufacturing technology, and integration; Future progress is full of hope and innovation. Applications and innovation efforts that overcome these challenges will pave the way for the continued development and use of polymer-based filters in many optical systems and systems.

	PVD	CVD
Process	Physical Vapor Deposition	Chemical Vapor Deposition
Deposition temperature	Low (<500°C)	High (>500 °C)
Coating thickness	Thinner (0.1-2 microns)	Thicker (1-10 microns)
Applications	Cutting tools, wear parts, jewelry, optics	Cutting tools, wear parts, medical implants

Fig. 6.7 Difference between PVD and CVD

Fig. 6.8 Optical coated mirrors (creative common by SA-3.0)

25 Optical Mirrors

Optical lenses made from polymer films have a long history and are important in a variety of fields, from telecommunications to aviation. Ancient Egyptians, Greeks, and Romans used mirrors for purposes such as cleaning and decoration. However, significant advances in mirror-making technology did not occur until the Middle Ages, with the advent of iron-clad glass (Fig. 6.8).

During the Renaissance, European craftsmen perfected the art of glassmaking and mirror-making, creating beautiful glass for research and personal use. Chemists such as Justus von Liebig revolutionized glass production by inventing silver plating in the nineteenth century, enabling large-scale glass production for optical instruments and architectural applications. It attracted people's attention due to its excellent performance. These glasses are widely used in industries where lightness and durability are important, such as automotive, aerospace, and electronics. By replacing glass lenses with a lightweight polymer alternative, automakers can reduce weight without having to do optical work, which plays an important role in this regard. These glasses enhance the user experience and functionality of electronic devices by providing an attractive appearance while reducing the size and weight of the product. These glasses can be used to transmit optical signals, for sensing, and for controlling outlook and position of moving vehicles and space crafts which help the scientific fraternity and other engineers to collect valuable data physically from earth.

Polymer glass is applied in cameras, smartphones as well as in devices such as displays where light and movement prevail. These glasses improve the aesthetic looks of electronic gadgets besides making to slim and lightweight.

26 Applications of Optical Mirrors

The glasses made with polymer films for the past few years have gained consumers' interest all across the world since they have a number of properties and uses in all kinds of industries. A thin layer of polymer is formed on the substrate and is lighter and more efficient than normal glass. Here, the use of optical glasses derived from polymer films is depicted and how it is vital in today's technological world is debated on.

27 Aerospace and Defense

In aviation or military industries, polymer film glass is commonly utilized when the item is light-weight and contains relevant data. These are spectacles which are optical devices to help see and perceive clearly and distinctly for observation detecting and purposes. The polymers which are used in window films are lightweight and thus can easily be incorporated into aircrafts, unmanned vehicles, and satellite systems in specific military operations to enhance situational awareness and operation efficiency.

28 Consumer Electronics

Polymer glass lenses are applied in more and more electronic appliances including smartphones, tablet PCs, and digital cameras. These glasses are used in elements of the optical systems used for vision, camera systems, and augmented reality innovations. The polymers used in window films are lightweight with flexibility that is pivotal in the mobility of the devices hence window films shall be associated with the production of capital in mobility electronic devices.

29 Solar Energy

Polymer film glass is installed within the solar energy systems to concentrate the solar energy to the photovoltaic cells or thermal receivers. This glass bends sunlight and focuses it on the material, thus allowing maximum the solar radiation to penetrate through it. Window film polymers are lightweight and cheaper hence suitability for large-scale solar projects and large-scale sustainable solutions to renewable resource generation.

30 Recent Advancements of Optical Mirrors

Recently, the application of polymer films in the manufacture of the optical glass has received some tremendous improvement. These glasses have been praised for their lightweight, adaptability, and practicality and are, therefore, finding acceptance in various areas such as imaging, sensing, graphics, and communicating.

31 Enhanced Optical Performance

Scientists who were involved in work with thin polymer glass had to increase its reflectivity, spectral selectivity, and smoothness of the surface. This is to enable the higher requirements and avant-garde nanostructured surface design to pave the way for diffraction drastically and to enhance the optical quality accompanied by the image quality. Spectroscopic ellipsometry, for example, can provide precise reflections about the thin-film properties of the polymer glasses so can atomic force microscopy.

32 Tailored Mechanical Properties

Newer innovations that have been made on the production of polymer glass have seen the production of polymer glass with enhanced characteristics that offer rigidity and stability in numerous circumstances. Polymer materials are strong, stronger, and more durable and, therefore, built to endure high temperatures, moisture, and stress. In the case of using polymers, attention is paid to the development of material with better characteristics in terms of composition, where along with multilayer polymer composites, hybrids are also studied containing the properties of different polymers for obtaining improved performance and relative durability.

33 Integration with Emerging Technologies

Having established the correlations between the polymer glasses and new technologies including AR/VR and electric vehicles, the aspects of visual experience and performance ability are enhanced. Accessory mirrors can also be incorporated into wearable devices, headrests, and car interiors as they improve the usability and interaction of the device or component it is installed. The polymer optics are utilized to offer the optical parts in carrying portable devices, UAVs, and medicinal devices.

34 Sustainable and Eco-Friendly Solutions

Concerning the two broader areas of sustainable environment, researchers have worked on green materials and polymer film glass technology was adopted. Lighting and illumination specialists are introducing bio-based polymers and recycled materials as well as green technologies to minimize the effect on the environment of optical products. This way the product performance can be enhanced particularly for items like glass polymer glass, boosting on the circular economy as well as cutting on expenses will be done by determining the lifespan and ecological design.

New developments in the area of polymer thin-film optics brought improvements in the area of optics, properties of the thin films, manufacturing processes, and integration of new technologies and eco-friendly solutions. These innovations open the prospects of the creation of new optical systems and devices with increased parameters of performance, stability, and eco-friendliness. Looking to the future of the industry, research and development of new material and engineering science innovation will enable thin polymer glasses to deliver the next generation of optical solutions.

35 Photovoltaics

Organic photovoltaics or polymer PV, polymer solar cells, and plastic solar cells are the other names for polymer PV. Conventional silicon-based solar cell devices lack flexibility, portability, and are expensive to produce in large quantities, while polymer-based solar cells are flexible, wearable, and likely to reduce manufacturing costs.

Scientists first discovered the potential of organic materials for solar energy conversion. However, it was not until the 1990s, thanks to the discovery of semiconductors, that significant progress was made in this area[16]. Composed of alternating single and double bonds, these polymers have excellent light absorption and flexibility, making them ideal for solar energy applications, as reported the first efficient solar cell based on a combination of polymers and fullerene derivatives. This achievement paves the way for further research and development in the field to improve the performance, stability, and energy efficiency of polymer solar cells.

Over, the years, many research institutes, universities, and companies have contributed to the development of photovoltaic systems, contributing to the commercialization of OPV technology and its integration into many uses (Fig. 6.9).

The importance of polymer photovoltaics lies in their ability to revolutionize the solar industry by providing lightweight, flexible, and efficient silicon-based solar cells [17]. Unlike rigid silicon materials, polymer solar cells can be produced using roll-to-roll printers, allowing for lower costs. This scalability and affordability make OPV particularly suitable for applications where traditional solar panels are not practical or economical.

Regulations often limit the installation of solar panels [18]. By integrating flexible solar panels into building materials, facades, and wearable devices, OPV can harness the power of the sun, innovate, reduce dependence on fossil fuels, and reduce the impact of climate change (Fig. 6.10).

This image shows three methods used in photovoltaic production: electroplating, inkjet printing, and casting. While coating uses a thin layer of material on the substrate, inkjet printing effectively deposits the material in a controlled pattern.

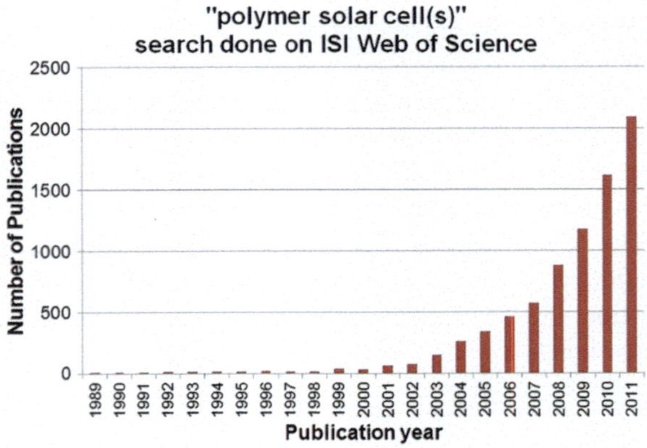

Fig. 6.9 Polymer solar cells publishing papers rate. (creative common by SA-3.0)

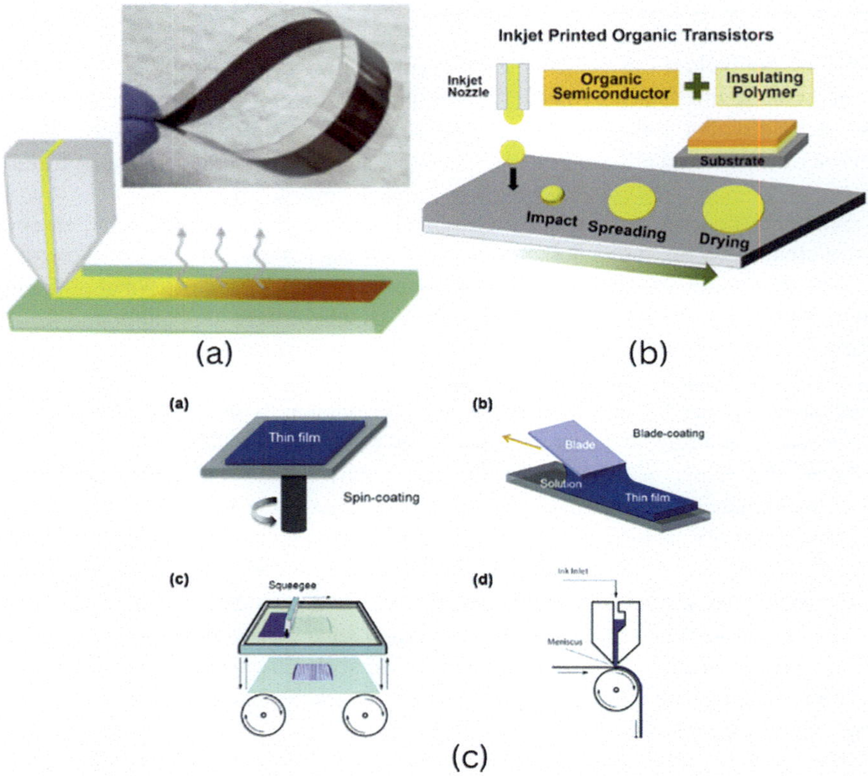

Fig. 6.10 Methods adopted in the production of photovoltaics. They are **a** coating, **b** ink-jet printing, and **c** casting

Foundries use molds to design and manufacture photovoltaic devices, providing scalability and versatility in production.

36 Applications of Photovoltaics

Polymer PhotoVoltaics or Organic PhotoVoltaics are very promising next-generation solar cells, being lightweight, flexible, and cheaper as compared to Silicon Solar cells. These solar panels have many uses in many fields and are taking the world to the future energy solution of renewables.

37 Building-Integrated Photovoltaics (BIPV)

The BIPV are designed to be a part of curtain, outer shell, window, roof, or skylight in a building and employ light to produce electricity. OPV is particularly suitable for BIPV applications due to its features as a simple, lightweight, appearing material. BIPV integrates solar cells into a building [18], providing new concepts for architectural design of a building and construction [19]. It lowers energy utilization, emission of carbon, and has green building certification like LEED and BREEAM.

38 Portable Electronics

The application of polymer photovoltaic cells would indeed be suitable to the electronic devices including smartphones, tablets, and other portable technologies. Owing to the lightweight and flexibility of OPVs, they can be easily integrated into the electronic manufacturing, therefore prolonging battery's life and decreasing frequency rates. Thus, using solar energy, it is possible to make equipment work for a longer time without involving electricity or batteries which will make the movement of the consumers easier [20].

39 Outdoor Recreation

Various devices with photovoltaic features may be employed for charging electric gadgets commonly used outdoors like backpacks, tents, umbrellas, or LED lights. These portable solar power solutions are convenient for outdoor lovers, campers, and adventurers, by providing them with a clean source of energy for outdoor and environmentally friendly living [21, 22]. It makes outdoor toys eco-friendly and also cuts down the use of electrical appliances which are used to power toys [21, 22].

40 Off-Grid Applications

Polymer photovoltaics are ideal for off-grid applications such as remote sensing, environmental monitoring, and rural energy. By providing simple solutions, OPV improves access to electricity in underserved communities, allowing individuals and communities to meet their electricity needs. Off-grid solar solutions produce energy, reduce energy poverty, and support economic development in remote and rural areas [23, 24].

Visits to parks and recreational facilities and the external power sources and supply. Thus, OPV has four areas of positive impact: sustainability, energy independence, and environmental protection due to the use of solar energy. Thus, technology development in polymer photovoltaic applications should increase and incorporate solar energy into the solution to the energy problem globally.

41 Optical Enhancement Techniques

These correcting accessories are very useful when enhancing the light quality and efficiency of photovoltaic elements particularly those involving polymer films. Some of these ideas are the incorporation of optical components on equipments to enhance the absorption of sunlight and enhance the efficiency of energy.

42 Nanostructuring and Texturing

Surface nanostructuring and surface texturing of polymer films have been identified as the possible ways for enhancing light trapping in optoelectronic applications. Therefore, when researchers incorporate nanopillars, nanowires, or nanogrid into the surface of the polymer film they are able to trap and diffract light more effectively, extending the light and supporting the vacuum. Techniques such as nanoimprint lithography, electron beam lithography, and many others allow determining size, shape, and contrast of the nanostructures on the polymer materials in order to provide better light trapping and light absorption. The photovoltaic application is one of the most important aspects of nanostructured polymer films because the films dramatically improve the performance of the organic solar cells, dye-sensitized solar cells, and perovskite solar cells in terms of energy conversion and stability of the device [25].

43 Plasmonic Nanoparticles

Conjugation of plasmonic nanoparticles like gold and silver nanoparticles in the polymer films can increase the light absorption through the plasmon resonance effect (LSPR). There are various ways on how gold nanoapertures can enhance the photovoltaic process, which are significantly scattering and capturing light for an increased absorption cross section and photon to electron conversion efficiency [26]. The right introduction, choice of the type of plasmonic nanoparticles, their controlled synthesis and dispersion into the polymer matrices can provide enhancement of the optical properties of organic, hybrid, and thin-film solar cells and enhance the photovoltaic characteristics [26].

44 Photonic Crystal Structure

One or more segments of the dielectric or metallic nanostructure can be incorporated in photonic crystal structures which can then be embedded in polymer films to control the propagation of light and to solve the interaction problems. IS scientists make new families of photonic crystals and change their optical responses; this way, the transmission or reflection or absorption of the particular wavelength of light can be controlled and therefore the absorption of photovoltaic devices can be optimized. There are also some drawbacks of photonic bandgap structures made from the polymer; they are cheap and lightweight and can be deposited on the flexible substrate [27].

By incorporating a polymer film for optical processing then advancements in photovoltaic devices' light standards and efficiency should be realized. Through application of nanostructured surfaces, plasmonic nanoparticles, construction of the photonic crystal structures, and utilization of lightweighted structures the scientists can control the optical characteristics of photovoltaic elements and enhance the process of energy conversion and the economic efficiency, and contribute to the future development in the field. For these end uses, the expanding field of research in data science and photonics will rely on sophisticated optical systems to optimize electronic devices and control photovoltaic devices, accommodating a multitude of applications.

45 Global Impact

Organic thin-film photovoltaics, or polymer thin-film photovoltaics, are an important player in today's transition to renewal energy sources. Thanks to the lightness, flexibility and relatively low cost, polymer thin-film solar cells may be useful in utilizing the solar energy and decreasing the usage of fossil fuel sources.

- Polymer thin-film photovoltaic system offers an opportunity to the society and people to harness solar energy since they are lightweight, portable, flexible, and which can be adopted in power, portable, and off-grid applications. The ability of constructing at low cost makes it possible for large-scale production and ability to produce more energy and decrease electricity to the remote areas and the developing nations.
- The widespread use of polymer thin-film photovoltaic power generation can replace fossil fuel generation with clean and renewable solar energy and help reduce carbon emissions. By converting sunlight into electricity without emitting greenhouse gases or pollutants, polymer thin-film solar cells help reduce the negative environmental impacts of burning fossil fuels, such as pollution, climate change, and environmental degradation.
- The growth of the polymer thin film optoelectronics industry is supporting business growth and job creation in the world's design, research and installation industries. Investing in solar infrastructure, manufacturing and R&D can support innovation, attract investment and create jobs for local communities and global businesses.
- Polymer thin-film photovoltaics help achieve sustainable development goals by providing everyone with access to affordable, reliable, sustainable and modern energy. The use of solar energy technology, including polymer thin-film solar energy, is consistent with global efforts to combat climate change, reduce energy poverty, and promote sustainable development as defined in the United Nations Sustainable Development Goals (SDGs).

46 Conclusion

The progress continues in optical and photonic applications, particularly in the fields of interference technology, filters and glasses, and photovoltaic applications (especially thin films). Antireflective coatings on polymer substrates are changing the industry by reducing reflections, increasing clarity, and improving optical performance. These layers are flexible, consistent, and can be easily changed. The combination of optical filters and polymer glasses allows control of light wavelengths that can be used for communications, photography, and science. Meanwhile, polymer thin-film solar cells have emerged as a lightweight, flexible, and scalable solar energy harvesting option. These cells are developed using technologies such as nanostructures and plasmonic nanoparticles to improve absorption and energy transfer to promote the development of stable energy. Innovations in these polymer films have had many impacts, from improving energy efficiency to revolutionizing communications and healthcare, and the benefits of continued research and development can create a brighter, better future.

References

1. Alias, M., Zabidi, A.N., Harun, A.M.: Optical characterization and properties of polymeric materials for optoelectronic and photonic applications; Alias, A.N., Zabidi, Z.M., Ali, A.M.M., Harun, M.K.: Faculty of applied sciences. Universiti Teknologi MARA 40450 Shah Alam, Selangor D. E. Int. J. Appl. Sci. Technol. **3**(5), 11–38 (2013)
2. Amiri, I.S., et al.: Introduction to photonics: Principles and the most recent applications of microstructures. Micromachines **9**(9) (2018). https://doi.org/10.3390/mi9090452
3. Nyachionjeka, K., Tarus, H., Langat, K.: Design of a photonic crystal fiber for optical communications application. Sci. Afr. **9**, e00511 (2020). https://doi.org/10.1016/j.sciaf.2020.e00511
4. Buskens, P., Burghoorn, M., Mourad, M.C.D., Vroon, Z.: Antireflective coatings for glass and transparent polymers. Langmuir **32**(27), 6781–6793 (2016). https://doi.org/10.1021/acs.langmuir.6b00428
5. Shan, D., et al.: Polymeric biomaterials for biophotonic applications. Bioact. Mater. **3**(4), 434–445 (2018). https://doi.org/10.1016/j.bioactmat.2018.07.001
6. Bilro, L., Alberto, N., Pinto, J.L., Nogueira, R.: Optical sensors based on plastic fibers. Sensors (Switzerland) **12**(9), 12184–12207 (2012). https://doi.org/10.3390/s120912184
7. Wei, T., et al.: Microstructure, corrosion resistance and oxidation behavior of Cr-coatings on Zircaloy-4 prepared by vacuum arc plasma deposition. Corros. Sci. **158** (2019). https://doi.org/10.1016/j.corsci.2019.06.029
8. Shah, S., Yu, C.N., Zheng, M., Kim, H., Eggleston, M.S.: Microparticle-based biochemical sensing using optical coherence tomography and deep learning. ACS Nano **15**(6), 9764–9774 (2021). https://doi.org/10.1021/acsnano.1c00497
9. Axtell, D.H., Parada, R.J., Morgan, D.: Polymer optics. Rev. Opt. Manuf. 2000 to 2020 (2021). https://doi.org/10.1117/3.2599940.ch7
10. Luo, D., Chen, Q., Liu, B., Qiu, Y.: Emergence of flexible white organic light-emitting diodes. Polymers (Basel) **11**(2), 1–28 (2019). https://doi.org/10.3390/POLYM11020384
11. Bernède, J.C.: Organic photovoltaic cells: history, principle and techniques. J. Chil. Chem. Soc. **53**(3), 1549–1564 (2008). https://doi.org/10.4067/S0717-97072008000300001
12. Hwang, S.H., Lee, K.K., Jung, J.C.: A novel organic bottom anti-reflective coating material for 193 nm excimer laser lithography. Polymer (Guildf) **41**(17), 6691–6694 (2000). https://doi.org/10.1016/S0032-3861(00)00010-0
13. Hossain, J., Mondal, B.K., Mostaque, S.K., Al Ahmed, S.R., Shirai, H.: Optimization of multi-layer anti-reflection coatings for efficient light management of PEDOT: PSS/c-Si heterojunction solar cells. Mater. Res. Express **7**(1) (2019). https://doi.org/10.1088/2053-1591/ab5ac7
14. Mousavi, B.K., Mousavu, A.K., Busani, T., Zadeh, M.H., Brueck, S.R.J.: Nanostructured anti-reflection coatings for enhancing transmission of light. J. Appl. Math. Phys. **07**(12), 3083–3100 (2019). https://doi.org/10.4236/jamp.2019.712217
15. Wang, P., Yan, X., Zeng, J., Luo, C., Wang, C.: Anti-reflective superhydrophobic coatings with excellent durable and self-cleaning properties for solar cells. Appl. Surf. Sci. **602**(May), 154408 (2022). https://doi.org/10.1016/j.apsusc.2022.154408
16. Zhang, Z., Pitwon, R.C.A., Feng, J.: Optics communications: special issue on polymer photonics and its applications. Opt. Commun. **362**(March 2016), 1–2 (2016). https://doi.org/10.1016/j.optcom.2015.11.010
17. Barkhouse, D.A.R., Gunawan, O., Gokmen, T., Todorov, T.K., Mitzi, D.B.: Yield predictions for photovoltaic power plants:empirical validation, recent advances and remaining uncertainties. Prog. Photovoltaics Res. Appl. **20**(1), 6–11 (2015). https://doi.org/10.1002/pip
18. Tripathy, M., Sadhu, P.K., Panda, S.K.: A critical review on building integrated photovoltaic products and their applications. Renew. Sustain. Energy Rev. **61**, 451–465 (2016). https://doi.org/10.1016/j.rser.2016.04.008
19. Dai, Y., Bai, Y.: Performance improvement for building integrated photovoltaics in practice: a review. Energies **14**(1) (2021). https://doi.org/10.3390/en14010178

20. Facchetti, A.: π-Conjugated polymers for organic electronics and photovoltaic cell applications. Chem. Mater. **23**(3), 733–758 (2011). https://doi.org/10.1021/cm102419z
21. Espinosa, N., Zimmermann, Y.S., Dos Reis Benatto, G.A., Lenz, M., Krebs, F.C.: Outdoor fate and environmental impact of polymer solar cells through leaching and emission to rainwater and soil. Energy Environ. Sci. **9**(5), 1674–1680 (2016). https://doi.org/10.1039/c6ee00578k
22. Shi, H., Xia, R., Zhang, G., Yip, H.L., Cao, Y.: Spectral engineering of semitransparent polymer solar cells for greenhouse applications. Adv. Energy Mater. **9**(5), 1–8 (2019). https://doi.org/10.1002/aenm.201803438
23. Emmott, C.J.M., et al.: In-situ, long-term operational stability of organic photovoltaics for off-grid applications in Africa. Sol. Energy Mater. Sol. Cells **149**, 284–293 (2016). https://doi.org/10.1016/j.solmat.2016.01.036
24. Kumar, R., Bag, M., Jain, S.M.: Dual-edged sword of ion migration in perovskite materials for simultaneous energy harvesting and storage application. iScience **26**(11), 108172 (2023). https://doi.org/10.1016/j.isci.2023.108172
25. Amalathas, A.P., Alkaisi, M.M.: Nanostructures for light trapping in thin film solar cells. Micromachines **10**(9), 1–18 (2019). https://doi.org/10.3390/mi10090619
26. Hamed, M.S.G., Ike, J.N., Mola, G.T.: Plasmonic nano-particles mediated energy harvesting in thin-film organic solar cells. J. Phys. D. Appl. Phys. **55**(1) (2021). https://doi.org/10.1088/1361-6463/ac24c8
27. Starczewska, A.: Photonic Crystal Structures for Photovoltaic Applications (2024)

Thin Films in Material Science

Muzamil Jalil Ahmed, Marwah Hamdan Aljohani, and Basheer A. Alshammari

Abstract Thin films have diverse applications in energy storage, electronic devices, biomedical implants, and optical coatings. Researchers are exploring novel materials, deposition techniques, and functionalisation strategies to unlock new opportunities. The chapter discusses substrate synthesis and surface modification, focussing on vinyl polymers, polyacrylamides, polyphthalamides, polystyrene, polyimides, polyketones, zwitterionic compounds, and celluloses. It also discusses the role of inorganic materials in thin film applications. Surface modification and functionalisation of inorganic thin films involve grafting from and grafting to approaches, providing precise control over surface chemistry and composition. Techniques like surface-initiated polymerisation, atom transfer radical polymerisation, and iCVD allow controlled chain growth and versatile surface functionalities. Superhydrophobic and superhydrophilic thin films are used in various fields, including multilayered packaging, membrane separation, sensing technologies, membrane separation, catalytic systems, and pharmaceutical delivery.

M. J. Ahmed (✉)
Department of Chemistry, NED University of Engineering and Technology, Main Campus, University Road, Karachi 75270, Sindh, Pakistan
e-mail: ahmed4000166@cloud.neduet.edu.pk

Thin-Film Lab, Advanced Materials and Sustainable Environment (AMSE) Group, NED University of Engineering and Technology, Main Campus, University Road, Karachi 75270, Sindh, Pakistan

Lehrstuhl des Holzwissenschaft, School of Engineering and Design, Holzforschungs München (Wood Research Institute), Technische Universität München, Winzererstr 45, 80797 Munich, Germany

M. H. Aljohani
Chemistry Department, College of Science, Imam Abdulrahman Bin Faisal University, Dammam, Kingdom of Saudi Arabia

B. A. Alshammari
National Centre of Petrochemicals, King Abdullah City of Science and Technology, Riyadh, Kingdom of Saudi Arabia

A. D. Chandio and I. A. Channa (eds.), *Thin Films and Coatings*, Engineering Materials,
https://doi.org/10.1007/978-981-96-0005-2_7

Abbreviations

A-*g*-B	A grafted on/to/from B polymer
AIBN	Azobisisobutyronitrile
ATRP	Atom-transfer radical polymerisation
CIGS	Copper indium gallium selenide
CSVT	Close spaced vapour transport
CTA	Chain transfer agent
CuAAC	Cu-catalysed Azide-Alkyne Cycloaddition
DMF	Dimethylformamide
PDADMAC	Poly(diallyldimethylammonium chloride
PMMA	Polymethyl methacrylate
PPA	Polyphthalamide
PS	Polystyrene
PVA	Polyvinyl alcohol
PVB	Polyvinyl butyral
PVDF	Olyvinyl difluoride
QVBC	Poly(quaternary 4-vinylbenzyl chloride)
RAFT	Reversible addition-fragmentation chain transfer, RIT (4,4-dimethyloxazoline-2,2,6,6-tetramethylpiperidin-1-yloxyl)
ROMP	Ring Opening Metathesis Polymerisation
ROP	Ring Opening Polymerisation
RuAAC	Ru-catalysed Azide-Alkyne Cycloaddition,
SI-	Surface initiated
SPAAC	Strain-promoted Azide-Alkyne Cycloaddition; TEMPO (2,2,6,6-tetramethylpiperidine-1-oxyl)

1 Introduction

The use of thin films in modern material science is indispensable due to their unique properties and diverse applications. The thin film market encompassing both organic and inorganic thin films, boasts a substantial value of US $12.45bn in 2021, growing at a rate of US $17.46bn or a compound annual growth rate (CAGR) of 4.4% from 2021 to 2029 [1]. The electrical and electronics industry is expected to be the largest market for said thin films, followed by optical coatings, photovoltaics, and other end-use industries (see Fig. 1a). Understanding the film-forming process is fundamental for designing thin films with desired properties. By controlling their composition, structure, and morphology, researchers can tailor thin films to exhibit desired functionalities, making them an essential component of advanced materials [2]. Various types of inorganic thin films exist to cater to different applications for which they are designed. Their average efficiency is typically 7–22% compared to other monocrystalline and polycrystalline materials 3–5, depending on thickness, effective area, and other properties. Organic thin films are a relatively recent, growing area of greener and more sustainable thin film fabrication than their inorganic counterparts. Various fatty acids, lipids, polymers, biopolymers, liquid crystals, and inorganic materials have been employed to investigate structure-induced property modifications and to achieve unique performance and versatility of the thin films at the nanoscale [3].

Lower efficiencies, coupled with material scarcity and processing complexities are key cost drivers for thin films [3]. By optimising temperature, pressure, and precursor composition, material scientists can precisely control the film's properties, such as thickness, uniformity, and crystallinity. Material choice and processing involved define the end-use performance of thin films, for which there has remained an unprecedented focus on the process-structure–property-performance (PSPP) relationship between the process and materials selected for end-use thin film fabrication. Cost, technological maturity, and sustainability of such processes and materials are also key considerations that impact the marketable value of the fabricated thin films, as shown in Fig. 1b, c.

2 Materials Design and Impact on End-Use Applications

Enhancing thin film functionality is crucial for expanding their applications. By introducing functional groups, nanoparticles, or coatings, the properties of thin films can be modified to improve adhesion, conductivity, optical properties, and chemical resistance, amongst others. Material design has focussed on leveraging the surface chemistry, and intrinsic properties of inorganic, organic, or hybrid/composite materials to allow for tunability and performance enhancement [6]. Control of film thickness during and after surface modification(s). In this regard, a core focus of recent innovations is dependent on numerous process-specific and material-specific factors [7]. The present chapter concerns the pre-fabrication aspect of the materials design, i.e., synthesis and surface modification of substrates. A focus on manufacturing methods is provided in subsequent chapters.

2.1 Choice of Materials and Implications for End-Use

Understanding the characteristics of these materials is essential for optimising thin film performance in diverse applications. The implications of the surface chemistry, crystalline properties, grain boundaries, monolayer formation, and interaction between the inorganic, organic or hybrid material and the substrate impact the intended application and desired properties [7–10]. Inorganic materials and various polymers (see Fig. 2), including vinyl polymers, polyacrylamides, polyphthalamides, polystyrene, aramids, polyketones, and celluloses, offer unique advantages in terms of mechanical strength, flexibility, chemical resistance, and biocompatibility. The adequate choice of the use of the filler/loading and the matrix materials gives way

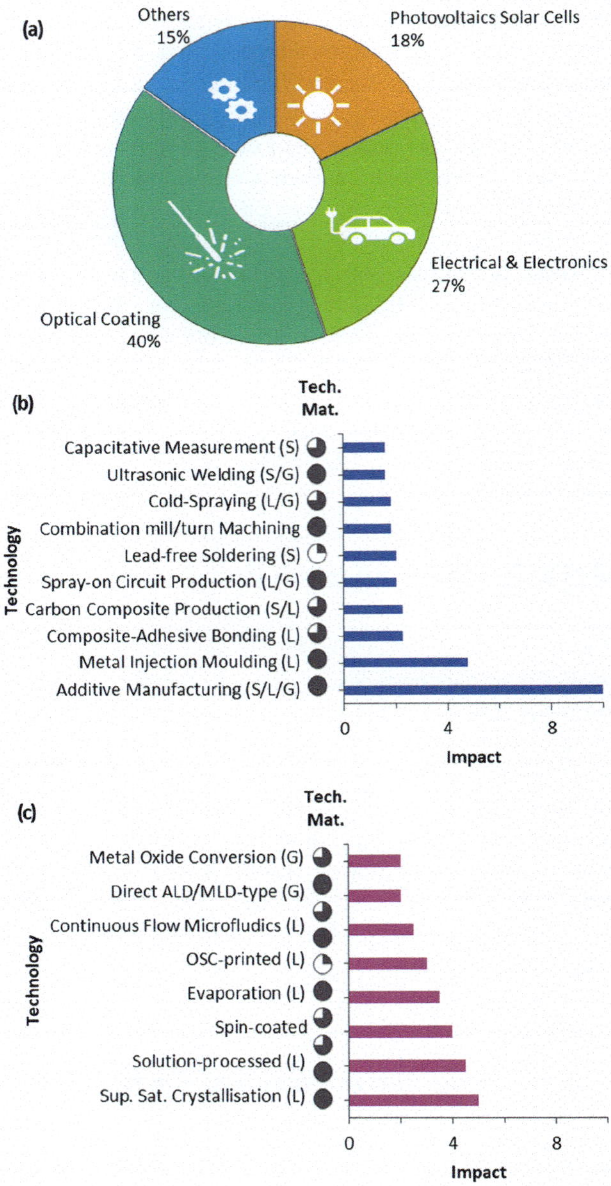

Fig. 1 **a** Market share of thin films by end-use Industry; **b** advanced manufacturing technologies by maturity and overall impact on manufacturing efficiency and cost-effectiveness, for **c** inorganic, and **d** organic thin films. 'S', 'L', and 'G' denote solid, liquid, and gaseous phase manufacturing processes. Data acquired from Ref(s). [1] (Fig. 1a), [3] and (Fig. 1c) [4, 5] (Fig. 1d). Scores for technological maturity and impact have been re-scaled on a five-tier scale based on reported performances and efficiencies for the thin films

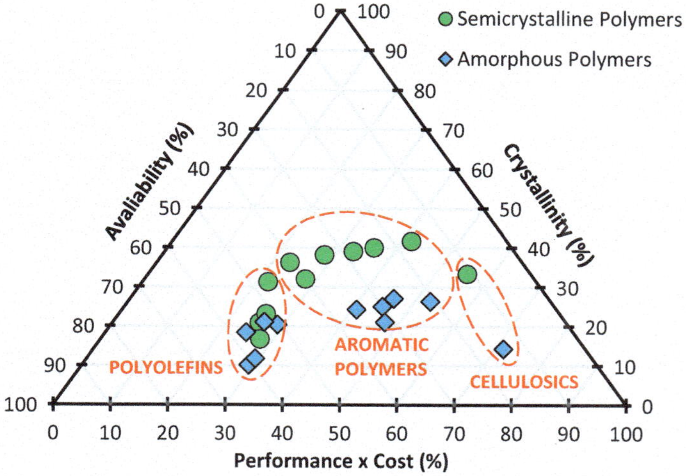

Fig. 2 The tradeoff between the performance time cost, the availability, and the crystallinity (herein, a measure of optimal property and microstructure) for organic thin film precursors. Where blue diamonds relate to semi-crystalline polymers and green circles relate to amorphous polymers. General polymer types are indicated in the figure

to versatile materials, though it is significant to consider the lifecycle aspect of such optimisation [11, 12]. In the context of the present chapter, the material choice herein implies the specific set of functionalities that could be provided to the final thin film for the desired application.

2.1.1 Amorphous Materials

Amorphous thin films are basically structures, with short-range order that result from variations in bond length and angle compared to a crystalline lattice. Typically, these films grow at substrate temperatures, where adatoms (adsorbed atoms) have limited mobility on the substrate surface [13]. At temperatures, adatoms reach an equilibrium with the substrate restricting energy available for diffusion across the surface. Moving these nearly immobile adatoms is essentially integrated into the film structure upon contact with the substrate surface. Another factor that encourages growth is a deposition rate as it hinders adatoms from migrating to more energetically favourable sites for equilibrium due to limited diffusion time. As a result, they become part of the film structure upon striking the substrate surface. Some studies have shown that incorporating gases like oxygen and nitrogen during deposition can lead to the formation of this disordered microstructure by impeding crystallite growth [14, 15]. Both PVD and CVD techniques are suitable for depositing structures. The deposition of the amorphous film(s) is essential in various applications such, as solar cells, transistors, optoelectronics, and dielectric films.

2.1.2 Polyolefins, Polyvinyls and Polyacrylates

Vinyl polymers, including polyvinyl chloride (PVC), polyvinyl alcohol (PVA), polyvinyl acetate (PVAc), polyacrylonitrile (PAN), and polyvinylidene fluoride (PVDF), are commonly employed in thin film applications due to their outstanding film-forming characteristics, flexibility, and chemical resistance [16]. Vinyl polymers are atactic materials that allow for high crystallinity. For PVA, the 1,3-diol linkages as well as some 1,2-diol linkages contribute to good film-forming, emulsifying, and adhesive properties. Similar effects are also observed for dicyano or diacetyl linkages in vinyl polymers. These polymers are easily processed utilising processes such as solvent casting, extrusion, and moulding, making them appropriate for a wide range of applications, including packaging, coatings, and biomedical devices [17–22]. Certain vinyl polymers can be readily functionalised with unique functionalities that can lead to versatile properties and end uses, e.g., divinylbenzene-cross-linked PAN for use in ion exchange resins (Amberlite) [23] or high-grade carbon fibres [24].

The crystallinity of PVA and the dehydration process during film formation can lead to unique fractal features in thin films, suggesting crystallinity owing to the simple linear chains, is a key factor in determining the final film structure [25]. Surface treatments, such as light plasma oxidation, can significantly enhance the wettability of PVA thin films on poly(dimethylsiloxane) (PDMS) substrates [26], demonstrating that surface modifications can be used to tailor thin film properties. The linear structure incurs low steric inhibitions and allows for versatile modification of PVA via the fine-tuning of the crystallisation rate, adsorption onto hydrophobic substrates (e.g., PDMS), cross-linking (by heating/dehydration) [17, 27] and tailoring hydrophobic interactions with other film components [28, 29]. This can allow for optimal thickness, functionality, and wettability control [28]. This is particularly much suited for multilayering applications, e.g., packaging.

Polyacrylamides, such as polyacrylamide (PAM) and poly(N-isopropylacrylamide) (PNIPAM), are commonly employed in thin film formulations for their high-water absorption capacity, biocompatibility, and stimuli-responsive behaviour [30]. These polymers can undergo significant volume changes in response to temperature, pH, or solvent composition changes, making them ideal for applications in drug delivery, tissue engineering, and smart coatings [31]. Adjacent hydrophilic polymer layers, such as poly(acrylic acid) and poly(styrenesulfonate), can alter the physical ageing of thin polymer films and their residual stress [32]. Photoreactivity of polyacrylamides also allows for improved control over film thickness and uniformity, which can affect dynamic contact angles and frictional coefficients in photoreactive PAM thin films [33]. PAM can also be grafted onto cellulosic (e.g., chitosan) films, resulting in surface reconstruction and polyacrylamide growth inside the soft substrate [34]. This allows for higher amounts of grafted polymer compared to traditional hard substrates.

Lipid monolayers and bilayers supported on polyacrylamide films can be utilised to create biocompatible receptive surfaces for biosensors or surfaces that are biofunctionalized [35]. Hydrogen bonds arise between the PAM amino groups and silicon substrates, causing side-group orientation leading to improved thickness control.

Surface morphology and separation performance of thin film composite polyamide membranes depend strongly on their porous substrate, with smaller pores inhibiting leaf-like features and larger pores diminishing surface roughness [36, 37].

2.1.3 Aromatic Polymers

Polyphthalamides (PPAs) are engineering thermoplastics known for their high heat resistance, mechanical strength, and chemical resistance. These polymers exhibit excellent dimensional stability and resistance to creep and fatigue, making them suitable for demanding applications in automotive, electrical, and industrial sectors, where durability and performance are critical [38, 39]. The polyphenol moiety in PPA allows for high thermal and mechanical resistance due to stiff rod-like chains and strong hydrogen bonding, leading to effective crystalline microdomains and high intermolecular packing [40]. The mechanical behaviour and application performance of semi-aromatic polyamide thin films are influenced by temperature-dependent microdeformation mechanisms. These structures typically come about via three known mechanisms: chain scission crazing, shear deformation, and chain disentanglement crazing [41]. Functionalisation of aromatic polyamide is relatively convenient and allows to develop ion exchange membranes with enhanced hydrophilicity, acid resistance, and reduced fouling [42]. Aromatic polyamides can be functionalised without compromising their performance, e.g., membrane permeability and selectivity [43, 44].

Polystyrene (PS) is a versatile thermoplastic polymer with a wide range of applications in thin film technology. PS films offer excellent optical clarity, stiffness, and dimensional stability, making them ideal for applications such as packaging, optical films, and microfluidic devices [45, 46]. Nanoporous PS thin films with homogeneously distributed carboxylic acid functions can be patterned using light, allowing for selective generation of porosity and fluorescent patterns [47]. The cohesion mechanisms of PS thin films are influenced by film preparation conditions, with annealed films preventing further cohesion unless partial dewetting occurs [48]. Nanometre-thin PS films demonstrate altered electrical properties when subjected to different annealing temperatures, which is crucial for their application in organic electronics [45]. PS thin films have high functionality and microstructure properties, with two different cohesion mechanisms observed: equilibrium desorption and nonlinear force-extension behaviour [49]. Guided dewetting of PS thin films on chemically patterned substrates can produce ordered patterns with applications in proteomics, biosensors, and single-cell studies [25].

Polyimide thin films possess excellent thermal stability, with high glass transition temperatures (T_g) exceeding 300 °C [50]. This property makes them suitable for use in high-temperature environments, such as aerospace, automotive, and electronic applications. Films are derived from polyimide polymers, which are synthesised through the reaction of dianhydride and diamine monomers. Polyimide thin films exhibit high tensile strength and modulus and good resistance to abrasion and tear [51]. Polyimide thin films have excellent electrical insulation properties, with low dielectric constants and high breakdown voltages [52]. Biphenyl polyamide thin films have a three-layer structure, with a dense middle layer responsible for separation and a loose surface layer for fluid transport. The aromaticity of the aromatic polyamides permits in-situ thickness and free volume control that can lead to uniform and tailored micro/nano-porous structures [53].

2.1.4 Polyketones and Polyetherketones

Polyketones, such as polyetherketone (PEK) and polyetheretherketone (PEEK), are high-performance engineering thermoplastics with excellent mechanical, thermal, and chemical properties [54, 55]. Polyketone thin films can be synthesised with tunable functionalities, allowing for micro- and nano-engineered materials to be designed. Polyketone thin films with favourable surface roughness can enhance surface area and achieve high-water permeance and salt rejection in reverse osmotic desalination of NaCl solution [56]. Noncovalent functionalisation using polyaromatic hydrocarbons (PAHs) such as 1-aminopyrene shows significant improvement in water vapour transmission rate and barrier performance [57]. Blending with hydrophilic materials such as vinyl alcohols leads to higher oxygen barrier properties. Polyketones exhibit either a tortuous diffusion path for O_2 and hydrogen bonding interactions that restrict O_2 diffusion [58]. Polyketone substrates with appropriate surface roughness can generate favourable "ridge-and-valley" structures, enhancing water permeance and salt rejection in reverse osmosis desalination processes. Thin films of PEEK blends show localised shear deformation at low temperatures, with high PEEK content suppressing disentanglement crazing and allowing for consistent bulk deformation behaviour. Polyketone films offer high-temperature resistance, low moisture absorption, and exceptional chemical resistance, making them suitable for demanding applications in automotive, aerospace, and oil and gas industries [59].

2.1.5 Zwitterionic Compounds

Zwitterionic polymer films such as polybetaines, sulphobetaines, and carboxybetaines; are characterised by the presence of both positive and negative charges within their structure [60, 61]. Zwitterionic polymers exhibit strong hydration effects and resistance to adsorption by amine-functionalised substrates (i.e., substrate) [62]. These properties are foundational to their exceptional properties such as lubrication at interfaces, promotion of protein stabilities, and antifreezing in solutions [60]. Zwitterionic films have higher water and salt permeability compared to non-zwitterionic counterparts due to interactions between zwitterions, water molecules, and salt ions, suggesting potential as antifouling coatings [61, 63]. The zwitterionic polymer is more hydrophilic when positive and negative charges are dispersed in the same side chain due to the higher density of the cationic and anionic groups, which affects the polymer's characteristics and uses [62]. In zwitterionic hydrogels, dipole–dipole and electrostatic interactions cause the zwitterionic chains to cross-link, resulting in a dynamic cross-linked network [64]. Cross-linked polymeric ionic liquid iongel films based on zwitterionic salts exhibit tunable conductivity and mechanical properties, indicating their potential as conducting materials. Grafted zwitterionic polymer surfaces exhibit superhydrophilicity, strong salt-affinity, and antifouling properties, with potential applications in biosystems, separation processes, and as "stealth" coatings for nanomedicines [65]. Zwitter-wettable nanostructured films demonstrate antifogging properties by allowing water vapour absorption while maintaining hydrophobicity to liquid water, showcasing the importance of water permeation rate in surface design [66].

2.1.6 Celluloses

One of the most remarkable properties of cellulose is its inherent strength and stiffness, which stems from its highly crystalline structure [67]. Cellulose molecules arrange themselves into long, linear chains that can form strong hydrogen bonds with neighbouring chains, resulting in robust, stable structures [68]. This strength makes cellulose thin films suitable for various applications where mechanical integrity is crucial, such as in packaging materials, membranes, and coatings. In addition to its mechanical properties, cellulose is also highly versatile regarding its surface chemistry [69, 70]. Functional groups present on the cellulose molecule, such as hydroxyl (–OH) groups, offer sites for chemical modification and functionalisation. Their transparency and high surface area-to-volume ratio make cellulose thin films promising candidates for use in flexible electronics, optical devices, and sensors. For example, surface modification techniques can introduce hydrophobic or hydrophilic properties, enhance adhesion [71–74], or incorporate bioactive molecules for biomedical applications [75].

Cellulose is a promising sustainable, greener material to develop thin films for versatile applications, though moisture-sensitivity, plasticity, and hydrophilicity have been critical in determining its widespread use in lieu of most petroleum-derived polymers [76–78]. Cellulose thin films also exhibit excellent optical transparency, making them suitable for optical and electronic applications [79, 80]. Cellulose thin films can also be engineered to degrade at controlled rates via surface hydrophobisation, allowing for sustained release of therapeutic agents or growth factors within the body [81]. Layer-by-layer (LbL) assembly, solvent casting, spin coating, and other deposition processes can all be used to create cellulose thin films, giving exact control over the composition, thickness, and shape of the films [82–84].

Sustainability entails choosing materials that minimise environmental impact throughout their lifecycle, from sourcing to disposal, while cost-optimisation focuses on maximising performance at minimal expense [85]. Green material design emphasises the use of renewable resources, eco-friendly processing methods, and recyclable materials. This figure illustrates how different materials incur varying costs against performance and properties, highlighting the importance of balancing efficiency, affordability, and environmental responsibility [86]. Additionally, crystallinity serves as a critical measure of material quality, with highly crystalline structures often

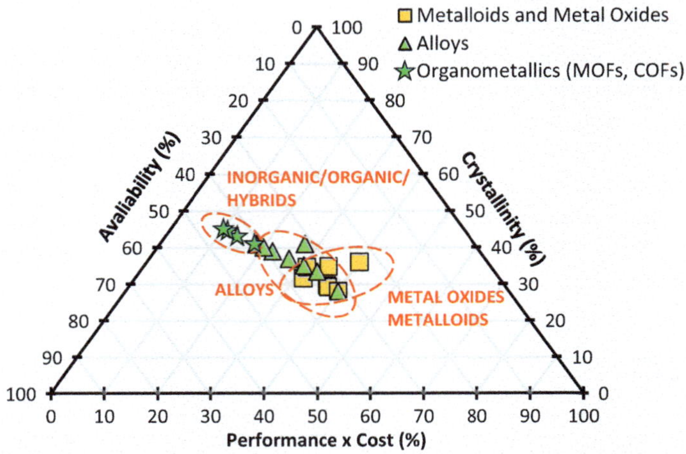

Fig. 3 The tradeoff between the performance time cost, the availability, and the crystallinity (herein, a measure of optimal property and microstructure) for inorganic thin film precursors. Where gold squares relate to metalloids and metal oxides, lime green triangles as alloys, and parrot green stars as organometallic polymers. General inorganic material types are indicated in the figure

exhibiting superior properties for specific applications, such as mechanical strength or electrical conductivity [22, 87–90]. Figures 2 and 3 map out how different inorganic/hybrid and organic thin film materials span in terms of the tradeoff between the categories (i) performance time cost, (ii) availability, and (iii) their crystallinity. The individual materials have been plotted on a five-tier ranking basis, where '5' indicates the highest score and '1' is the lowest score possible for each of the three categories.

2.1.7 Inorganic Materials (Organometallics)

Inorganic materials play a crucial role in thin film applications owing to their distinct properties, including high-temperature stability, optical transparency, electrical conductivity, and mechanical strength. These materials are utilised across diverse industries, ranging from electronics and optoelectronics to coatings, sensing, and energy storage [91]. Silicon dioxide (SiO_2) stands out as one of the most prevalent inorganic materials in thin film technology. Renowned for its excellent thermal stability, chemical inertness, and electrical insulation properties, SiO_2 thin films are commonly employed as passivation layers in semiconductor devices, dielectric layers in capacitors, and protective coatings in optics [6]. Titanium dioxide (TiO_2) is valued for its high refractive index, photocatalytic activity, and UV absorption capabilities [92]. Moreover, TiO_2 thin films are integral to gas sensors, electrochromic devices, and biomedical coatings.

Aluminium oxide (Al_2O_3) exhibits exceptional mechanical strength, thermal stability, and chemical resistance. These attributes make Al_2O_3 thin films ideal for use as barrier coatings, insulating layers, and protective coatings in electronics, packaging, and aerospace applications [93]. Additionally, they serve as dielectric layers in capacitors, passivation layers in microelectronic devices, and protective coatings in optical and MEMS (Microelectromechanical Systems) devices. A transparent conductive oxide (TCO) material with exceptional optical transparency and electrical conductivity is indium tin oxide (ITO). ITO thin films are frequently utilised as transparent electrodes in solar cells, optoelectronic devices, touch screens, and displays, providing efficient light transmission and electrical conductivity [94]. Zinc oxide (ZnO) has a wide bandgap, high electron mobility, and piezoelectric properties. ZnO thin films are applied in UV photodetectors, gas sensors, thin film transistors, and piezoelectric devices [95]. Moreover, they play a crucial role in energy harvesting applications, including dye-sensitised solar cells and piezoelectric generators (Fig. 4).

Copper indium gallium selenide (CIGS) and Cadmium telluride (CdTe) are common semiconductor materials employed in thin film solar cells. CIGS thin films offer high absorption coefficients, excellent photovoltaic efficiency, and potential for low-cost solar energy generation [104]. These films are deposited onto flexible substrates, enabling light-weight and flexible solar panels. The active layers of the CdTe-based thin film solar cells are only a few microns thick, or one-tenth the diameter of a human hair. CdTe thin films are deposited using a variety of methods, such as sputtering, chemical spray pyrolysis, electrodeposition, and close-spaced vapour transport (CSVT) [105]. These cells are the second most common photovoltaic technology globally, representing approximately 5% of the world market.

2.2 *Surface Modification and Functionalisation Approaches*

Grafting from and grafting to approaches involve attaching functional groups or polymers directly onto the surface of thin films [106]. In *grafting from* approach, polymerisation reactions are initiated from the surface of the film, leading to the growth of polymer chains directly attached to the substrate [107]. Conversely, in *grafting to* approach, pre-synthesised polymers or functionalised molecules are chemically linked to the surface of the film. These approaches offer precise control over surface chemistry and composition [108].

2.2.1 '*Grafting from*' Approaches

Grafting from approach offers precise control over the density, length, and composition of grafted chains, enabling the tailoring of surface properties for specific applications. Surface-initiated polymerization (SIP) is a widely used technique that starts the formation of polymer chains from the substrate surface by immobilising polymerization initiators on the surface of inorganic thin films [109]. This technique allows

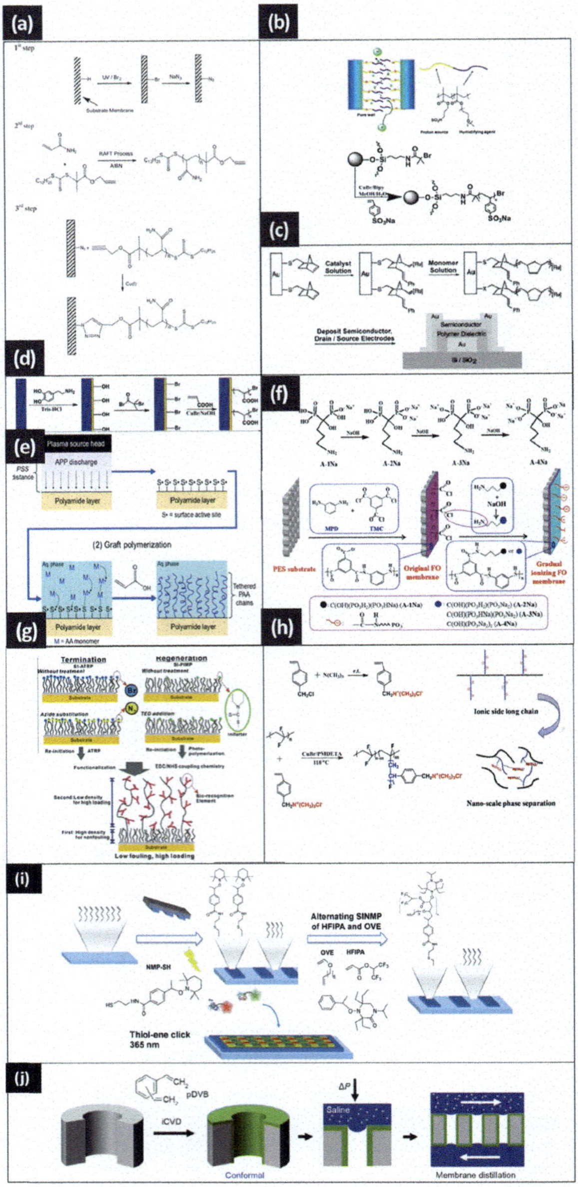
(a)
(b)
(c)
Catalyst Solution
Monomer Solution
Deposit Semiconductor, Drain / Source Electrodes
Semiconductor
Polymer Dielectric
Si / SiO2
(d)
(e)
Plasma source head
APP discharge
Polyamide layer
S• = surface active site
(2) Graft polymerization
Tethered PAA chains
M = AA monomer
(f)
PES substrate
Original FO membrane
Gradual ionizing FO membrane
(g)
Termination
Regeneration
Low fouling, high loading
(h)
Ionic side long chain
Nano-scale phase separation
(i)
Alternating SINMP of HFIPA and OVE
OVE
HFIPA
NMP-SH
Thiol-ene click 365 nm
(j)
pDVB
iCVD
Saline
Conformal
Membrane distillation

◀**Fig. 4** **a** Three-step click coupling modification for PPMM membranes. Adapted with permission from Ref. [96]. Copyrights 2014 Elsevier B.V. **b** pore-filling surface and inorganic surface grafting polymerisations using ATRP. Adapted with permission from Ref. [97]. Copyrights 2014 Elsevier B.V. **c** Dielectric polymer thin films based on Pentacene-Si/SiO_2 from SI-ROP approach. Adapted with permission from Ref. [98]. Copyrights 2004 American Chemical Society. **d** ATRP-initiated monolayer deposition over nylon surfaces. Adapted with permission from Ref. [96]. Copyrights 2014 Elsevier B.V. **e** Plasma surface-activated graft polymerisation of SWRO membranes with tethered PAAs. Adapted with permission from Ref. [99]. Copyrights 2014 Elsevier B.V. **f** Alendronates and Alendronate-grafting on polyamide surfaces of PES membranes. Adapted with permission from Ref. [100]. Copyrights 2018 Elsevier B.V. **g** ultra-low-fouling first layer and high-loading second layer achieved by both ATRP and NMP approach. Adapted with permission from Ref. [101]. Copyrights 2016 Elsevier B.V **h** PVDF-g-QVBC grafter polymer membranes for nanoscale phase separation prepared by ATRP. Adapted with permission from Ref. [96]. Copyrights 2014 Elsevier B.V. **i** Microprinting biochip preparation using NMP approach. Adapted from Ref. [102]. Copyrights 2021 Elsevier B.V. **j** Solvent-less fluorine-free hydrophobic conformal coating based on iCVD for membrane distillation. Adapted with permission from Ref. [103]. Copyrights 2016 Elsevier B.V

for the synthesis of polymer brushes with controlled thickness and composition, providing versatile surface functionalities such as enhanced lubricity, biocompatibility, or stimuli-responsiveness. Table 1 presents the process, microstructure, and property outcomes associated with various grafting from approaches used to develop thin films for various applications.

Another prevalent 'grafting from' approach for surface modification of inorganic thin films is atom transfer radical polymerisation (ATRP). Functional monomers are polymerised in the presence of transition metal catalysts immobilised onto the substrate surface, enabling the controlled growth of polymer chains with well-defined molecular weights and architectures [97, 110]. ATRP-based surface modification techniques find applications in various fields, where tailored surface properties are essential for device performance and functionality [108]. This method offers excellent control over polymer chain growth and end-group functionality, allowing for the precise tuning of surface properties such as adhesion, wettability, and surface energy [109]. ATRP-based surface modification techniques are widely used in applications such as surface patterning, biomaterials, and responsive coatings.

Additionally, initiated chemical vapour deposition (iCVD) is a versatile 'grafting from' approach utilised for surface modification of inorganic thin films. In iCVD, monomer vapour is polymerised on the substrate surface using thermal or plasma-initiated reactions, forming conformal and pinhole-free polymer coatings [111, 112]. This technique offers advantages such as solvent-free processing, uniform film deposition, and the ability to incorporate a wide range of functional groups into the polymer coatings. iCVD-based surface modification methods are employed in various applications, including barrier coatings, antifouling surfaces, and flexible electronics, where precise control over surface properties and film thickness is crucial [113, 114]. On complicated surfaces with micro- and nanoscale features, well-defined thin polymeric films can be fabricated using oxidative chemical vapour deposition (oCVD) and initiated chemical vapour deposition (iCVD) processes, avoiding wetting and solution effects and allowing for the creation of conformal films [115].

Table 1 PPSP outcomes for various *'Grafting To'* and *'Grafting From'* approaches for thin film surface modification

	Process parameters *f*(T, P, Cat., Solv.)	Impact on property	Impact on microstructure	Application/Performance outcomes	References
Radical Polymerisation (e.g., ATRP, RAFT)	Elevated temperatures (e.g., 80 °C)	Enhanced chemical and thermal stability Improved mechanical properties	Densely grafted polymer chains on the surface Increased surface roughness and area	Surface modification of membranes for water purification Improved fouling resistance and separation efficiency	[131]
	Typically ambient pressure	Increased surface wettability	Homogeneous layer of grafted chains Controlled thickness and density	Enhanced colloidal stability Controlled release applications possible (thin film drug delivery)	[132]
	Often requires radical initiators (e.g., alkyne-CTA, AIBN)	Increased surface energy	Well-defined grafting density and orientation Increased surface chemistry and reactivity	Stimuli-responsive properties for sensing and drug delivery	[119]
	Selection depends on the compatibility with the substrate (e.g., DMF for PDADMAC)	Improved adhesion to the substrate Reduced aggregation of grafted chains	Tailored interpenetrating network structure Higher mechanical strength and flexibility	Enhanced adhesion to different substrates Improved surface properties	[133]
Click Chemistry (e.g., CuAAC, RuAAC, SPAAC)	Varies based on the reactivity of functional groups (e.g., 25–100°C)	Preserves the properties of the pristine polymer, facilitates post-modification	Monolayer or multilayer formation Functionalisation of surface polymers affects the surface chemistry	Biosensing Bioimaging applications	[134, 135]
	Often under ambient pressure	Retention of the bulk properties of the substrate Compatibility with sensitive materials	Well-defined surface layer Controlled thickness and chemical composition	Controlled drug release Tissue engineering applications	[136]

(continued)

Table 1 (continued)

	Process parameters f(T, P, Cat., Solv.)	Impact on property	Impact on microstructure	Application/Performance outcomes	References
	May require coupling agents	Enhanced chemical versatility, preservation of the substrate's structural integrity	Uniform and highly functionalised surface layer Changes in surface energy and adhesion	Functional coatings for corrosion protection Improved surface activity for protein rejection	[96, 137]
	Depends on the compatibility with both substrates	Improved efficiency of grafting Reduced side reactions	Well-controlled interfacial layer, Changes in interfacial properties and stability	Enhanced biocompatibility Antimicrobial properties	[43]
Condensation Polymerisation (e.g., Polycondensation)	Varies based on the reactivity of functional groups (80–100 °C)	Preserves the properties of the pristine polymer Facilitates post-modification	Monolayer or multilayer formation Changes in surface energy and adhesion	High dielectric constants achieved via the surface treatment	[138]
	Typically conducted under ambient pressure	Retention of the bulk properties of the substrate Compatibility with sensitive materials	Well-defined surface layer Controlled thickness and chemical composition	Oriented aromatic polyesters which improve mechanical strength and surface activity	[139]
	May require coupling agents	Enhanced chemical versatility Retention of substrate's structural integrity	Uniform and highly functionalised surface layer Affect on the surface energy and adhesion	Mixed matrix membranes for gas separation with high efficiency	[140]

(continued)

Table 1 (continued)

	Process parameters *f*(T, P, Cat., Solv.)	Impact on property	Impact on microstructure	Application/Performance outcomes	References
	Depends on the compatibility with both substrates	Improved efficiency of grafting Reduced side reactions	Well-controlled interfacial layer Stability		
Ring-opening Polymerisation (e.g., polysiloxanes, polycarbonates, polyester, polyamide)	Elevated temperatures (e.g., ≤ 100°C)	Enhanced chemical and thermal stability, improved mechanical properties	Formation of densely grafted polymer chains on the surface, leading to increased surface roughness and area	Water purification, resulting in improved anti-chlorine resistance Separation efficiency	[142]
	Typically under ambient pressure	Increased surface wettability, altered surface energy	Formation of a homogeneous layer of grafted chains with controlled thickness and density	Enhanced colloidal stability Improved hydrophilicity control	[143]
	Often requires radical initiators	Increased surface activity Improved functionality	Well-defined grafting density Increased orientation, influencing the surface chemistry and reactivity	Heightened thermoresponsiveness Improved nanorod shape control	[144, 154]
	Selection depends on the compatibility with substrate	Improved adhesion to the substrate Reduced aggregation of grafted chains	Tailored interpenetrating network structure Improved mechanical strength and flexibility	Polymer films for increased barrier performance Improved surface properties	[146]
Atom Transfer Radical Polymerisation (ATRP)	Typically at elevated temperatures (e.g., ~ 100°C)	Enhanced chemical and thermal stability, improved mechanical properties Increased surface roughness and area	Monolayer or multilayer formation on the surface Densely grafted polymer chains on the surface	Chemical-resistant semiconductor thin films with controlled bulk morphology, uniform molecular organisation, and stability	[147]

(continued)

Table 1 (continued)

	Process parameters f(T, P, Cat., Solv.)	Impact on property	Impact on microstructure	Application/Performance outcomes	References
	Typically under ambient pressure	Increased surface wettability, altered surface energy	Homogeneous layer of grafted chains Well-defined surface layer Controlled thickness and density	Reduced cell adhesion for developing antifoulant metal surfaces	[148]
	Often requires radical initiators (e.g., alkyl halides)	Tailored interpenetrating network structure	Well-defined orientation Higher grafting density Increased surface chemistry and reactivity	Thermo-responsive films with significant hydrophobic property control	[149]
	Selection depends on the compatibility with substrate	Improved adhesion to the substrate Reduced aggregation of grafted chains	Improved mechanical strength and flexibility	Area selective control and improved etch resistance	[150]
Initiated Chemical Vapour Deposition (iCVD)	Typically at moderate to elevated temperatures (40–100°C)	Enhanced chemical and thermal stability, improved mechanical properties	Densely surface grafted polymer chains Increased surface roughness and area	Successful coating of soft material substrates with highly durable superhydrophobic coating layers	[115]
	Typically conducted under vacuum or low pressure	Increased surface wettability, altered surface energy	Conformal and pinhole-free polymer coating on the surface	Flexible polymer buffer layer in moisture barrier systems for electronic devices	[151]
	None is required as it is a solvent-free process	Introduction of functional groups for further chemical modification and attachment	Uniform and highly functionalised surface layer Improved the surface energy and adhesion	Cross-linked polymers with high chemical and thermal stability	[152]

(continued)

Table 1 (continued)

	Process parameters *f*(T, P, Cat., Solv.)	Impact on property	Impact on microstructure	Application/Performance outcomes	References
	Solvent-free process	Improved efficiency of grafting reactions, reduced side reactions	Well-controlled interfacial layer Optimised interfacial properties and stability	Thin film electrolyte for solid-state battery applications with high energy storage capacity and thermal stability	[153]
Nitroxide-Mediated Polymerisation (NMP)	Typically conducted at elevated temperatures (e.g., ~ 100°C)	Enhanced chemical and thermal stability, improved mechanical properties	Densely grafted polymer chains on the surface Increased surface roughness and area	Improved wear resistance in PS thin films	[154]
	Typically conducted under ambient pressure	Increased surface wettability Increased surface energy	Homogeneous layer of grafted chains Controlled thickness and density	Nanostructured materials with stimuli-responsiveness, and improved mechanical or barrier properties	[102]
	Often requires radical initiators (e.g., TEMPO, RIT)	Increased surface energy	Well-defined grafting density and orientation Enhanced dispersibility, adjustable surface wettability,	Resistance to dissolution during subsequent layer deposition Electrochemical behaviour remained consistent	[155]
	Selection depends on the compatibility with substrate	Improved adhesion to the substrate Reduced aggregation of grafted chains	Tailored interpenetrating network structure Increased mechanical strength and flexibility		

Note AIBN—Azobisisobutyronitrile, ATRP—Atom-transfer radical polymerisation, CTA—Chain transfer agent, CuAAC—Cu-catalysed Azide-Alkyne Cycloaddition, DMF—Dimethylformamide, PDADMAC—Poly(diallyldimethylammonium chloride, RAFT—Reversible addition-fragmentation chain transfer, RIT (4,4-dimethyloxazoline-2,2,6,6-tetramethylpiperidin-1-yloxyl), RuAAC—Ru-catalysed Azide-Alkyne Cycloaddition, and SPAAC—Strain-promoted Azide-Alkyne Cycloaddition; TEMPO (2,2,6,6-tetramethylpiperidine-1-oxyl) iCVD utilises a thermally labile initiator to enhance deposition rates while maintaining chemical functionalities, making it suitable for delicate substrates, whereas oCVD forms infusible, electrically conductive films directly on substrates by introducing oxidant and monomer simultaneously, collectively representing versatile platform technologies for thin film deposition. The use of iCVD has also demonstrated conformal and partially conformal surfaces that allow for complete or partial control of water vapour permeance and pH-dependence—useful in protein separations [103].

Radical (co)polymerisation involves the initiation and propagation of polymerisation reactions using free radicals generated either on the surface of thin films or in solution. This method enables the direct production of polymer coatings onto the substrate with regulated functionality, content, and thickness [116]. Moreover, 'grafting from' methods for surface modification of inorganic thin films include surface-initiated controlled radical polymerization processes including addition-fragmentation chain transfer (RAFT) polymerization and nitroxide-mediated polymerization (NMP). These methods allow molecular weight control and end-group functionality to be preserved while allowing for the controlled development of polymer chains from the substrate surface [117, 118]. Surface-initiated controlled radical polymerisation methods offer versatility in designing functional coatings with tailored properties, including adhesion, biocompatibility, and surface roughness. These techniques find applications in diverse fields, including coatings, sensors, and biomedical devices, where precise control over surface properties is essential for performance optimisation.

2.2.2 '*Grafting to*' Approaches

Various 'grafting to' approaches are utilised to functionalise or modify the surface of inorganic thin films, enabling the attachment of specific functional groups or molecules onto the substrate [119]. These approaches offer precise control over surface chemistry and properties, enhancing the functionality and performance of thin film materials across diverse applications. One common method involves the use of self-assembled monolayers (SAMs) composed of molecules with tailored functional groups [120, 121]. SAMs can be formed through chemisorption or physisorption onto the surface of inorganic thin films, providing a highly ordered and densely packed molecular layer [122]. By selecting appropriate SAM components, such as alkyl thiols or silanes with desired functional groups, researchers can tailor the surface properties of the thin film, including wettability, adhesion, and biocompatibility.

Alternatively, the use of CVD for polymer thin film deposition preserves the functionalities and enables integration into micro- and nanodevices for improved performance [123]. Grain growth in the process of film formation or post-deposition annealing can significantly influence the mechanical and microstructural features of polycrystalline thin films with a thickness of less than 1 μm [19]. iCVD-based surface modification methods are employed where precise control over surface properties and film thickness is crucial for device performance and reliability [124]. Substrate characteristics that promote wetting and charge conduction are beneficial for the formation of well-defined, ultra-thin films, emphasising the influence of substrate selection on film formation [125].

Another prevalent 'grafting to' approach for surface modification of inorganic thin films is the covalent immobilisation of functional molecules using linker molecules or coupling agents [126]. Linker molecules possess reactive functional groups at both ends, facilitating the attachment of the thin film surface and the desired functional molecules [125, 127, 128]. This method allows for precise control over the surface orientation and density, enabling the creation of customised surfaces with specific properties like enhanced chemical reactivity, biocompatibility, or antimicrobial activity. This method allows for precise control over the surface orientation and grafting density, and surface functionalisation with specific properties like enhanced chemical reactivity, biocompatibility, or antimicrobial activity [129]. The reactive functional groups on the thin film's surface react with complementary functional groups on the target molecule by the use of coupling agents or click chemistry [130]. This approach enables precise control over the attachment of functional groups and ensures high surface coverage with minimal side reactions.

Photoinitiators or photoinitiating systems are employed to initiate cross-linking reactions, leading to the formation of a network structure within the thin film [156]. Photoinitiated cross-linking finds applications in the fabrication of coatings, adhesives, and optical materials. Comparably, functional molecules can be covalently bound to the surface of inorganic thin films by coupling agents such as phosphonic acids or organosilanes. This creates a durable and reliable interface for a range of uses, including surface patterning, sensing, and catalysis [157]. Photopolymerisation can be incorporated into LbL assembly processes to cross-link or stabilise deposited layers, enhancing the mechanical integrity and stability of the thin films [158]. This approach allows for the fabrication of multilayered thin films with tailored properties and functionalities.

In order to create multilayered thin films with regulated thickness and composition, oppositely charged polyelectrolytes or functional molecules are alternately adsorbed onto the substrate surface during LbL deposition [159, 160]. LbL assembly allows for the sequential deposition of oppositely charged polyelectrolytes, including zwitterionic polymers, to form multilayer thin films with tailored properties [60]. Surface-initiated polymerisation techniques enable the grafting of zwitterionic polymer brushes onto substrate surfaces, providing enhanced control over film thickness and surface coverage [65]. LbL assembly techniques offer advantages such as scalability, versatility, and compatibility with a wide range of substrates, making them suitable for various applications in optics, electronics, and biotechnology.

2.3 *Superhydrophobic and Superhydrophilic Thin Films*

Superhydrophobic films are characterised by their extreme water repellency, often referred to as the "lotus effect" due to their resemblance to the water-repellent properties of lotus leaves. These films typically exhibit high contact angles (>150°) and low contact angle hysteresis, causing water droplets to form nearly spherical shapes and roll off the surface effortlessly [161, 162]. The key to achieving superhydrophobicity lies in creating rough micro- or nanostructures on the surface, combined with low surface energy coatings. Superhydrophilic films, on the other hand, exhibit extreme water affinity, causing water droplets to spread rapidly across the surface and form a thin, continuous film. These films typically possess low contact angles (<10°) and high contact angle hysteresis, promoting rapid wetting and spreading of liquids [163]. Superhydrophilic surfaces are often achieved by modifying the surface chemistry to increase surface energy or introducing hydrophilic functional groups.

RAFT polymerisation can result in uniform coatings with excellent mechanical properties and high stability. The controlled nature of RAFT polymerisation enables the incorporation of specific functional groups, facilitating surface modification for enhanced hydrophobicity [96, 133]. Studies have shown that RAFT-synthesised polymers can form densely packed, rough surfaces ideal for superhydrophobicity. ATRP-derived polymers typically exhibit well-defined architectures with controlled chain lengths and end functionalities. ATRP offers versatility and the ability to incorporate hydrophobic side chains, enhancing water repellency and surface roughness [97, 109, 110]. Arguably, RAFT polymerisation excels in providing precise control over polymer structure and composition, leading to uniform coatings with excellent mechanical properties [164] (Fig. 5).

Conventional chemical functionalisation of thin film surfaces has leveraged the Schiff-base reaction. Patil et al. [166] grafted only 0.5% of the amine-modified kaolin clay Si-HNT with long-chain silane onto the dialdehyde cellulose [165]. Then the C_7 long-chain fatty acid was further esterified with the cellulose hydroxyl to obtain the superhydrophobic cotton fabric with a maximum wet contact angle. of 153° [166]. Superhydrophobic coatings prepared via ATRP often demonstrate excellent water repellency due to the presence of hydrophobic side chains. Additionally, the tunable nature of ATRP allows for the incorporation of various additives or nanoparticles to further enhance surface roughness and hydrophobicity [116]. NMP enables the formation of well-defined polymer architectures with tailored properties, contributing to the development of robust and durable superhydrophobic coatings. Furthermore, the compatibility of NMP with a variety of monomers allows for the design of polymers with specific functionalities to optimise surface hydrophobicity [117]. NMP combines controlled polymerisation with functional group tolerance, allowing for the creation of tailored polymers with optimised surface properties.

3 Thin Films and Ultrathin Film Coatings for Versatile Applications

Different designs and fabrication processes are used to deposit thin films precisely controlling their composition, shape, and thickness. For the manufacture of thin films, chemical deposition techniques like as spin coating, dip coating, and LbL assembly

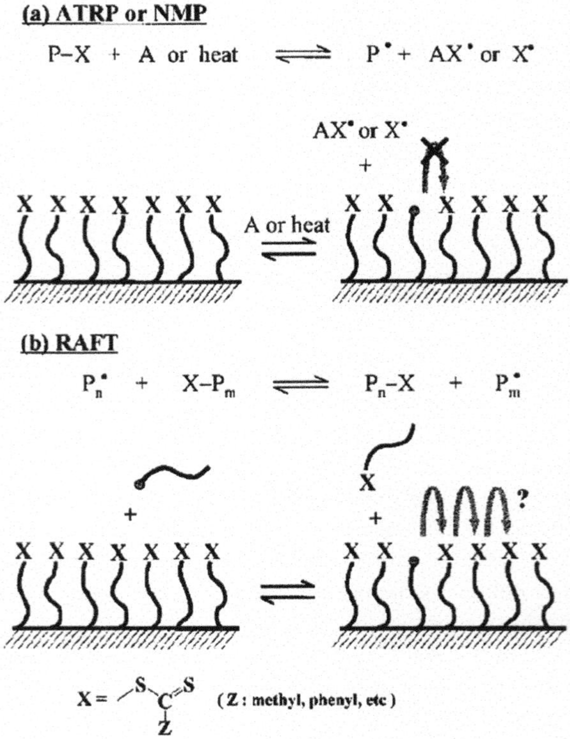

Fig. 5 Schematic representations of the RAFT-based superhydrophobisation of fluoropolymer surface, with substantial hydrophobicity visible. Adapted from Ref. [164] with permission. Copyrights 2001 American Chemical Society

provide great throughput and consistency [7]. Physical deposition techniques such as sputtering, evaporation, and molecular beam epitaxy provide excellent control over film properties and are suitable for depositing thin films on diverse substrates [167]. The various applications that are possible as a result of the versatile syntheses that have been discussed prior.

3.1 Multilayered Packaging

Multilayered packaging systems incorporating thin films offer superior barrier properties against oxygen, moisture, and light, preserving the quality and freshness of packaged goods [168]. By optimising film composition and structure, researchers can tailor packaging materials to meet specific requirements for food packaging, pharmaceuticals, and electronics, reducing food waste and extending product shelf life.

3.2 Extreme Environments—Corrosion Protection and Tribological Coatings

Thin film coatings play a vital role in protecting materials and components operating in extreme environments, including aerospace, automotive, and marine applications [2]. These coatings provide resistance against corrosion, oxidation, and mechanical wear, ensuring the reliability and durability of critical infrastructure and equipment exposed to harsh conditions. Thin film coatings offer effective corrosion protection for metals and alloys by acting as a barrier against corrosive environments. By controlling film composition and morphology, researchers can design coatings with enhanced adhesion and durability, prolonging the lifespan of metallic components in infrastructure, transportation, and manufacturing industries.

Tribological coatings play a crucial role in reducing friction, wear, and surface damage in mechanical systems, thereby enhancing efficiency, reliability, and lifespan [16]. These offer excellent lubricating properties and resistance to abrasion, making them ideal for applications in automotive components, cutting tools, and biomedical implants [168]. Researchers can tailor tribological coatings to meet specific requirements for load-bearing, temperature, and environmental conditions, facilitating advancements in transportation, manufacturing, and healthcare industries.

3.3 Sensing

Thin films play a crucial role in sensing technologies for detecting gases, chemicals, and biological analytes with high sensitivity and selectivity [122, 170, 171]. Functionalised thin films enable the development of portable, rapid diagnostic devices for healthcare, environmental monitoring, and security applications, facilitating early detection and mitigation of potential hazards. The use of sensing devices frequently involves the deposition of electrochemically active nanoparticles such as graphene oxide over substrate surfaces for point of care testing or in vitro pharmacological studies [172].

3.4 Membrane Separation and Purification Processes

Functionalised thin films and specifically, membranes, find applications in membrane separation and purification processes for water treatment, gas separation, and desalination [56, 96, 173]. By controlling pore size, surface chemistry, and morphology, researchers can design membranes with high flux, selectivity, and stability, addressing global challenges related to water scarcity and environmental pollution. Resource recovery and chemical recycling processes have also been proposed based on functional membrane-based systems, yielding significant advantages over conventional chemical separation processes [174].

3.5 *Unit Processes (Syntheses, Reactions and Catalytic Systems)*

Thin films serve as active components in catalytic systems for promoting chemical reactions and synthesis processes with high efficiency and selectivity [175, 176]. By immobilising catalysts on thin film supports, researchers can enhance catalytic activity, recyclability, and durability, enabling sustainable production processes with reduced energy consumption and waste generation.

3.6 *Pharmaceuticals*

Thin film drug delivery systems have gained significant attention in the pharmaceutical industry due to their potential to improve drug efficacy, patient compliance, and therapeutic outcomes. These systems consist of thin polymeric films that can be applied topically, orally, or transdermally to deliver drugs at controlled rates [19, 156]. By modulating film properties and drug release mechanisms, researchers can achieve precise control over drug release rates, minimising fluctuations in plasma drug concentrations and improving therapeutic outcomes. Thin films provide a protective barrier against environmental factors, such as light, moisture, and oxygen, preserving the stability and efficacy of sensitive drugs.

3.7 *Miscellaneous*

Quantum well structures consist of thin film layers with nanoscale thicknesses, enabling the confinement and manipulation of electrons and photons [177]. Magnetic multilayers comprise alternating thin film layers of ferromagnetic and non-magnetic materials, exhibiting unique magnetic properties, such as giant magnetoresistance (GMR) and exchange bias [98]. These multilayers are employed in magnetic storage devices, such as hard disk drives (HDDs) and magnetic random-access memory (MRAM), enabling high-density data storage and fast data access speeds.

4 Challenges and Perspectives in Thin Film Applications

The wide range of applications, for films necessitates physical, mechanical, and chemical properties that are influenced by the structure and morphology resulting from the deposition techniques and parameters used. There are still challenges to overcome in understanding how different phases interact during thin film fabrication. Both CVD and PVD methods have variations that result in the final products lack of reproducibility, poor adhesion to substrates high deposition temperatures limiting substrate choices, and limited control over final properties. Therefore, advancements in deposition technology are crucial for microstructure control and increased deposition rates for cost large-scale applications. While nanostructured films hold promise for fields like microelectronics, optics, photovoltaics, and biomedicine some require transfer to specific substrates, for optimal operation. Improving the transferring techniques is essential to harness the potential of film technology.

Despite the variety of methods to analyse aspects of thin films challenges arise when the film thickness is reduced to a few nanometers. Techniques, like Raman spectroscopy and X-ray diffraction, may be limited by interference from the substrate in the obtained data. This makes using models to extract information necessary. Fu and colleagues developed a modelling technique to study the wrinkling and restabilization of soft films made of various compressible and incompressible hyperplastic materials. This technique combines a path-following continuation method using the asymptotic numerical method (ANM) with spatial discretization using a spectral collocation method [178].

Developing more advanced in vitro models can expedite the identification of films. Despite advancements in deposition and characterisation techniques predicting film properties based on microstructure remains challenging due to factors such as defects, grain boundaries, material phases quantum effects, in thin films, and interface interactions. Therefore, it is essential to develop models that encompass these structural elements to effectively link microstructure, with film characteristics and drive advancements in thin film technology [22].

5 Conclusion

Thin films find diverse applications in various other fields, including energy storage and conversion, electronic devices, biomedical implants, and optical coatings. By continuously exploring novel materials, deposition techniques, and functionalisation strategies, researchers can unlock new opportunities for thin films in addressing emerging societal and technological challenges. The chapter covers the synthesis and surface modification of substrates, focussing on materials like vinyl polymers, polyacrylamides, polyphthalamides, polystyrene, polyimides, polyketones, zwitterionic compounds, and celluloses, each offering unique advantages and applications. It details the characteristics, synthesis methods, and applications of these materials in diverse fields such as packaging, coatings, biomedical devices, and electronics. The discussion also highlights the importance of sustainability, cost-optimisation, and crystallinity in material selection, as well as the trade-offs involved in choosing between different materials based on their performance, availability, and crystallinity. Additionally, it delves into the role of inorganic materials in various thin film applications such as semiconductor devices, gas sensors, solar cells, and optoelectronic devices. A focus on organic preparative methods is retained as they are potential platforms for innovations to come.

Surface modification and functionalisation of inorganic thin films involve grafting from and grafting to approaches, offering precise control over surface chemistry and composition. Grafting from methods initiate polymerisation reactions directly from the film surface, enabling the growth of polymer chains tailored to the substrate. Techniques like surface-initiated polymerisation (SIP) and atom transfer radical polymerisation (ATRP) allow for controlled chain growth and versatile surface functionalities. Initiated chemical vapour deposition (iCVD) offers advantages like solvent-free processing and uniform film deposition. Grafting to approaches involve attaching pre-synthesised polymers or functionalised molecules to the film surface, providing control over surface properties through self-assembled monolayers (SAMs), chemical vapour deposition (CVD), and covalent immobilisation methods. Superhydrophobic and superhydrophilic thin films, achieved through rough micro- or nanostructures and tailored surface chemistry, find applications in various fields. Thin films play crucial roles in multilayered packaging, extreme environment protection, sensing technologies, membrane separation, catalytic systems, pharmaceutical delivery, and advanced material structures like quantum wells and magnetic multilayers.

References

1. Fortune Business Insights.: The global thin film material market size was valued at $12.45 billion in 2021 & is projected to grow from $12.91 billion in 2022 to $17.46 billion by 2029. Web report. https://www.fortunebusinessinsights.com/thin-film-material-market-107350. Accessed 04 March 2024
2. Sathish, M., Radhika, N., Saleh, B.: Current status, challenges, and future prospects of thin film coating techniques and coating structures. J. Bio- Tribo-Corrosion **9**(2), 35 (2023). https://doi.org/10.1007/s40735-023-00754-9
3. Sheppard, B., Shaw, C., Williams, B., Persaud, J., Bromberge, J.: Factory of the Future: Advanced Manufacturing Technologies, Including 3-D Printing, Will Disrupt How We Manufacture. Are You Ready to Implement Them? (2014). https://www.mckinsey.com/~/media/mckinsey/ourpeople/bensheppard/factoryofthefuture-advancedmanufacturingtechnologies.ashx.
4. Guo, X., Xu, Y., Ogier, S., Ng, T.N., Caironi, M., Perinot, A., Li, L., Zhao, J., Tang, W., Sporea, R.A., Nejim, A., Carrabina, J., Cain, P., Yan, F.: Current status and opportunities of organic thin-film transistor technologies. IEEE Trans. Electron Dev. **64**(5), 1906–1921 (2017). https://doi.org/10.1109/TED.2017.2677086

5. Crivello, C., Sevim, S., Graniel, O., Franco, C., Pané, S., Puigmartí-Luis, J., Muñoz-Rojas, D.: Advanced technologies for the fabrication of MOF thin films. Mater. Horizons **8**(1), 168–178 (2021). https://doi.org/10.1039/D0MH00898B
6. Eslamian, M.: Inorganic and organic solution-processed thin film devices. Nano-Micro Lett. **9**(1), 3 (2017). https://doi.org/10.1007/s40820-016-0106-4
7. Abu-Thabit, N.Y., Makhlouf, A.S.H.: Fundamental of smart coatings and thin films: synthesis, deposition methods, and industrial applications. In: Advances in Smart Coatings and Thin Films for Future Industrial and Biomedical Engineering Applications; Elsevier, pp. 3–35 (2020). https://doi.org/10.1016/B978-0-12-849870-5.00001-X
8. Wang, X., Li, B., Xu, X., Han, Y.: Surface roughening, premelting and melting of monolayer and bilayer crystals. Soft Matter **17**(3), 688–693 (2021). https://doi.org/10.1039/D0SM01589J
9. Dehm, G., Cairney, J.: Implication of grain-boundary structure and chemistry on plasticity and failure. MRS Bull. **47**(8), 800–807 (2022). https://doi.org/10.1557/s43577-022-00378-3
10. Xie, W., Deng, Z., Liu, Z., Famprikis, T., Butler, K.T., Canepa, P.: Effects of grain boundaries and surfaces on electronic and mechanical properties of solid electrolytes. Adv. Energy Mater. (2024). https://doi.org/10.1002/aenm.202304230
11. Bustamante-Torres, M., Romero-Fierro, D., Arcentales-Vera, B., Pardo, S., Bucio, E.: Interaction between filler and polymeric matrix in nanocomposites: magnetic approach and applications. Polymers (Basel) **13**(17), 2998 (2021). https://doi.org/10.3390/polym13172998
12. Zafar, M.F., Siddiqui, M.A.: Effect of filler loading and size on the mechanical and morphological behaviour of natural fibre-reinforced polystyrene composites. Adv. Mater. Process. Technol. **7**(4), 647–659 (2021). https://doi.org/10.1080/2374068X.2020.1793261
13. Pelliccione, M., Lu, T.-M.: Evolution of Thin Film Morphology, Materials Science, vol. 108. Springer, New York (2008). https://doi.org/10.1007/978-0-387-75109-2
14. Kim, I., Shim, C., Kim, S.W., Lee, C., Kwon, J., Byun, K., Jeong, U.: Amorphous carbon films for electronic applications. Adv. Mater. 35(43) (2023). https://doi.org/10.1002/adma.202204912
15. Choy, K.: Chemical vapour deposition of coatings. Prog. Mater. Sci. **48**(2), 57–170 (2003). https://doi.org/10.1016/S0079-6425(01)00009-3
16. Whiteley, K.S., Heggs, T.G., Koch, H., Mawer, R.L., Immel, W.: Polyolefins. In :Ullmann's Encyclopedia of Industrial Chemistry. Wiley-VCH Verlag GmbH & Co. KGaA, Weinheim, Germany (2000). https://doi.org/10.1002/14356007.a21_487
17. Peppas, N.A.: Crystallization of polyvinyl alcohol-water films by slow dehydration. Eur. Polym. J. **12**(8), 495–498 (1976). https://doi.org/10.1016/0014-3057(76)90004-5
18. Nuruddin, M., Chowdhury, R.A., Szeto, R., Howarter, J.A., Erk, K.A., Szczepanski, C.R., Youngblood, J.P.: Structure-property relationship of cellulose nanocrystal-polyvinyl alcohol thin films for high barrier coating applications. ACS Appl. Mater. Interfaces **13**(10), 12472–12482 (2021). https://doi.org/10.1021/acsami.0c21525
19. Khlyustova, A., Cheng, Y., Yang, R.: Vapor-deposited functional polymer thin films in biological applications. J. Mater. Chem. B **8**(31), 6588–6609 (2020). https://doi.org/10.1039/D0TB00681E
20. Zhu, X., Cheng, X., Luo, X., Liu, Y., Xu, D., Tang, X., Gan, Z., Yang, L., Li, G., Liang, H.: Ultrathin thin-film composite polyamide membranes constructed on hydrophilic Poly(Vinyl Alcohol) decorated support toward enhanced nanofiltration performance. Environ. Sci. Technol. **54**(10), 6365–6374 (2020). https://doi.org/10.1021/acs.est.9b06779
21. Mavukkandy, M.O., McBride, S.A., Warsinger, D.M., Dizge, N., Hasan, S.W., Arafat, H.A.: Thin film deposition techniques for polymeric membranes–a review. J. Memb. Sci. **610**, 118258 (2020). https://doi.org/10.1016/j.memsci.2020.118258
22. Acosta, E.: Thin films/properties and applications. In: Thin Films. IntechOpen (2021). https://doi.org/10.5772/intechopen.95527
23. de Dardel, F., Arden, T.V.: Ion exchangers. In: Ullmann's Encyclopedia of Industrial Chemistry. Wiley (2008). https://doi.org/10.1002/14356007.a14_393.pub2

24. Serkov, A.T., Radishevskii, M.B.: Status and prospects for production of carbon fibres based on polyacrylonitrile. Fibre Chem. **40**(1), 24–31 (2008). https://doi.org/10.1007/s10692-008-9012-y
25. Reiter, G.: Probing Properties of Polymers in Thin Films Via Dewetting, pp. 29–63 (2012). https://doi.org/10.1007/12_2012_174
26. Karki, A., Nguyen, L., Sharma, B., Yan, Y., Chen, W.: Unusual morphologies of Poly(Vinyl Alcohol) Thin films adsorbed on Poly(Dimethylsiloxane) substrates. Langmuir **32**(13), 3191–3198 (2016). https://doi.org/10.1021/acs.langmuir.6b00470
27. Sau, S., Pandit, S., Kundu, S.: Crosslinked Poly (Vinyl Alcohol): structural, optical and mechanical properties. Surf. Interfaces **25**, 101198 (2021). https://doi.org/10.1016/j.surfin.2021.101198
28. Jayaraman, K., Hsu, S.L., McCarthy, T.J.: Versatile multilayer thin film preparation using hydrophobic interactions, crystallization, and chemical modification of Poly(Vinyl Alcohol). Langmuir **23**(6), 3260–3264 (2007). https://doi.org/10.1021/la063451r
29. Kozlov, M., Quarmyne, M., Chen, W., McCarthy, T.J.: Adsorption of Poly(Vinyl Alcohol) onto hydrophobic substrates. a general approach for hydrophilizing and chemically activating surfaces. Macromolecules **36**(16), 6054–6059 (2003). https://doi.org/10.1021/ma021681g
30. Bashir, S., Hina, M., Iqbal, J., Rajpar, A.H., Mujtaba, M.A., Alghamdi, N.A., Wageh, S., Ramesh, K., Ramesh, S.: Fundamental concepts of hydrogels: synthesis, properties, and their applications. Polymers (Basel) **12**(11), 2702 (2020). https://doi.org/10.3390/polym12112702
31. Funaki, M., Janmey, P.A.: Technologies to engineer cell substrate mechanics in hydrogels. In: Biology and Engineering of Stem Cell Niches, pp. 363–373. Elsevier (2017). https://doi.org/10.1016/B978-0-12-802734-9.00023-8
32. Lewis, E.A., Stafford, C.M., Vogt, B.D.: Effect of adjacent hydrophilic polymer thin films on physical aging and residual stress in thin films of Poly(Butylnorbornene- Ran -hydroxyhexafluoroisopropyl Norbornene). J. Polym. Sci. Part B Polym. Phys. **57**(15), 992–1000 (2019). https://doi.org/10.1002/polb.24855
33. Smithson, R.L.W., Evans, D.F., Monfils, J.D., Guire, P.E.: Ultrathin film characterization: photoreactive polyacrylamide. Colloids Surf. B Biointerfaces **1**(6), 349–355 (1993). https://doi.org/10.1016/0927-7765(93)80029-X
34. Chu, E., Sidorenko, A.: Surface reconstruction by a "Grafting Through" approach: polyacrylamide grafted onto chitosan film. Langmuir **29**(40), 12585–12592 (2013). https://doi.org/10.1021/la402609w
35. Kühner, M., Tampé, R., Sackmann, E.: Lipid mono- and bilayer supported on polymer films: composite polymer-lipid films on solid substrates. Biophys. J. **67**(1), 217–226 (1994). https://doi.org/10.1016/S0006-3495(94)80472-2
36. Peng, L.E., Yao, Z., Yang, Z., Guo, H., Tang, C.Y.: Dissecting the role of substrate on the morphology and separation properties of thin film composite polyamide membranes: seeing is believing. Environ. Sci. Technol. **54**(11), 6978–6986 (2020). https://doi.org/10.1021/acs.est.0c01427
37. Lu, X., Mi, Y.: Characterization of the interfacial interaction between polyacrylamide and silicon substrate by fourier transform infrared spectroscopy. Macromolecules **38**(3), 839–843 (2005). https://doi.org/10.1021/ma0486896
38. García, J.M., García, F.C., Serna, F., de la Peña, J.L.: High-performance aromatic polyamides. Prog. Polym. Sci. **35**(5), 623–686 (2010). https://doi.org/10.1016/j.progpolymsci.2009.09.002
39. Gonzalez de Gortari, M., Misra, M., Mohanty, A.K.: Polyphthalamide polymers: a review on synthesis, properties, and advance manufacturing and emerging applications. J. Appl. Polym. Sci. **139**(40) (2022). https://doi.org/10.1002/app.52965
40. Chong, J.Y., Wang, R.: From micro to nano: polyamide thin film on microfiltration ceramic tubular membranes for nanofiltration. J. Memb. Sci. **587**, 117161 (2019). https://doi.org/10.1016/j.memsci.2019.06.001
41. Brûlé, B., Kausch, H., Monnerie, L., Plummer, C.J., Halary, J.: Microdeformation mechanisms in thin films of amorphous semi-aromatic polyamides. Polymer (Guildf). **44**(4), 1181–1192 (2003). https://doi.org/10.1016/S0032-3861(02)00862-5

42. Zhang, X., Lv, Y., Yang, H.-C., Du, Y., Xu, Z.-K.: Polyphenol coating as an interlayer for thin-film composite membranes with enhanced nanofiltration performance. ACS Appl. Mater. Interfaces **8**(47), 32512–32519 (2016). https://doi.org/10.1021/acsami.6b10693
43. Goh, P.S., Ismail, A.F.: Chemically functionalized polyamide thin film composite membranes: the art of chemistry. Desalination **495**, 114655 (2020). https://doi.org/10.1016/j.desal.2020.114655
44. Yu, S., Ma, M., Liu, J., Tao, J., Liu, M., Gao, C.: Study on polyamide thin-film composite nanofiltration membrane by interfacial Polymerization of Polyvinylamine (PVAm) and Isophthaloyl Chloride (IPC). J. Memb. Sci. **379**(1–2), 164–173 (2011). https://doi.org/10.1016/j.memsci.2011.05.061
45. Prime, D., Paul, S., Josephs-Franks, P.: Electrical properties of nanometre thin film polystyrene for organic electronic applications. IEEE Trans. Dielectr. Electr. Insul. **15**(4), 905–909 (2008). https://doi.org/10.1109/TDEI.2008.4591208
46. Ye, C., Pan, Z., Shen, Y.: Facile conversion of polystyrene waste into an efficient sorbent for water purification. Polymers (Basel) **14**(21), 4477 (2022). https://doi.org/10.3390/POLYM14214477/S1
47. Schumers, J., Vlad, A., Huynen, I., Gohy, J., Fustin, C.: Functionalized nanoporous thin films from photocleavable block copolymers. Macromol. Rapid Commun. **33**(3), 199–205 (2012). https://doi.org/10.1002/marc.201100739
48. Balzer, B.N., Gallei, M., Sondergeld, K., Schindler, M., Müller-Buschbaum, P., Rehahn, M., Hugel, T.: Cohesion mechanisms of polystyrene-based thin polymer films. Macromolecules **46**(18), 7406–7414 (2013). https://doi.org/10.1021/ma401173y
49. Ghezzi, M., Thickett, S.C., Neto, C.: Early and intermediate stages of guided dewetting in polystyrene thin films. Langmuir **28**(27), 10147–10151 (2012). https://doi.org/10.1021/la301773h
50. Yu, Y., Zhu, W., Li, H., Long, J., Huang, W., Li, J., Chen, L., Zhang, Y.: Transparent semi-aliphatic polyimide films with excellent thermal stabilities for applications in the display technology. Eur. Polym. J. **203**, 112653 (2024). https://doi.org/10.1016/j.eurpolymj.2023.112653
51. Wang, L., Yu, X., Wang, D., Zhao, X., Yang, D., UrRehman, S., Chen, C., Zhou, H., Dang, G.: High modulus and high strength ultra-thin polyimide films with hot-stretch induced molecular orientation. Mater. Chem. Phys. **139**(2–3), 968–974 (2013). https://doi.org/10.1016/j.matchemphys.2013.02.063
52. Zhao, K., Wei, S., Cao, M., Wang, M., Li, P., Li, H., Zhang, X., Zhang, Y., Chen, Y.: Dielectric polyimide composites with enhanced thermal conductivity and excellent electrical insulation properties by constructing 3D oriented heat transfer network. Compos. Sci. Technol. **245**, 110323 (2024). https://doi.org/10.1016/j.compscitech.2023.110323
53. Mulhearn, W.D., Oleshko, V.P., Stafford, C.M.: Thickness-dependent permeance of molecular layer-by-layer polyamide membranes. J. Memb. Sci. **618**, 118637 (2021). https://doi.org/10.1016/j.memsci.2020.118637
54. Peters, E.N.: Engineering thermoplastics—materials, properties, trends. In: Applied Plastics Engineering Handbook, pp. 3–26. Elsevier (2017). https://doi.org/10.1016/B978-0-323-39040-8.00001-8
55. Veazey, D., Hsu, T., Gomez, E.D.: Next generation high-performance carbon fiber thermoplastic composites based on polyaryletherketones. J. Appl. Polym. Sci. **134**(6) (2017). https://doi.org/10.1002/app.44441
56. Guan, K., Sasaki, Y., Jia, Y., Gonzales, R.R., Zhang, P., Lin, Y., Li, Z., Matsuyama, H.: Interfacial polymerization of thin film selective membrane layers: effect of polyketone substrates. J. Memb. Sci. **640**, 119801 (2021). https://doi.org/10.1016/j.memsci.2021.119801
57. Cho, J., Jeon, I., Kim, S.Y., Lim, S., Jho, J.Y.: Improving dispersion and barrier properties of polyketone/graphene nanoplatelet composites via noncovalent functionalization using aminopyrene. ACS Appl. Mater. Interfaces **9**(33), 27984–27994 (2017). https://doi.org/10.1021/acsami.7b10474

58. Kim, J., Oh, S., Cho, S.M., Jun, J., Kwak, S.: Oxygen barrier properties of polyketone/EVOH blend films and their resistance to moisture. J. Appl. Polym. Sci. **137**(47) (2020). https://doi.org/10.1002/app.49537
59. Alshammari, B.A., Alsuhybani, M.S., Almushaikeh, A.M., Alotaibi, B.M., Alenad, A.M., Alqahtani, N.B., Alharbi, A.G.: Comprehensive review of the properties and modifications of carbon fiber-reinforced thermoplastic composites. Polymers (Basel) **13**(15), 2474 (2021). https://doi.org/10.3390/polym13152474
60. Qu, K., Yuan, Z., Wang, Y., Song, Z., Gong, X., Zhao, Y., Mu, Q., Zhan, Q., Xu, W., Wang, L.: Structures, properties, and applications of zwitterionic polymers. ChemPhysMater **1**(4), 294–309 (2022). https://doi.org/10.1016/j.chphma.2022.04.003
61. Blackman, L.D., Gunatillake, P.A., Cass, P., Locock, K.E.S.: An introduction to zwitterionic polymer behavior and applications in solution and at surfaces. Chem. Soc. Rev. **48**(3), 757–770 (2019). https://doi.org/10.1039/C8CS00508G
62. Li, Q., Wen, C., Yang, J., Zhou, X., Zhu, Y., Zheng, J., Cheng, G., Bai, J., Xu, T., Ji, J., Jiang, S., Zhang, L., Zhang, P.: Zwitterionic biomaterials. Chem. Rev. **122**(23), 17073–17154 (2022). https://doi.org/10.1021/acs.chemrev.2c00344
63. Li, D., Wei, Q., Wu, C., Zhang, X., Xue, Q., Zheng, T., Cao, M.: Superhydrophilicity and strong salt-affinity: zwitterionic polymer grafted surfaces with significant potentials particularly in biological systems. Adv. Colloid Interface Sci. **278**, 102141 (2020). https://doi.org/10.1016/j.cis.2020.102141
64. Zhang, Z., Chao, T., Jiang, S.: Physical, chemical, and chemical−physical double network of zwitterionic hydrogels. J. Phys. Chem. B **112**(17), 5327–5332 (2008). https://doi.org/10.1021/jp710683w
65. Jensen, M.J., Peel, A., Horne, R., Chamberlain, J., Xu, L., Hansen, M.R., Guymon, C.A.: Antifouling and mechanical properties of photografted zwitterionic hydrogel thin-film coatings depend on the cross-link density. ACS Biomater. Sci. Eng. **7**(9), 4494–4502 (2021). https://doi.org/10.1021/acsbiomaterials.1c00852
66. Lee, H., Gilbert, J.B., Angilè, F.E., Yang, R., Lee, D., Rubner, M.F., Cohen, R.E.: Design and fabrication of zwitter-wettable nanostructured films. ACS Appl. Mater. Interfaces **7**(1), 1004–1011 (2015). https://doi.org/10.1021/am508157m
67. Hubbe, M.A., Rojas, O.J., Lucia, L.A., Sain, M.: Cellulosic nanocomposites: a review. BioResources **3**(3), 929–980 (2008). https://doi.org/10.15376/biores.3.3.929-980
68. Michelin, M., Gomes, D.G., Romaní, A., Polizeli, M. de L.T.M., Teixeira, J.A.: Nanocellulose production: exploring the enzymatic route and residues of pulp and paper industry. Molecules **25**(15), 3411 (2020). https://doi.org/10.3390/molecules25153411
69. Negro, C., Pettersson, G., Mattsson, A., Nyström, S., Sanchez-Salvador, J.L., Blanco, A., Engstrand, P.: Synergies between fibrillated nanocellulose and hot-pressing of papers obtained from high-yield pulp. Nanomaterials **13**(13), 1931 (2023). https://doi.org/10.3390/nano13131931
70. Mahmud, M.A., Anannya, F.R.: Sugarcane bagasse—a source of cellulosic fiber for diverse applications. Heliyon **7**(8), e07771 (2021). https://doi.org/10.1016/J.HELIYON.2021.E07771
71. Vaca-Garcia, C., Thiebaud, S., Borredon, M.E., Gozzelino, G.: Cellulose esterification with fatty acids and acetic anhydride in lithium chloride/ N, N -dimethylacetamide medium. J. Am. Oil Chem. Soc. **75**(2), 315–319 (1998). https://doi.org/10.1007/s11746-998-0047-2
72. Peng, S.X., Chang, H., Kumar, S., Moon, R.J., Youngblood, J.P.: A comparative guide to controlled hydrophobization of cellulose nanocrystals via surface esterification. Cellulose **23**(3), 1825–1846 (2016). https://doi.org/10.1007/s10570-016-0912-3
73. Terrié, C., Mahieu, A., Lequart, V., Martin, P., Leblanc, N., Joly, N.: Towards the hydrophobization of thermoplastic starch using fatty acid starch ester as additive. Molecules **27**(19), 6739 (2022). https://doi.org/10.3390/molecules27196739
74. Rodríguez-Fabià, S., Torstensen, J., Johansson, L., Syverud, K.: Hydrophobization of lignocellulosic materials part II: chemical modification. Cellulose **29**(17), 8957–8995 (2022). https://doi.org/10.1007/s10570-022-04824-y

75. Seddiqi, H., Oliaei, E., Honarkar, H., Jin, J., Geonzon, L.C., Bacabac, R.G., Klein-Nulend, J.: Cellulose and its derivatives: towards biomedical applications. Cellul. **28**(4), 1893–1931 (2021). https://doi.org/10.1007/S10570-020-03674-W
76. Shen, R., Guo, Y., Wang, S., Tuerxun, A., He, J., Bian, Y.: Biodegradable electrospun nanofiber membranes as promising candidates for the development of face masks. Int. J. Environ. Res. Public Heal. **20**(2), 1306 (2023). https://doi.org/10.3390/IJERPH20021306
77. Yang, J., Ching, Y., Chuah, C.: Applications of lignocellulosic fibers and lignin in bioplastics: a review. Polymers (Basel) **11**(5), 751 (2019). https://doi.org/10.3390/polym11050751
78. Pacheco, A., Evangelista-Osorio, A., Muchaypiña-Flores, K.G., Marzano-Barreda, L.A., Paredes-Concepción, P., Palacin-Baldeón, H., Dos Santos, M.S.N., Tres, M.V., Zabot, G.L., Olivera-Montenegro, L.: Polymeric materials obtained by extrusion and injection molding from lignocellulosic agroindustrial biomass. Polymers (Basel) **15**(20), 4046 (2023). https://doi.org/10.3390/polym15204046
79. Aulin, C., Flodberg, G., Ström, G., Lindström, T.: Enhanced mechanical and gas barrier performance of plasticized cellulose nanofibril films. Nord. Pulp Pap. Res. J. **37**(1), 138–148 (2022). https://doi.org/10.1515/npprj-2021-0061
80. Sharifi Zamani, E., Ahadian, H., Maloney, T.: Twin-roll forming, a novel method for producing high-consistency microfibrillated cellulosic films. Cellulose **29**(18), 9627–9636 (2022). https://doi.org/10.1007/s10570-022-04884-0
81. Acharya, S., Liyanage, S., Parajuli, P., Rumi, S.S., Shamshina, J.L., Abidi, N.: Utilization of cellulose to its full potential: a review on cellulose dissolution, regeneration, and applications. Polymers (Basel) **13**(24), 4344 (2021). https://doi.org/10.3390/polym13244344
82. Nadeem, H., Athar, M., Dehghani, M., Garnier, G., Batchelor, W.: Recent advancements, trends, fundamental challenges and opportunities in spray deposited cellulose nanofibril films for packaging applications. Sci. Total. Environ. **836**, 155654 (2022). https://doi.org/10.1016/j.scitotenv.2022.155654
83. Koppolu, R., Lahti, J., Abitbol, T., Aulin, C., Kuusipalo, J., Toivakka, M.: Tailoring the performance of nanocellulose-based multilayer-barrier paperboard using biodegradable-thermoplastics, pigments, and plasticizers. Cellulose **30**(11), 6945–6958 (2023). https://doi.org/10.1007/s10570-023-05281-x
84. Kallakas, H., Kattamanchi, T., Kilumets, C., Tarasova, E., Krasnou, I., Savest, N., Ahmadian, I., Kers, J., Krumme, A.: Tensile and surface wettability properties of the solvent cast cellulose fatty acid ester films. Polymers (Basel) **15**(12), 2677 (2023). https://doi.org/10.3390/polym15122677
85. Ho, D., Kim, J., Lee, M., Kim, C.: Sustainable approaches in the design of dielectric materials for organic thin-film transistors. In: Sustainable Strategies in Organic Electronics, pp. 179–208. Elsevier (2022). https://doi.org/10.1016/B978-0-12-823147-0.00005-7
86. Isotta, E., Andrade-Arvizu, J., Syafiq, U., Jiménez-Arguijo, A., Navarro-Güell, A., Guc, M., Saucedo, E., Scardi, P.: Towards low cost and sustainable thin film thermoelectric devices based on quaternary chalcogenides. Adv. Funct. Mater. **32**(32) (2022). https://doi.org/10.1002/adfm.202202157
87. Ahmed, M.J., Ashfaq, J., Sohail, Z., Channa, I.A., Sánchez-Ferrer, A., Ali, S.N., Chandio, A.D.: Lignocellulosic bioplastics in sustainable packaging–recent developments in materials design and processing: a comprehensive review. Sustain. Mater. Technol. 1–77 (2024)
88. Izawa, S., Nakano, K., Suzuki, K., Chen, Y., Kikitsu, T., Hashizume, D., Koganezawa, T., Nguyen, T.-Q., Tajima, K.: Crystallization and polymorphism of organic semiconductor in thin film induced by surface segregated monolayers. Sci. Rep. **8**(1), 481 (2018). https://doi.org/10.1038/s41598-017-18881-y
89. Özdal, T., Chtouki, T., Kavak, H., Figa, V., Guichaoua, D., Erguig, H., Mysliwiec, J., Sahraoui, B.: Effect of annealing temperature on morphology and optoelectronics properties of spin-coated CZTS thin films. J. Inorg. Organomet. Polym. Mater. **31**(1), 89–99 (2021). https://doi.org/10.1007/s10904-020-01646-y
90. Zhou, K., Qi, Z., Zhao, B., Lu, S., Wang, H., Liu, J., Yan, H.: The influence of crystallinity on the electrochromic properties and durability of NiO thin films. Surf. Interfaces **6**, 91–97 (2017). https://doi.org/10.1016/j.surfin.2016.12.004

91. Baisnab, D.K., Mukherjee, S., Das, S.: A short review on inorganic thin films from device perspective. In: Chemical Solution Synthesis for Materials Design and Thin Film Device Applications, pp 231–275. Elsevier (2021). https://doi.org/10.1016/B978-0-12-819718-9.00007-8
92. Calisi, N., Galvanetto, E., Borgioli, F., Martinuzzi, S.M., Bacci, T., Caporali, S.: Thin films deposition of fully inorganic metal halide perovskites: a review. Mater. Sci. Semicond. Process. **147**, 106721 (2022). https://doi.org/10.1016/j.mssp.2022.106721
93. Benelmekki, M., Erbe, A.: Nanostructured Thin Films–Background, Preparation and Relation to the Technological Revolution of the 21st Century, pp. 1–34 (2019). https://doi.org/10.1016/B978-0-08-102572-7.00001-5
94. Eshaghi, A., Graeli, A.: Optical and Electrical Properties of Indium Tin Oxide (ITO) nanostructured thin films deposited on polycarbonate substrates "Thickness Effect." Optik (Stuttg) **125**(3), 1478–1481 (2014). https://doi.org/10.1016/j.ijleo.2013.09.011
95. Chen, X., Zhou, Z., Lin, Y.-H., Nan, C.: Thermoelectric thin films: promising strategies and related mechanism on boosting energy conversion performance. J. Mater. **6**(3), 494–512 (2020). https://doi.org/10.1016/j.jmat.2020.02.008
96. Wu, X.-M., Wang, L.-L., Wang, Y., Gu, J.-S., Yu, H.-Y.: Surface modification of polypropylene macroporous membrane by marrying RAFT polymerization with click chemistry. J. Memb. Sci. **421–422**, 60–68 (2012). https://doi.org/10.1016/j.memsci.2012.06.033
97. Ran, J., Wu, L., Zhang, Z., Xu, T.: Atom Transfer Radical Polymerization (ATRP): a versatile and forceful tool for functional membranes. Prog. Polym. Sci. **39**(1), 124–144 (2014). https://doi.org/10.1016/j.progpolymsci.2013.09.001
98. Rutenberg, I.M., Scherman, O.A., Grubbs, R.H., Jiang, W., Garfunkel, E., Bao, Z.: Synthesis of polymer dielectric layers for organic thin film transistors via surface-initiated ring-opening metathesis polymerization. J. Am. Chem. Soc. **126**(13), 4062–4063 (2004). https://doi.org/10.1021/ja035773c
99. Kim, S., Rahardianto, A., Walker, J.S., Wolfe, T., Coleman, K., Cohen, Y.: Upgrading polyamide TFC BWRO and SWRO membranes to higher swro membrane performance via surface nano-structuring with tethered poly(Acrylic Acid). J. Memb. Sci. **597**, 117736 (2020). https://doi.org/10.1016/j.memsci.2019.117736
100. Xie, G., Xu, W., Ge, Q.: Controlling membrane ionization with bifunctional alendronates to benefit desalination through forward osmosis. Desalination **447**, 147–157 (2018). https://doi.org/10.1016/j.desal.2018.09.014
101. Liu, B., Liu, X., Shi, S., Huang, R., Su, R., Qi, W., He, Z.: Design and mechanisms of antifouling materials for surface plasmon resonance sensors. Acta Biomater. **40**, 100–118 (2016). https://doi.org/10.1016/j.actbio.2016.02.035
102. Beyou, E., Bourgeat-Lami, E.: Organic-inorganic hybrid functional materials by nitroxide-mediated polymerization. Prog. Polym. Sci. **121**, 101434 (2021). https://doi.org/10.1016/j.progpolymsci.2021.101434
103. Servi, A.T., Kharraz, J., Klee, D., Notarangelo, K., Eyob, B., Guillen-Burrieza, E., Liu, A., Arafat, H.A., Gleason, K.K.: A systematic study of the impact of hydrophobicity on the wetting of MD membranes. J. Memb. Sci. **520**, 850–859 (2016). https://doi.org/10.1016/j.memsci.2016.08.021
104. Liu, F., Zeng, Q., Li, J., Hao, X., Ho-Baillie, A., Tang, J., Green, M.A.: Emerging inorganic compound thin film photovoltaic materials: progress, challenges and strategies. Mater. Today **41**, 120–142 (2020). https://doi.org/10.1016/j.mattod.2020.09.002
105. Arce-Plaza, A., Sánchez-Rodriguez, F., Courel-Piedrahita, M., Vigil Galán, O., Hernandez-Calderon, V., Ramirez-Velasco, S., Ortega López, M.: CdTe thin films: deposition techniques and applications. In: Coatings and Thin-Film Technologies. IntechOpen (2019). https://doi.org/10.5772/intechopen.79578
106. Purohit, P., Bhatt, A., Mittal, R.K., Abdellattif, M.H., Farghaly, T.A.: Polymer grafting and its chemical reactions. Front. Bioeng. Biotechnol. **10** (2023). https://doi.org/10.3389/fbioe.2022.1044927

107. Olson, R.A., Korpusik, A.B., Sumerlin, B.S.: Enlightening advances in polymer bioconjugate chemistry: light-based techniques for grafting to and from biomacromolecules. Chem. Sci. **11**(20), 5142–5156 (2020). https://doi.org/10.1039/D0SC01544J
108. Minko, S.: Grafting on solid surfaces: "Grafting to" and "Grafting from" methods. In: Polymer Surfaces and Interfaces, pp. 215–234. Springer, Berlin (2008). https://doi.org/10.1007/978-3-540-73865-7_11
109. Khabibullin, A., Mastan, E., Matyjaszewski, K., Zhu, S.: Surface-Initiated Atom Transfer Radical Polymerization, pp. 29–76 (2015). https://doi.org/10.1007/12_2015_311
110. Coessens, V., Pintauer, T., Matyjaszewski, K.: Functional polymers by atom transfer radical polymerization. Prog. Polym. Sci. **26**(3), 337–377 (2001). https://doi.org/10.1016/S0079-6700(01)00003-X
111. Yu, S.J., Pak, K., Kwak, M.J., Joo, M., Kim, B.J., Oh, M.S., Baek, J., Park, H., Choi, G., Kim, D.H., Choi, J., Choi, Y., Shin, J., Moon, H., Lee, E., Im, S.G.: Initiated chemical vapor deposition: a versatile tool for various device applications. Adv. Eng. Mater. **20**(3) (2018). https://doi.org/10.1002/adem.201700622
112. Gleason, K.K.: Fluoropolymers by Initiated Chemical Vapor Deposition (ICVD). In: Opportunities for Fluoropolymers, pp 113–135. Elsevier (2020). https://doi.org/10.1016/B978-0-12-821966-9.00005-5
113. Lee, B., Jiao, A., Yu, S., You, J.B., Kim, D.-H., Im, S.G.: Initiated chemical vapor deposition of thermoresponsive Poly(N-Vinylcaprolactam) thin films for cell sheet engineering. Acta Biomater. **9**(8), 7691–7698 (2013). https://doi.org/10.1016/j.actbio.2013.04.049
114. Song, Q., Zhu, M., Chen, X., Liu, T., Xie, M., Mao, Y.: Flexible membranes fabricated by initiated chemical vapor deposition for water treatment, battery, and drug delivery. Chem. Eng. J. **477**, 146911 (2023). https://doi.org/10.1016/j.cej.2023.146911
115. Tenhaeff, W.E., Gleason, K.K.: Initiated and oxidative chemical vapor deposition of polymeric thin films: ICVD and OCVD. Adv. Funct. Mater. **18**(7), 979–992 (2008). https://doi.org/10.1002/adfm.200701479
116. van Herk, A.M.: Introduction to radical (Co)Polymerisation. In: Chemistry and Technology of Emulsion Polymerisation, pp. 23–42. Wiley (2013). https://doi.org/10.1002/9781118638521.ch2
117. Lamontagne, H.R., Lessard, B.H.: Nitroxide-mediated polymerization: a versatile tool for the engineering of next generation materials. ACS Appl. Polym. Mater. **2**(12), 5327–5344 (2020). https://doi.org/10.1021/acsapm.0c00888
118. Titirici, M.-M., Sellergren, B.: Thin molecularly imprinted polymer films via reversible addition−fragmentation chain transfer polymerization. Chem. Mater. **18**(7), 1773–1779 (2006). https://doi.org/10.1021/cm052153x
119. Mazloomi-Rezvani, M., Salami-Kalajahi, M., Roghani-Mamaqani, H.: "Grafting to" approach for surface modification of AuNPs with RAFT-mediated synthesized smart polymers: stimuli-responsive behaviors of hybrid nanoparticles. J. Phys. Chem. Solids **123**, 183–190 (2018). https://doi.org/10.1016/j.jpcs.2018.08.002
120. Wink, T., van Zuilen, S., Bult, A., van Bennekom, W.: Self-assembled monolayers for biosensors. Analyst **122**(4), 43R-50R (1997). https://doi.org/10.1039/A606964I
121. Schierbaum, K.D., Weiss, T., van Veizen, E.U.T., Engbersen, J.F.J., Reinhoudt, D.N., Gopel, W.: Molecular recognition by self-assembled monolayers of cavitand receptors. Science (80-) **265** (5177), 1413–1415 (1994). https://doi.org/10.1126/science.265.5177.1413
122. Mandler, D., Kraus-Ophir, S.: Self-Assembled Monolayers (SAMs) for electrochemical sensing. J. Solid State Electrochem. **15**(7–8), 1535–1558 (2011). https://doi.org/10.1007/s10008-011-1493-6
123. Gleason, K.K.: Nanoscale control by chemically vapour-deposited polymers. Nat. Rev. Phys. **2**(7), 347–364 (2020). https://doi.org/10.1038/s42254-020-0192-6
124. Gleason, K.K.: Organic surface functionalization by initiated <scp>CVD</Scp> (<scp>iCVD</Scp>). In: Surface Modification of Polymers, pp. 107–134. Wiley (2019). https://doi.org/10.1002/9783527819249.ch4

125. Kaymaz, S.V., Nobar, H.M., Sarıgül, H., Soylukan, C., Akyüz, L., Yüce, M.: Nanomaterial surface modification toolkit: principles, components, recipes, and applications. Adv. Colloid Interface Sci. **322**, 103035 (2023). https://doi.org/10.1016/j.cis.2023.103035
126. Kiliona, K.P.S., Lwal, A.L.J., Tao, H., Lin, N.: Surface Modification with Grafting Functional Molecules on Nanopolysaccharides, pp. 55–85 (2019). https://doi.org/10.1007/978-981-15-0913-1_2
127. Lin, J., Huang, J., Wang, J., Yu, J., You, X., Lin, X., Van der Bruggen, B., Zhao, S.: High-performance porous anion exchange membranes for efficient acid recovery from acidic wastewater by diffusion dialysis. J. Memb. Sci. **624**, 119116 (2021). https://doi.org/10.1016/J.MEMSCI.2021.119116
128. Halali, M.A., de Lannoy, C.-F.: The effect of cross-linkers on the permeability of electrically conductive membranes. Ind. Eng. Chem. Res. **58**(9), 3832–3844 (2019). https://doi.org/10.1021/acs.iecr.8b05691
129. Arya, S.K., Solanki, P.R., Datta, M., Malhotra, B.D.: Recent advances in self-assembled monolayers based biomolecular electronic devices. Biosens. Bioelectron. **24**(9), 2810–2817 (2009). https://doi.org/10.1016/j.bios.2009.02.008
130. Hilf, S., Kilbinger, A.F.M.: Functional end groups for polymers prepared using ring-opening metathesis polymerization. Nat. Chem. **1**(7), 537–546 (2009). https://doi.org/10.1038/nchem.347
131. Kochameshki, M.G., Marjani, A., Mahmoudian, M., Farhadi, K.: Grafting of diallyldimethylammonium chloride on graphene oxide by RAFT polymerization for modification of nanocomposite polysulfone membranes using in water treatment. Chem. Eng. J. **309**, 206–221 (2017). https://doi.org/10.1016/j.cej.2016.10.008
132. Asem, H., Zheng, W., Nilsson, F., Zhang, Y., Hedenqvist, M.S., Hassan, M., Malmström, E.: Functional nanocarriers for drug delivery by surface engineering of polymeric nanoparticle post-polymerization-induced self-assembly. ACS Appl. Bio Mater. **4**(1), 1045–1056 (2021). https://doi.org/10.1021/acsabm.0c01552
133. Fromel, M., Pester, C.W.: Polycarbonate surface modification via aqueous SI-PET-RAFT. Macromolecules **55**(12), 4907–4915 (2022). https://doi.org/10.1021/acs.macromol.2c00714
134. Shellaiah, M., Chen, Y.-T., Thirumalaivasan, N., Aazaad, B., Awasthi, K., Sun, K.W., Wu, S.-P., Lin, M.-C., Ohta, N.: Pyrene-based AIEE active nanoprobe for Zn 2+ and tyrosine detection demonstrated by DFT, bioimaging, and organic thin-film transistor. ACS Appl. Mater. Interfaces **13**(24), 28610–28626 (2021). https://doi.org/10.1021/acsami.1c04744
135. Chen, M., Li, Y., Han, R., Chen, Q., Jiang, L., Luo, X.: Click reaction-assisted construction of antifouling immunosensors for electrochemical detection of cancer biomarkers in human serum. Sens. Actuators B Chem. **363**, 131810 (2022). https://doi.org/10.1016/j.snb.2022.131810
136. Salunkhe, S., Dheeraj, Basak, M., Chitkara, D., Mittal, A.: Surface functionalization of exosomes for target-specific delivery and in vivo imaging and tracking: strategies and significance. J. Control. Release **326**, 599–614 (2020). https://doi.org/10.1016/j.jconrel.2020.07.042
137. Yang, S., Zou, P., Wang, Y., Song, R., Chen, Z.: Novel method for protecting copper: an in-situ click-assembly film on copper surface. Colloids Surf A Physicochem Eng Asp **677**, 132355 (2023). https://doi.org/10.1016/j.colsurfa.2023.132355
138. Mandal, S., Hou, Y., Wang, M., Anthopoulos, T.D., Choy, K.L.: Surface modification of hetero-phase nanoparticles for low-cost solution-processable high-k dielectric polymer nanocomposites. ACS Appl. Mater. Interfaces **15**(5), 7371–7379 (2023). https://doi.org/10.1021/acsami.2c19559
139. Matsubara, R., Aida, S., Fukushima, S., Kubono, A.: Preparation of vertically oriented aromatic polyester thin films by thermal chemical vapor deposition. Jpn. J. Appl. Phys. **63**(1), 015505 (2024). https://doi.org/10.35848/1347-4065/ad1938
140. Baniasadi, H., Seppälä, J.: Novel long-chain aliphatic polyamide/surface-modified silicon dioxide nanocomposites: in-situ polymerization and properties. Mater. Today Chem. **20**, 100450 (2021). https://doi.org/10.1016/j.mtchem.2021.100450

141. Zhang, Q., Luo, S., Weidman, J., Guo, R.: Surface modification of <scp>ZIF</Scp> -90 with triptycene for enhanced interfacial interaction in <scp>mixed-matrix</Scp> membranes for gas separation. J. Polym. Sci. **58**(18), 2675–2687 (2020). https://doi.org/10.1002/pol.20200123
142. Kwon, Y.-N., Hong, S., Choi, H., Tak, T.: Surface modification of a polyamide reverse osmosis membrane for chlorine resistance improvement. J. Memb. Sci. **415–416**, 192–198 (2012). https://doi.org/10.1016/j.memsci.2012.04.056
143. Li, G., Sampson, N.S.: Alternating Ring-Opening Metathesis Polymerization (AROMP) of hydrophobic and hydrophilic monomers provides oligomers with side-chain sequence control. Macromolecules **51**(11), 3932–3940 (2018). https://doi.org/10.1021/acs.macromol.8b00562
144. Deng, Y., Zou, T., Tao, X., Semetey, V., Trepout, S., Marco, S., Ling, J., Li, M.-H.: Poly(ε-Caprolactone)-block-polysarcosine by ring-opening polymerization of sarcosine N-Thiocarboxyanhydride: synthesis and thermoresponsive self-assembly. Biomacromol **16**(10), 3265–3274 (2015). https://doi.org/10.1021/acs.biomac.5b00930
145. Wieringa, R.H., Schouten, A.J.: Oriented thin film formation by surface graft polymerization of γ-Methyl<scp>l</Scp>-Glutamate *N*-carboxyanhydride in the melt. Macromolecules **29**(8), 3032–3034 (1996). https://doi.org/10.1021/ma9508358
146. Carey, M., Hinton, Z., Sokol, M., Alvarez, N.J., Barsoum, M.W.: Nylon-6/Ti 3 C 2 T z MXene nanocomposites synthesized by in situ ring opening polymerization of ε-caprolactam and their water transport properties. ACS Appl. Mater. Interfaces **11**(22), 20425–20436 (2019). https://doi.org/10.1021/acsami.9b05027
147. Youm, S.G., Hwang, E., Chavez, C.A., Li, X., Chatterjee, S., Lusker, K.L., Lu, L., Strzalka, J., Ankner, J.F., Losovyj, Y., Garno, J.C., Nesterov, E.E.: Polythiophene thin films by surface-initiated polymerization: mechanistic and structural studies. Chem. Mater. **28**(13), 4787–4804 (2016). https://doi.org/10.1021/acs.chemmater.6b01957
148. Fan, X., Lin, L., Dalsin, J.L., Messersmith, P.B.: Biomimetic anchor for surface-initiated polymerization from metal substrates. J. Am. Chem. Soc. **127**(45), 15843–15847 (2005). https://doi.org/10.1021/ja0532638
149. Jennings, G.K., Brantley, E.L.: Physicochemical properties of surface-initiated polymer films in the modification and processing of materials. Adv. Mater. **16**(22), 1983–1994 (2004). https://doi.org/10.1002/adma.200400810
150. Pattison, T.G., Hess, A.E., Arellano, N., Lanzillo, N., Nguyen, S., Bui, H., Rettner, C., Truong, H., Friz, A., Topuria, T., Fong, A., Hughes, B., Tek, A.T., DeSilva, A., Miller, R.D., Qiao, G.G., Wojtecki, R.J.: Surface initiated polymer thin films for the area selective deposition and etching of metal oxides. ACS Nano **14**(4), 4276–4288 (2020). https://doi.org/10.1021/acsnano.9b09637
151. Seok, J.-H., Kim, S.H., Cho, S.M., Yi, G.-R., Lee, J.Y.: Crosslinked organosilicon-acrylate copolymer moisture barrier thin film fabricated by Initiated Chemical Vapor Deposition (ICVD). Macromol. Res. **26**(13), 1257–1264 (2018). https://doi.org/10.1007/s13233-019-6149-2
152. Yılmaz, K., Şakalak, H., Gürsoy, M., Karaman, M.: Vapor deposition of stable copolymer thin films in a batch <scp>iCVD</Scp> reactor. J. Appl. Polym. Sci. **138**(13) (2021). https://doi.org/10.1002/app.50119
153. Chen, N., Reeja-Jayan, B., Liu, A., Lau, J., Dunn, B., Gleason, K.K.: ICVD cyclic polysiloxane and polysilazane as nanoscale thin-film electrolyte: synthesis and properties. Macromol. Rapid Commun. **37**(5), 446–452 (2016). https://doi.org/10.1002/marc.201500649
154. Matsuno, R., Otsuka, H., Takahara, A.: Polystyrene-grafted titanium oxide nanoparticles prepared through surface-initiated nitroxide-mediated radical polymerization and their application to polymer hybrid thin films. Soft Matter **2**(5), 415 (2006). https://doi.org/10.1039/b601197g
155. Melville, O.A., King, B., Imperiale, C., Lessard, B.H.: Orthogonally processable carbazole-based polymer thin films by nitroxide-mediated polymerization. Langmuir **32**(51), 13640–13648 (2016). https://doi.org/10.1021/acs.langmuir.6b03920

156. Becker, A.L., Johnston, A.P.R., Caruso, F.: Layer-by-layer-assembled capsules and films for therapeutic delivery. Small **6**(17) (2010). https://doi.org/10.1002/smll.201000379
157. Mallakpour, S., Madani, M.: A review of current coupling agents for modification of metal oxide nanoparticles. Prog. Org. Coatings **86**, 194–207 (2015). https://doi.org/10.1016/j.porgcoat.2015.05.023
158. Guzmán, E., Mateos-Maroto, A., Ruano, M., Ortega, F., Rubio, R.G.: Layer-by-layer polyelectrolyte assemblies for encapsulation and release of active compounds. Adv. Colloid Interface Sci. **249**, 290–307 (2017). https://doi.org/10.1016/j.cis.2017.04.009
159. Pavinatto, A., Mercante, L.A., Leandro, C.S., Mattoso, L.H.C., Correa, D.S.: Layer-by-layer assembled films of chitosan and multi-walled carbon nanotubes for the electrochemical detection of 17α-ethinylestradiol. J. Electroanal. Chem. **755**, 215–220 (2015). https://doi.org/10.1016/j.jelechem.2015.08.002
160. Cao, R., Shi, S., Cao, H., Li, Y., Duan, F., Li, Y.: Surface composite modification of anion exchange membrane by electrodeposition and self-polymerization for improved antifouling performance. Colloids Surf. Physicochem. Eng. Asp. **648**, 129402 (2022). https://doi.org/10.1016/J.COLSURFA.2022.129402
161. Alimohammadzadeh, R., Sanhueza, I., Córdova, A.: Design and fabrication of superhydrophobic cellulose nanocrystal films by combination of self-assembly and organocatalysis. Sci. Rep. **13**(1), 3157 (2023). https://doi.org/10.1038/s41598-023-29905-1
162. Elzaabalawy, A., Meguid, S.A.: Advances in the development of superhydrophobic and icephobic surfaces. Int. J. Mech. Mater. Des. **18**(3), 509–547 (2022). https://doi.org/10.1007/s10999-022-09593-x
163. Drelich, J., Chibowski, E.: Superhydrophilic and superwetting surfaces: definition and mechanisms of control. Langmuir **26**(24), 18621–18623 (2010). https://doi.org/10.1021/la1039893
164. Tsujii, Y., Ejaz, M., Sato, K., Goto, A., Fukuda, T.: Mechanism and kinetics of RAFT-mediated graft polymerization of styrene on a solid surface. 1. experimental evidence of surface radical migration. Macromolecules **34**(26), 8872–8878 (2001). https://doi.org/10.1021/ma010733j
165. Wang, Y., Zhang, H., Lin, L., Wu, R., Wang, X., Ren, J., Liu, C., Wang, X.: Modification of cellulose by hydrophobic long-chain molecules: advances and prospects. Pap. Biomater. **5**(3), 31–43 (2020). https://doi.org/10.12103/j.issn.2096-2355.2020.03.003
166. Patil, N.V., Netravali, A.N.: Direct assembly of silica nanospheres on halloysite nanotubes for "Green" ultrahydrophobic cotton fabrics. Adv. Sustain. Syst. **3**(8) (2019). https://doi.org/10.1002/adsu.201900009
167. Oke, J.A., Jen, T.-C.: Atomic layer deposition and other thin film deposition techniques: from principles to film properties. J. Mater. Res. Technol. **21**, 2481–2514 (2022). https://doi.org/10.1016/j.jmrt.2022.10.064
168. Shao, H., Niu, D., Liu, B., Xu, P., Yang, W., Lemstra, P.J., Bastiaansen, C.W.M., Wang, Z., Wang, C., Ma, P.: Mono-layer films with superior barrier properties and full recyclability: the system of Poly(Ethylene Terephthalate)/Poly(Glycolic Acid). Polymer (Guildf) **265**, 125594 (2023). https://doi.org/10.1016/j.polymer.2022.125594
169. Ghadai, R.K., Singh, K., Sharma, A., Roy, M.K., Swain, B.P.: Mechanical and Tribological Properties of Metal Incorporated DLC Thin Film, pp. 229–263 (2021). https://doi.org/10.1007/978-981-15-8307-0_12
170. Ma, Y., Shen, X.-L., Zeng, Q., Wang, H.-S., Wang, L.-S.: A multi-walled carbon nanotubes based molecularly imprinted polymers electrochemical sensor for the sensitive determination of HIV-P24. Talanta **164**, 121–127 (2017). https://doi.org/10.1016/j.talanta.2016.11.043
171. Oliveira, T., Morais, S.: New Generation of Electrochemical Sensors Based on Multi-walled Carbon Nanotubes, p .124 (2019). https://doi.org/10.3390/books978-3-03921-230-9
172. Ahmed, M.J., Perveen, S., Hussain, S.G., Khan, A.A., Ejaz, S.M.W., Rizvi, S.M.A.: Design of a facile, green and efficient graphene oxide-based electrochemical sensor for analysis of acetaminophen drug. Chem. Pap. **77**(4), 2275–2294 (2023). https://doi.org/10.1007/s11696-022-02628-9

173. Nagarale, R.K., Bavdane, P.P., Sreenath, S., Pawar, C.M., Dave, V., Satpati, A.K.: Polyaniline derivatized anion exchange membrane for acid recovery. J. Polym. Res. **29**(7), 1–12 (2022). https://doi.org/10.1007/S10965-022-03151-X/TABLES/3
174. Perveen, S., Hussain, S.G., Ahmed, M.J., Khawar, R., Siraj, T.B., Saleem, M.: A viable and sustainable flat-membrane plate-and-frame module for spent acid regeneration and metal ion recovery. Heliyon **9**(8), e18344 (2023). https://doi.org/10.1016/j.heliyon.2023.e18344
175. Guizard, C., Princivalle, A.: Preparation and characterization of catalyst thin films. Catal. Today **146**(3–4), 367–377 (2009). https://doi.org/10.1016/j.cattod.2009.05.012
176. Li, W., Mukherjee, S., Ren, B., Cao, R., Fischer, R.A.: Open framework material based thin films: electrochemical catalysis and state-of-the-art technologies. Adv. Energy Mater. **12**(4) (2022). https://doi.org/10.1002/aenm.202003499
177. Riahi, Z., Rhim, J.-W., Bagheri, R., Pircheraghi, G., Lotfali, E.: Carboxymethyl cellulose-based functional film integrated with chitosan-based carbon quantum dots for active food packaging applications. Prog. Org. Coatings **166**, 106794 (2022). https://doi.org/10.1016/j.porgcoat.2022.106794
178. Fu, C., Wang, T., Xu, F., Huo, Y., Potier-Ferry, M.: A modeling and resolution framework for wrinkling in hyperelastic sheets at finite membrane strain. J. Mech. Phys. Solids **124**, 446–470 (2019). https://doi.org/10.1016/j.jmps.2018.11.005

Nanotechnology and Thin Films: Explores Nanoscale Thin Films, Nanocomposites, Nanoparticles, and Their Application

Abdul Ghaffar Memon, Abdulmoseen Segun Giwa, and Jabran Ahmad

Abstract The chapter "Nanotechnology and Thin Films" introduces the reader to the world of nanoscale thin films and the importance of thin film technology in the current and further growth of several high-tech industries. This chapter discusses the growth and use of nanoscale thin films, nanocomposites, and nanoparticles their features, and opportunities. Nanoscale thin film fabrication has remarkable mechanical, electrical, and optical properties, placing them in high-tech applications in electronics, photonics, biotechnology, etc. Multifunctional nanocomposites incorporating nanoscale particles and thin films have optimal performance and efficiency. The chapter further examines the involvement of nanoparticles in thin film technology with a particular focus on enhancing film properties and establishing new techniques in the advancement of film technology. The chapter also demonstrates that nanotechnology remains irreplaceably significant for further development of thin film science to open new frontiers and prospects for other fields and industries. The various detailed discussions presented here serve to enrich the appreciation of nanotechnology-composite thin films, and the exciting opportunities they offer to transform the face of contemporary engineering and material science.

A. G. Memon (✉)
Department of Environmental Engineering, NED University of Engineering and Technology, Karachi 75270, Pakistan
e-mail: abdulghaffar@neduet.edu.pk

A. G. Memon · A. S. Giwa (✉) · J. Ahmad
State Key Joint Laboratory of ESPC, School of Environment, Tsinghua University, Beijing 100084, China
e-mail: giwaabdulmoseensegun@ncpu.edu.cn

A. S. Giwa
School of Civil and Environmental Engineering, Nanchang Institute of Science and Technology, Nanchang 330000, China

A. D. Chandio and I. A. Channa (eds.), *Thin Films and Coatings*, Engineering Materials,
https://doi.org/10.1007/978-981-96-0005-2_8

1 Introduction

In the ever-developing realm of modern science and technology, new innovative creations in the fields of thin film technology and nanotechnology have caused a fusion. This shift in material technology can be considered as creating new prospective and diverse discoveries for important breakthroughs of radical importance. Scientists and engineers embarked on a revolutionary process by exploring the potential of nanoscale thin films, nanocomposites, and nanoparticles.

This chapter consequently sets its focus on explaining the relationship between nanotechnology and thin films. This book looks at its formation, characteristics, and remarkable uses. Ranging from improvements in the mechanical density or elasticity of materials to greater control over surface texture and characteristics. The incorporation of nanotechnology in thin films has instigated great hazards in various fields. These sectors include electronics, energy, healthcare, and many more. By using the information available regarding modern events and tendencies as the basis. Thus, this chapter aims to provide readers with a nuanced understanding of the multifaceted role played by nanotechnology. In the future, technology will propel us toward unique inventions and more discoveries.

Nanostructured thin films have emerged as a specific focus in the scientific era. It draws considerable attention due to its distinctive physicochemical properties. The properties vary intricately with the size measurement of nanoparticles. These properties render them highly appealing and applications have benefited greatly. Conventionally defined as material layers spanning from a few nanometers to several micrometers in thickness, thin films have historically served as coatings for surface cover or modification [1]. However, recent technological strides have initiated the development of standalone thin layers. This has substantially broadened their scope.

Thin films, at the nanoscale, are defined as having a thickness of a few nanometers or even less. Due to their very small size, these films exhibit characteristics qualities such as enhanced mechanical, electrical, physical, chemical, or optical capabilities, a high surface area-to-volume ratio, and quantum effects. It is crucial to understand that thin films produced at the nanoscale have remarkable properties that are usually gathered with special and modern technology [2].

Nanostructured films can be produced via a variety of deposition methods and are very flexible. These films are precisely crafted for their intended uses. Due to their adjustable atomic-molecular arrangement in two or three dimensions, thickness, crystallinity, porosity, optical, mechanical, catalytic, and conductive capabilities, they offer unparalleled flexibility [3]. These characteristics make nanostructured films suitable for a variety of domains and applications, such as electronic devices (including superconductors, semiconductors, magnets, heat-resistant materials, and insulators.), energy reservoir and power transformation (nanostructured films are utilized in batteries, solar cells, and fuel cells), protective, transparent, and optical coatings (these films enhance the performance and durability of optical components and surface modification), and functional tools (nanostructured films serve as catalysts for chemical reactions and enable the detection of specific substances through sensors and biosensors) [4, 5]. The versatility of nanostructured films makes them invaluable in advancing technology and meeting a range of societal objectives.

Modern technology development focuses on the development of materials through nanotechnology, an area that has attracted the attention of numerous researchers searching for novel materials for a range of applications [6]. As these nanoparticles enhance specific qualities like strength, heat resistance, electrical conductivity, and moisture resistance, nanofillers (small particles) are added to various polymer materials [7]. Since the twentieth century, nanofillers have been used in composite materials. The goal is to use a component to hold the main material (matrix) and any additional strengthening ingredients (reinforcement) together. Since these minuscule particles have a considerable surface area relative to their volume, nanofillers become significant when considering the concept of nanotechnology. As a result, they will be able to create stronger bonds with the primary material, improving performance all around. The uniform distribution of nanofillers in the composite is a critical factor in defining its mechanical characteristics. By adding nanofillers, the polymer matrix, and fillers become more firmly connected with it, preventing premature failure of the composites [8].

2 History

Nanotechnology, particularly thin film technology, is a fascinating story of scientific discovery, technological growth, and industrial applications in the modern world. The field of nanotechnology in thin films has evolved over time, with new developments and advancements made possible by each passing era.

2.1 The First Decade (1950s–1970s)

The 1980s and 1990s featured a growth in nanotechnology. Nanotechnology began to take off in the 1980s and 90 s because of K. Eric Drexler's breakthrough concepts in his 1986 book. Advances in microscopy methods that allow scientists to observe and interact with materials at the nanoscale, such as the atomic force microscope and scanning tunneling microscope advances in chemical vapor deposition techniques that provide precise control [9].

2.2 The Emergence of Nanotechnology (1980s–1990s)

The 1980s and 1990s featured a growth in nanotechnology. Nanotechnology began to take off in the 1980s and 90 s because of K. Eric Drexler's breakthrough concepts in his 1986 book. Advances in microscopy methods that allow scientists to observe and interact with materials at the nanoscale, such as the atomic force microscope and scanning tunneling microscope advances in chemical vapor deposition techniques that provide precise control [9].

2.3 The Nanotechnology Revolution (2000s-Present)

Research and development in nanotechnology have increased dramatically in the twenty-first century, as it has the integration of nanotechnology into thin films. Continuous great research is conducted at this time on the characteristics of nanomaterials and how they might be used in a variety of industries, such as electronics, optics, energy, and packaging. Interaction between several fields, including engineering, chemistry, and materials science, to advance thin film technologies. Expanding the use of nano thin films in innovative domains including improved coatings and flexible electronics [10].

2.4 The Future of Nanotechnology

Nowadays, thin films enabled by nanotechnology are essential to many high-tech industries. However there are still some obstacles to be addressed; such as increasing the scale of nanomanufacturing procedures to satisfy market demands, increasing the sustainability and cost-effectiveness of nanotechnology, and addressing the potential impacts of nanoparticles on the environment. Through ongoing study and innovation, scientists and engineers have been working to solve these problems while achieving the full potential of nanotechnology in the future [11].

The history of thin film nanotechnology reflects an inventive problem-solving mindset, an unwavering quest for useful applications with noteworthy results, and an attitude of curiosity.

3 Impact of Nanoparticles on Diffusion Behavior in Thin Films

Within thin film technology, the enrichment of nanoparticles has a significant impact on the complex dynamics of dispersion in these films [12]. Nanoparticles provide an array of applications due to their precise dimensions and distinctive surface features. These influences have a complex shaping effect on diffusion processes. Diffusion in nanostructured thin films is the mechanism via which atoms or particles migrate inside these incredibly thin material layers, as seen in (Fig. 8.1b). Usually, these layers are nanoscale in nature. They introduce a range of effects when integrated into

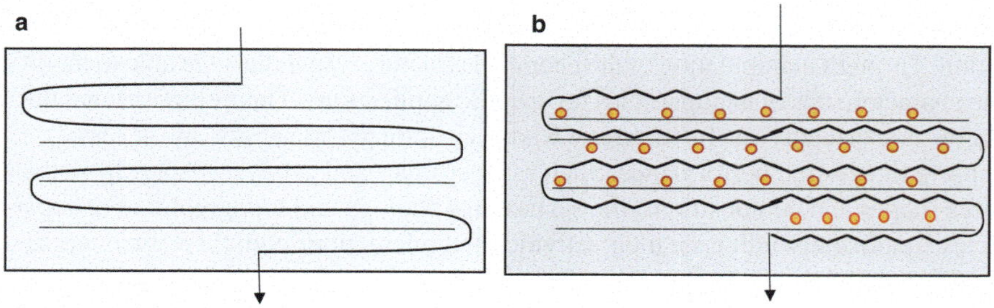

Fig. 8.1 **a** Diffusion in thin films. **b** Diffusion in nanostructured thin film

thin films. These outcomes comprise an improvement in a thin film. They also include hindrance of diffusion processes. Atomic motion can be accelerated or hindered by nanoparticles due to their modified atomic configurations and larger surface area. Furthermore, they possess the ability to selectively control diffusion. They hinder some species while favoring individuals providing a certain level of control over the movement of materials. Diffusion kinetics can be engineered to meet specific application requirements by fine-tuning nanoparticle parameters like size, shape, and composition. The complex interaction between thin films and nanoparticles offers opportunities to modify the behavior of materials and maximizing efficiency in a variety of technological fields [13].

4 Important Characteristic of Nanoparticles

Nanoparticles because they are extremely small have a lot of special qualities that make them really important for many different things. At the heart of their uniqueness lies their size. This nano-size is a defining characteristic. At least one dimension measures less than 100 nm. This small size gives nanoparticles some unique properties. These properties are different from those of larger or bulk materials. At the core of nanoparticles' exceptional characteristics lies their remarkable surface area-to-volume ratio [14].

A considerable portion of particles' atoms are located on their surfaces. As a result, nanoparticles react more strongly. They are more likely to interact with other substances. This unique characteristic makes nanoparticles valuable. They are important in fields like catalysis and drug delivery. Furthermore, the key factor in defining a nanoparticle's characteristics and potential applications is its shape. In bulk materials, shape changes are frequently insignificant. Nanoparticles may take on a variety of geometries. Each contributes unique optical, magnetic, and chemical properties. Controlling the form of nanoparticles is essential for both their production and their use. The spherical and rod-shaped nanoparticles might have dissimilar characteristics.

The adaptability and utility of nanoparticles are further enhanced by their composition. Through manipulation of the chemical structure researchers are able to modify the characteristics of nanoparticles for specific applications. This compositional flexibility is especially useful in areas where customized materials are necessary to achieve the desired performance. Such as electronics and catalysis. Apart from their sizes shapes and compositions, the surface morphology and topography of nanoparticles significantly influence their activities and interactions with their surroundings. Surface engineering is a crucial element of nanoparticle design in biomedical applications. It's also important in environmental applications. Complex surface characteristics of nanoparticles control their biological activity. They also affect cellular absorption and adherence to surfaces despite numerous advantages nanoparticles have some limitations [15].

There are several types of nanoparticles based on materials including gold (AuNP), silver (AgNP), copper (CuNP), iron (FeNP), and other metals. Their synthesis mechanisms are also diverse from chemical to thermal and green synthesis. Gold nanoparticles are proven to be stable with diverse applications from healthcare, and material to environment. Likewise, silver nanoparticles have applications in antimicrobial, wastewater treatment, and other applications [16–22].

The key concern is that they tend to agglomerate or cluster together. This can alter the characteristics and behavior of individual particles. Decreasing aggregation tendencies requires rigorous control of synthesis conditions [23]. It also needs surface modifications. To ensure the intended performance of nanoparticles. Moreover, the solubility properties of nanoparticles are important factors. These must be taken into consideration. This is especially true in biological systems [24]. Here interactions with physiological fluids and cellular components take place. Ineffectively soluble nanoparticles can highlight the problems for medication administration. Biomedical imaging applications also face challenges. This is due to their low bioavailability [25].

5 The Effects of Nanoparticles Vary Depending on Their Different Sizes

Nanoparticles exhibit a diverse range of properties that are influenced by their size, and this influence becomes more pronounced as their size decreases. Let's delve into the impact of nanoparticle size on their properties and applications in greater detail (Table 8.1).

Table 8.1 Effect of different sizes of nanoparticles [26]

Nanoparticle size range	Surface area to volume ratio	Quantum confinement	Properties	Applications
Large (10–100 nm)	Moderate	No	Improved mechanical properties, stability	Catalysis, drug delivery
Medium (1–10 nm)	High	Moderate	Unique optical, electronic, and magnetic properties	Electronics, sensing, biomedical implants
Small (<1 nm)	Very High	Strong	Tunable electronic, and optical properties, discrete energy level	Quantum dots (displays, solar cells), enhanced catalysis
Ultrafine (sub-nanometer)	Extremely high	Very strong	Molecular-like behavior, discrete electronic states	Fundamental research, single atom catalysis, molecular electronics

6 Classification of Nanostructured Thin Films

A systematic classification of nanostructured thin films is offered, defining different criteria for classification. This includes their intended applications, properties, fabrication techniques, and structural arrangement composition. Structural classification encompasses various arrangements. These include layered thin films [7]. Additionally, it also includes engineered thin films and graded nanocomposite. Based on their fundamental ingredients, such as organic materials, dielectrics, and metallic semiconductors, the composition basis classifies them. A film's functions are identified by its property foundation. These consist of electrical, mechanical, magnetic, and optical properties. Thin film production processes are categorized by the fabrication method. It includes physical deposition of vapors, atomic layer deposition, chemical vapor deposition, and sol–gel techniques. Ultimately, a variety of purposes are outlined via application-based categorization. Among these are biological implants, optical coatings, and sensors for electrical devices [27]. Figure 8.2 illustrates a comprehensive framework providing a structured approach to the Categorization of nanostructured thin films. The figure combines fabrication techniques including PVD, CVD, sol–gel, and ALD thin film. Properties of nanostructured films, composition, and structural classification. This helps in understanding the diverse nature and potential applications of nanostructured thin films.

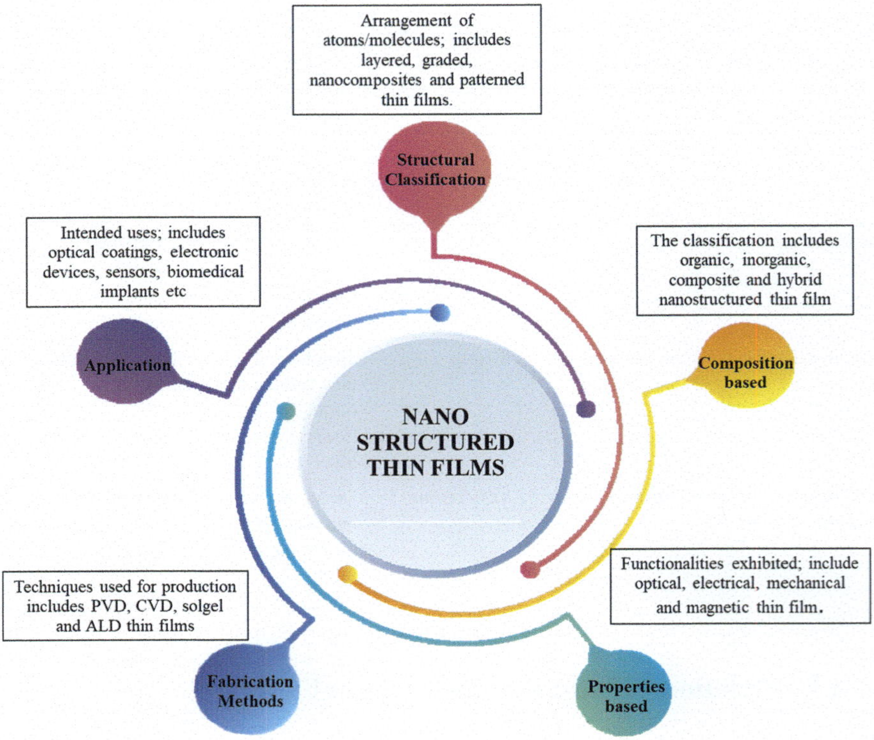

Fig. 8.2 Categorization of nanostructured thin films

7 Nanocomposite Thin Films

Nanocomposite thin films offer an innovative field of study in which polymers integrate with minuscule particles to produce thin films with extraordinary characteristics. A multitude of potentials has been explored by researchers by combining polymers and nanoparticles at the nanoscale. They produce thin layers of material [28]. These films feature exceptional electrical and mechanical properties. Impressive optical characteristics as well. We will explore synthesis methods in this chapter. We'll talk about distinctive qualities. We will also examine the diverse applications of nanocomposite thin films. Nanocomposite thin films offer an innovative field of study in which polymers integrate with minuscule particles to produce thin films with extraordinary characteristics. A multitude of potentials has been explored by researchers by combining polymers and nanoparticles at the nanoscale. They produce thin layers of material. These films feature exceptional electrical and mechanical properties. Impressive optical characteristics as well. We will explore synthesis methods in this chapter. We'll talk about distinctive qualities. We will also look at a variety of uses for thin films made of nanocomposite materials [29].

The potential of nanocomposite thin films to improve the characteristics of traditional materials makes them attractive. Polymer matrixes with nanoparticles introduced exhibit increased mechanical strength. They become more electrically conductible. There is an enhancement in thermal stability. Optical transparency is another feature of them [8].

In Fig. 8.3, a set of images illustrates the bio-nanocomposite thin films under various different conditions. Each image provides a visual representation of distinct compositions and concentrations of nanoparticles within the films. (a) Glycerol Film + 0% ZnO Nanoparticles: The glycerol base film without the incorporation of ZnO nanoparticles serves as the reference point, showcasing the film's characteristics in the absence of nanomaterials. (b) Glycerol Film + 3% ZnO Nanoparticles: Fig. 8.1b displays the glycerol film with a 3% concentration of ZnO nanoparticles, highlighting the impact of nanoparticle inclusion on the film's visual appearance. (c) Glycerol Film + 5% ZnO Nanoparticles: As seen in Fig. 8.1c, the glycerol film is enhanced with a ZnO nanoparticle concentration of 5%, providing information on potential morphological changes in the film as well as visual changes. (d) Sorbitol-Plasticized Film + 0% ZnO Nanoparticles: This image, represented in Fig. 8.1d, showcases the sorbitol-plasticized film in its pristine state without the addition of ZnO nanoparticles. (e) Sorbitol-Plasticized Film + 3% ZnO Nanoparticles: Fig. 8.1e presents the sorbitol-plasticized film integrated with a 3% concentration of ZnO nanoparticles, providing a visual comparison against the reference. (f) Sorbitol-Plasticized Film + 5% ZnO Nanoparticles: Lastly, Fig. 8.1f displays the sorbitol-plasticized film with a 5% concentration of ZnO nanoparticles, allowing for an examination of the visual impact associated with higher nanoparticle content.

These images offer a comprehensive view of the bio-nanocomposite films, allowing for qualitative assessments of their visual characteristics and potential morphological changes induced by varying concentrations of ZnO nanoparticles.

8 Fabrication Technique for Thin Films

Fabrication techniques for nanostructured thin films encompass a broad array of methods each offering unique advantages and applications.

8.1 CVD Fabrication Technique

Chemical vapor deposition (CVD) emerges as a robust technology for fabricating nanostructured thin films and coatings. It continually evolves to accommodate new materials and applications within modern industries. This primer explores the precision of CVD synthesis in creating nanostructured thin films. It emphasizes its application to inorganic 2D materials and high-purity polymeric thin films. It offers conformal deposition on diverse substrates. Using examples like graphene and 2D

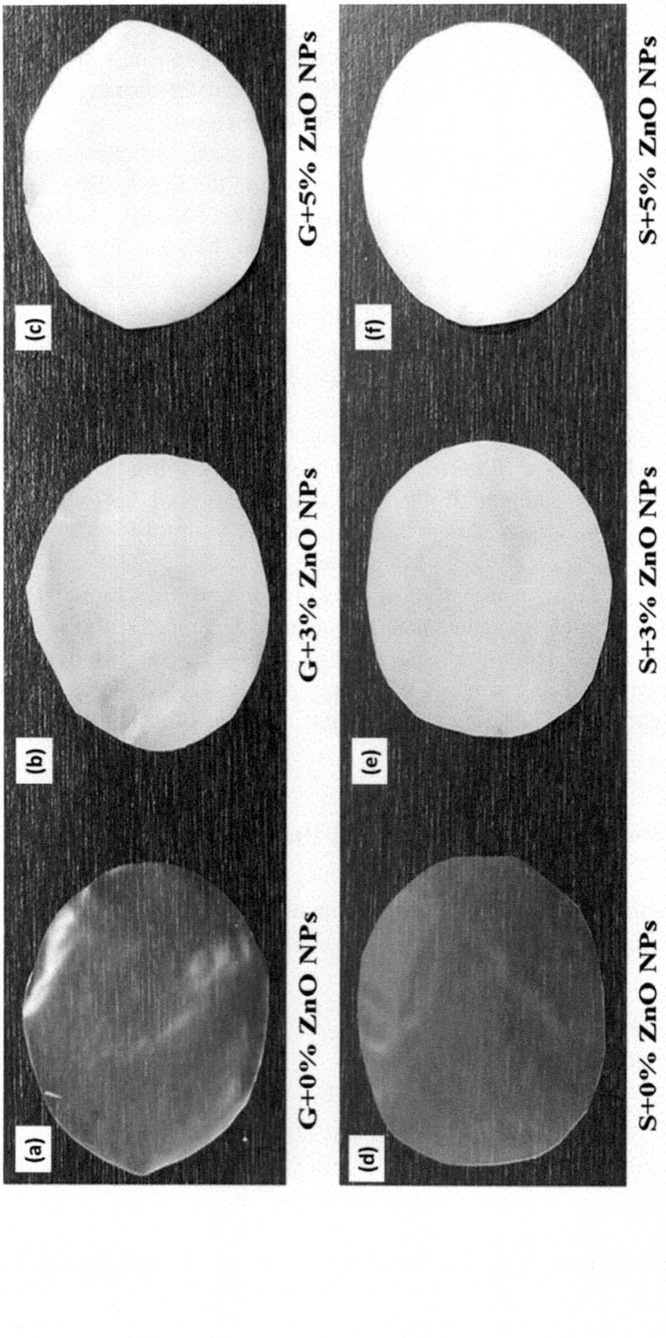

Fig. 8.3 Photographs of the bio-nanocomposite films. CC by 4.0 [30]

transition metal dichalcogenides (TMDs). And polymeric thin films. The primer elucidates optimal experimental practices. This includes substrate pretreatment high-temperature growth and post-growth processes.

For instance, in the production of graphene nanostructured thin films via CVD the process involves introducing hydrocarbon precursor gas. This could be something such as methane. It goes into a high-temperature chamber containing a catalytic substrate, typically copper at temperatures exceeding 1000 °C. Carbon atoms from precursor gas adsorb onto the catalytic surface. They form a graphene layer through the process of nucleation.

Similarly, the synthesis of 2D transition metal dichalcogenides (TMDs) nanostructured thin films via CVD follows a similar procedure. Metal and chalcogenide precursors are introduced into the reaction chamber at lower temperatures. This leads to the formation of layered structures like molybdenum disulfide (MoS2) or tungsten dieseline (WSe2). Post-growth processes may be employed. These include annealing or functionalization, to enhance the properties of deposited nanostructured thin films. Recent advancements and scaling-up challenges in nanostructured thin films are thoroughly examined. This sheds light on current limitations. It also explores optimization strategies. Furthermore, insights into potential future directions for CVD methods in nanostructured thin films are discussed. This emphasizes reactor design advancements. High-throughput and low-temperature growth are key focuses.

High-purity polycrystalline, amorphous, uniaxially oriented polycrystalline, and epitaxial layers can be produced by using CVD's flexibility. This demonstrates its wide applicability in the field of nanostructured thin film synthesis across several material production domains.

Functional nanomaterials are commonly created via the chemical vapor deposition (CVD) technique. It is frequently utilized to produce effective, high-quality thin films on enormous wafers or substrates with intricate designs. Originating from its effective use in the 1970s in the semiconductor production industry, CVD has made significant progress. It has developed into a sophisticated method for creating a variety of coatings. In addition to its application in producing thin films, CVD techniques have been advanced to manufacture materials featuring nanostructures. These include carbon nanotubes and nanotube fibers. Also ZnO nanowires and nanorods. While there is a lack of documented instances regarding the utilization of CVD for fabricating BFO nanomaterials, insights from analogous materials are anticipated to furnish the necessary direction for BFO. Undoubtedly this represents an area necessitating swift advancement in the foreseeable future [31] (Fig. 8.4).

8.2 PVD Fabrication Technique

PVD, or physical vapor deposition, is an effective technique for introducing nanostructured materials or nanoadditives to thin films. It comprises precisely flowing material under a vacuum at the atomic level.

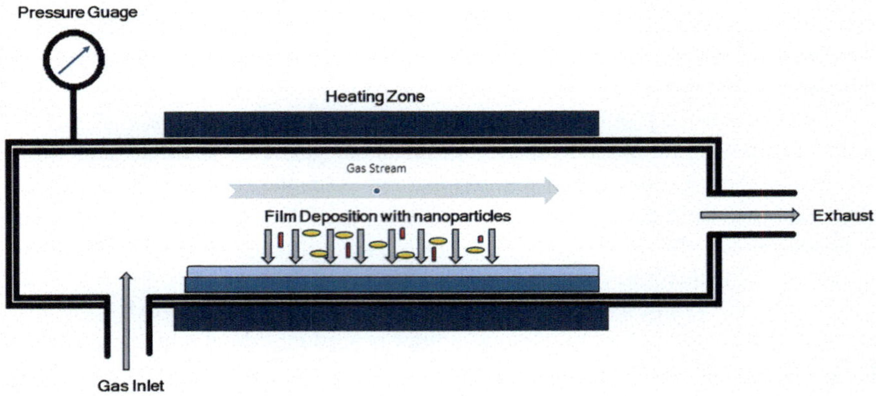

Fig. 8.4 Schematic of CVD technique for nanofilm modified CC BY 4.0

PVD uses solid precursors, which sets it apart from chemical vapor deposition (CVD). High-energy sources, like as ions or electron beams, evaporate them. The vapor is subsequently transferred to the substrate, where it condenses in this spot. Reactions with reactive gases, such as oxygen, nitrogen, or methane, help to expedite this process. PVD has a number of advantages. It includes the potential for coatings with enhanced properties compared to the substrate material. It also has the versatility to utilize various nanostructured inorganic materials or select organic nano additives. Moreover, its environmentally friendly nature sets it apart from alternative processes like electroplating.

Physical vapor deposition methods like sputtering and evaporation provide ways to add nanostructured materials or nano additives to thin films that are customized for a given need. The deposition of nanostructured materials or nano additives is rendered possible by these procedures. Thin film production is facilitated by it. These possess special qualities and abilities. They are appropriate for a wide range of sophisticated materials science and nanotechnology applications.

Applications of PVD include the deposition of precision tools and aluminized film for food packaging, as well as thin film solar panels. However, problems like expensive processing costs and difficult-to-coat forms still exist. PVD is nevertheless widely used in several sectors. These include aircraft and semiconductors. This promotes creativity and advancement [32] (Fig. 8.5).

8.3 *Laser Ablation Deposition Technique*

Laser ablation stands as a distinguished method for fabricating nanostructured thin films. It provides unparalleled accuracy. Impressive versatility, especially in academic and professional contexts, particularly when using nano additions. In particular,

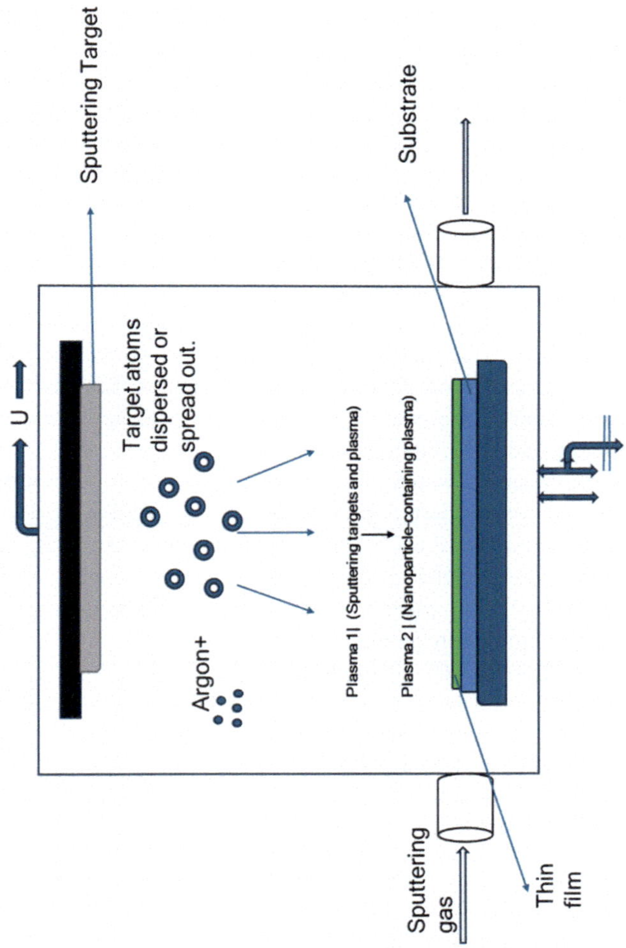

Fig. 8.5 Schematic of PVD technique for thin films modified CC BY 4.0

pulsed laser deposition (PLD) outperforms other deposition techniques. Its versatility is impressive. Providing the ability to transmit stoichiometrically and precisely control growth rate. Furthermore, PLD is remarkably flexible. Especially in the geometry of ablation.

The inception of nanostructured thin film deposition via laser ablation dates back to the emergence of pulsed laser deposition (PLD) in 1965. Seminal advancements were achieved by Dijjkamp et al. In 1987 they achieved stoichiometric transfer of complex molecular structures. A feat previously unattainable with conventional deposition methods. The evolution of PLD has led to the development of various laser ablation methods. This includes Combinatorial PLD (CPLD). It allows for the synthesis of complex material libraries with variable composition over the surface.

At the core of laser ablation lies the principle. The principle is instant local vaporization. This is induced by a high-intensity laser beam. A plasma plume is generated. It consists of photons electrons and ions. It also contains atoms molecules clusters and solid particles. This phenomenon is known as laser ablation. Laser ablation forms the basis for numerous applications. These applications include precise cutting. Hole drilling. Compositional analysis. Nanostructured thin film deposition.

Notably, PLD enables collection. The collection of ablated material onto a substrate. This facilitates the incorporation of nano additives into the thin film structure. This process allows the creation of nanostructured thin films. These have tailored properties. Their functionalities are suitable for various advanced applications. These include nanotechnology and materials science.

Factors influencing the laser ablation process include deposition conditions laser beam parameters and material properties of the target. These contribute to uniformity. Also the quality of deposited thin films. While PLD offers simplicity in experimental assembly, the underlying laser-material interaction involves a complex interplay of processes. These include optical energy coupling, surface melting, vaporization, photon absorption, and plume propagation.

The versatility of CPLD enables the synthesis of thin film libraries with diverse compositions. This is achieved in relatively short timeframes. It offers significant advantages for materials research. However, challenges such as plasma spread variability. Additionally, compositional testing complexities underscore the need for meticulous characterization. Optimization in CPLD applications remains crucial [33] (Fig. 8.6).

8.4 SOL–GEL Deposition Technique

The sol–gel process is considered a cornerstone in the field of nanostructured thin film deposition. It provides an adaptable approach to producing a wide range of materials, from catalysts to glass fibers. Atomic-scale thin films show incredible properties, especially when nanoadditions are included in them. It is renowned for its economic viability and offers procedural simplicity. Sol–gel deposition presents a compelling alternative to more intricate techniques. This includes physical vapor deposition and

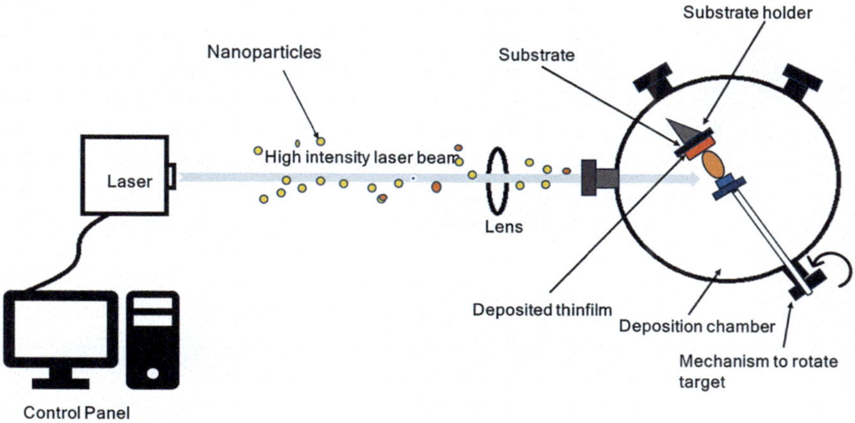

Fig. 8.6 Schematic diagram of laser ablation deposition technique modified CC BY 4.0

chemical vapor deposition. These often necessitate sophisticated infrastructure and resources.

Leveraging methodologies including dip coating, spin coating, electrochemical coating, and spray coating. Sol–gel deposition showcases remarkable adaptability. This spans various substrate types. It enables the fabrication of nanostructured thin films. These are tailored to a myriad of applications. They span the semiconductor industry in catalysis and photocatalysis at the nanoscale. The historical lineage of sol–gel processing traces back to Ebelman and Graham. They conducted pioneering investigations in the mid-1800s. These researchers delved into the properties of silica gels. Their seminal work laid the foundation for subsequent developments in the field. This culminated in the refinement of the hydrolysis process. This was specifically for metal alkoxide precursors. These precursors were employed for nanoscale applications.

In the context of optical fiber sensors at the nanoscale. Sol–gel coatings emerge as the linchpin. They facilitate the integration of fluorescent dyes. Mainly for temperature sensing applications at ultra-fine resolutions. By immersing optical fibers into sol–gel solutions containing precisely tuned precursor compositions controlled film deposition ensues. This results in coatings with finely calibrated properties optimized for nanoscale interactions. Film thickness and properties may be carefully adjusted to satisfy certain nanoscale performance requirements by carefully adjusting sol–gel parameters, such as withdrawal speed and solution composition. Moreover, the characteristics of sol–gel coatings may be further improved by post-treatment procedures, highlighting the flexibility and adaptability of this deposition method in the context of nanostructured thin films.

The sol–gel deposition method for thin films commonly employs two techniques: spin coating and dip coating. Spin coating is the process of dispensing a solution containing the necessary precursor chemicals onto a revolving substrate. Centrifugal force equally distributes the solution throughout the surface of the spinning substrate.

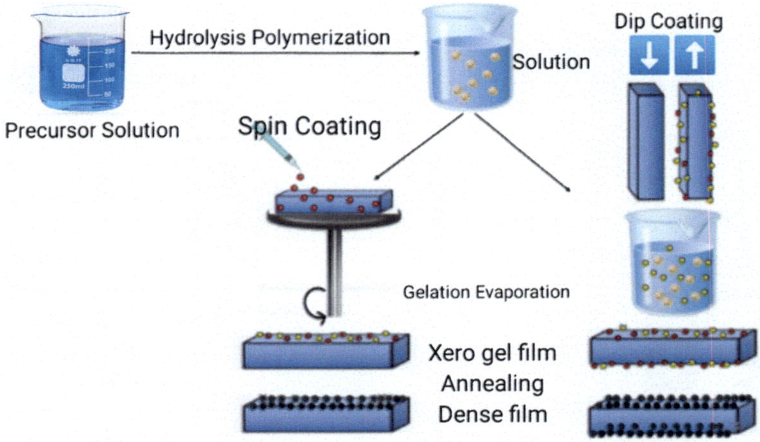

Fig. 8.7 A visual representation of the sol–gel deposition method, illustrating both spin coating and dip coating processes

A thin film develops as a result. The resultant film thickness is determined by rotational speed and duration. For simplicity's sake, this approach is preferred. It creates uniform thin sheets that can be precisely controlled in thickness. On the other hand, dip coating entails submerging the substrate in a precursor chemical solution. Withdrawal occurs at a regulated rate following it. Capillary action causes a thin layer to develop on the substrate's surface as it is removed from the solution. Film formation is also aided by the evaporation of solvents. There are several benefits to dip coating for massive or asymmetrical substrates. When contrasted with spin coating, it may be used to produce thicker coatings. However, optimization could be necessary to achieve uniformity in film thickness. Important variables include solution viscosity and withdrawal speed. For sol–gel thin film deposition, spin coating and dip coating each have particular advantages. For thinner films, spin coating excels in accuracy and homogeneity, while dip coating has its own advantages for coating larger or irregularly shaped substrates. This method is particularly useful for thicker films [34] (Fig. 8.7).

8.5 ALD Technique

The most advanced technique for thin film deposition is atomic layer deposition or ALD. Highly esteemed in the semiconductor industry and related fields, requiring precise control over the composition and thickness of films down to the atomic level. The ALD process is sequential and involves substrate preparation, which requires a thorough cleaning to get rid of contaminants. It also involves the sequential introduction of chemical precursors onto the substrate's surface. These orchestrate direct chemical reactions. These reactions form sub-monolayers of film. The method is inherently atomic and it ensures unparalleled precision in film deposition onto the substrate surface at the nanoscale.

Genesis of ALD can be traced back to the pioneering work of Prof. V.B. Aleskovskii in 1952. His doctoral research laid the foundation for this technique. Further advancements were made in collaboration with Prof. Kolt'sov in 1960. This culminated in the elucidation of ALD principles. Under the moniker 'Molecular Layering' the Evolution of ALD saw the emergence of distinct process modes such as Thermal ALD and plasma-enhanced ALD. PEALD being the most prevalent technique.

Thermal ALD is primarily surface-driven and operates exclusively through surface reactions. It offers exceptional thickness control and conformality irrespective of substrate geometry or reactor design. However, the requirement of relatively high temperatures (typically 150–350 °C) limits its applicability. A challenge addressed by PEALD which is also known as plasma-assisted ALD capitalizes on heightened reactivity of plasma species. This lowers deposition temperatures achieved without compromising film quality. This mode facilitates the use of a wider array of precursors and enables deposition on temperature-sensitive materials. Thereby broadening ALD's applicability, the industrialization of ALD gained momentum in 1970. Dr. Tuomo Suntola and collaborators pioneered the development of atomic layer epitaxy ALE further solidifying its position. Today it is the leading method for nanostructured thin film deposition.

Atomic layer deposition (ALD) represents the pinnacle in thin film deposition methodologies. Particularly revered in the semiconductor industry and allied domains necessitating meticulous control over film thickness and composition at the atomic level. The ALD procedure proceeds methodically, starting with substrate preparation that requires thorough cleaning to remove impurities. Substrate loading into an evacuated chamber proved to be successful. This creates a spotless environment for deposition. Inert gas (argon and nitrogen) purging sequentially removes any remaining moisture or air. This makes precursor adsorption more easier. The substrate surface interacts with introduced precursor gases initiating monolayer formation. Subsequently purging steps mitigate unwanted reactions ensuring uniform thin film deposition through precursor adsorption cycles. Post-deposition annealing may further enhance film properties. Culminating in rigorous characterization to confirm adherence to predefined specifications.

ALD's systematic orchestration confers unparalleled control over film thickness and composition. This is the peak of revolutionary progress. Atomic-scale engineering is on the verge of redefining technical boundaries [35].

Preparation and Deposition

Precursor preparation and precise vapor delivery are cornerstone elements in ALD processes. The production of precursor compounds is laborious. Guarantees volatility, stability, and purity. Precursor's vapor-phase introduction is crucial. Each precursor pulse introduces controlled amounts necessary for the reaction. The reactor's environment is often tailored. Temperature. Pressure. Surface characteristics fine-tuned. Ensures optimal kinetics and precursor-surface interaction.

Nucleation and Growth

Nucleation is a foundational step precursor adsorption initiates film growth. The growing layer consists of discrete nuclei growth that evolves from nucleation. Deposited materials coalesce into continuous layers. Subsequent cycles increase film thickness. With atomic precision, the surface must remain reactive. Ensuring uniformity and isolating unintended reactions are critical.

Film Uniformity and Composition Control

ALD techniques result in films with exquisite uniformity. It is achieved through self-limiting reactions ensuring layer-by-layer deposition. Each cycle contributes fractional film thickness, minimizing defects and high conformality will be achieved. Adjusting precursor types and sequences allows intricate multi-component structures, precise stoichiometry, and composition gradients achievable.

Applications in Advanced Technologies

Owing to precise control, ALD is essential in the creation of semiconductor devices, biotechnology, and catalysis. As the field of nanomaterials develops, ALD procedures are essential to innovation. Dielectrics with high K (high dielectric constant), applications include 3D nanostructures and encapsulation layers. The research is still ongoing and the usefulness of ALD is still being expanded with emerging technology paradigms [35] (Fig. 8.8).

9 Role and Effect of Nanofillers in Thin Films

Nanofillers are essential for improving thin film performance in a variety of fields owing to their unique characteristics and applications. Nanofillers are incredibly beneficial when used in thin films, which are frequently used in electronics, packaging, and medicinal equipment. In addition to improving barrier qualities, they also enhance their mechanical, electrical, and thermal characteristics. This comprehensive review explores the many kinds of nanofillers used in thin films. It also looks at significant effects on the functioning of films.

9.1 Carbon-Based Nanomaterials: Enhancing Mechanical and Thermal Properties

Carbon-based nanomaterials are among the most remarkable and in-demand materials in the engineering world. These materials include graphene and carbon

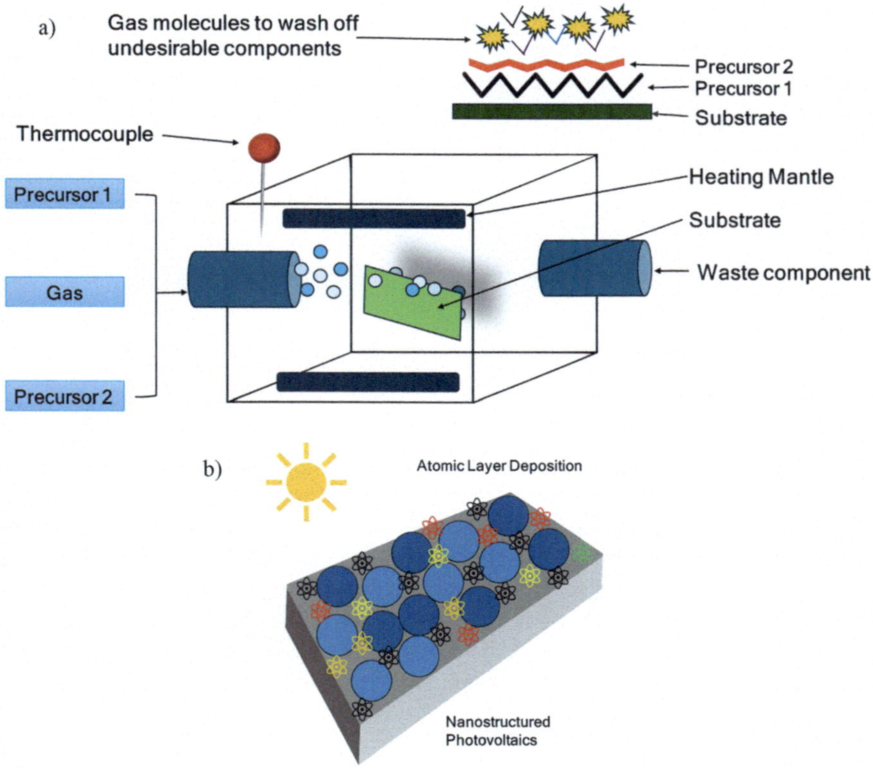

Fig. 8.8 Illustration of ALD methods for the fabrication of nanostructured thin films

nanotubes (CNTs) which are remarkable materials with outstanding mechanical strength, electrical conductivity and thermal stability are also impressive traits. These nanoparticles improve mechanical qualities including tensile strength and toughness. By significantly strengthening films' structural integrity when included in thin films. Furthermore, carbon-based nanomaterials and graphene have exceptional heat-dissipation properties. This property makes them indispensable for electrical and electronics applications where effective thermal control is necessary. They are excellent additions as reinforcement due to their distinct structural characteristics and high aspect ratio [36].

9.2 Nanoparticles of Metal Oxide: Improving Barrier, Optical, and Antimicrobial Characteristics

Incorporating metal oxide nanoparticles into thin films offers several benefits including zinc oxide (ZnO) and titanium dioxide (TiO_2). TiO_2 nanoparticles are widely recognized for potent photocatalytic characteristics. These properties enable them to enhance UV filtering. This also preserves optical transparency. As such they are highly suitable for applications where transparency is crucial. Furthermore, the barrier qualities of thin films are significantly enhanced by metal oxide nanoparticles. This offers defense against organic vapors and protection against oxygen and moisture. Adding ZnO nanoparticles to the film gives it strong antibacterial features. This successfully extends packed items' shelf lives. It also improves suitability for food packaging [37].

9.3 Clay Nanoparticles: Reinforcing Mechanical and Gas Barrier Properties

Since montmorillonite clay nanoparticles may significantly improve the mechanical and gas barrier characteristics of thin films. Their inclusion is very desirable. Their distinctively layered structure has a strengthening effect. This increases the film's tensile strength and lowers its gas permeability. Maintaining product freshness and prolonging shelf life are top concerns in the food packaging industry. Clay nanoparticles have therefore been widely used in this sector. The performance of thin films has been shown to be enhanced by the addition of clay nanoparticles, making them more appropriate for a variety of packaging applications [38].

9.4 Nanoparticles: Enhancing Mechanical Strength and Transparency

Thin film mechanical strength and transparency can be enhanced by silica nanoparticles. These nanoparticles are well known for their properties. Nanoparticles strengthen the film matrix's modulus and increase tensile strength. This results in increased resistance to mechanical forces. Furthermore, even after being altered, the film's transparency remains maintained by silica nanoparticles while optical clarity is preserved. This quality proves helpful in industries like display technology and optical coatings, where aestheticness and surface uniqueness are essential [39].

9.5 Natural Biopolymers: Cellulose and Chitosan for Biodegradability and Antimicrobial Activity

Thin films made of natural biopolymers such as chitosan and cellulose, provide special benefits. These biopolymers derived from renewable resources, provide thin films with the ability to decompose naturally over time. This renders them more eco-friendly than conventional plastics, reducing greenhouse emissions. The addition of chitosan enhances the biodegradability factor. It confers remarkable antibacterial qualities. Such characteristics make it an ideal choice for industries that place a premium on food safety and cleanliness. Researchers can improve the performance of thin films while also lessening their environmental effects. Using natural biopolymers it opens the door for environmentally friendly packaging options [40].

9.6 Synthetic Polymers: Polyethylene and Polystyrene for Mechanical Reinforcement and Barrier Properties

Synthetic polymers have become standard matrix materials for improving the performance of thin films in a range of applications including electronics and food packaging. Manufacturers can ensure their thin films have appropriate mechanical strength, also have thermal stability and barrier qualities by adding polymers like polystyrene and polyethylene. These man-made polymers provide a strong barrier against environmental factors such as moisture. Also oxygen and light. This gives packed goods optimal protection. Synthetic polymers such as polystyrene and polyethylene are essential for producing dependable. They produce high-quality packaging solutions due to their extensive applications and unique characteristics.

An extensive summary of the numerous types of nanofillers utilized in thin films is given in the following sections. It describes how each of them improved the qualities of the films. Nanofillers have an influence on films by improving mechanical strength, electrical properties, thermal durability, and optical transparency. They also increase the effectiveness of antibiotics. These nanofillers are essential for enhancing thin film properties and applications across a wide range of industries [41].

9.7 Microscopic Examination

The distinctive properties and behaviors of nanostructured thin films must be comprehended through microscopic analysis. Energy dispersive X-ray spectroscopy (EDS), scanning electron microscopy (SEM), transmission electron microscopy (TEM), and advanced (scanning) transmission electron microscopy ((S)TEM) are a few sophisticated techniques that may be used for this purpose and ensuring the quality and uniqueness of thin films.

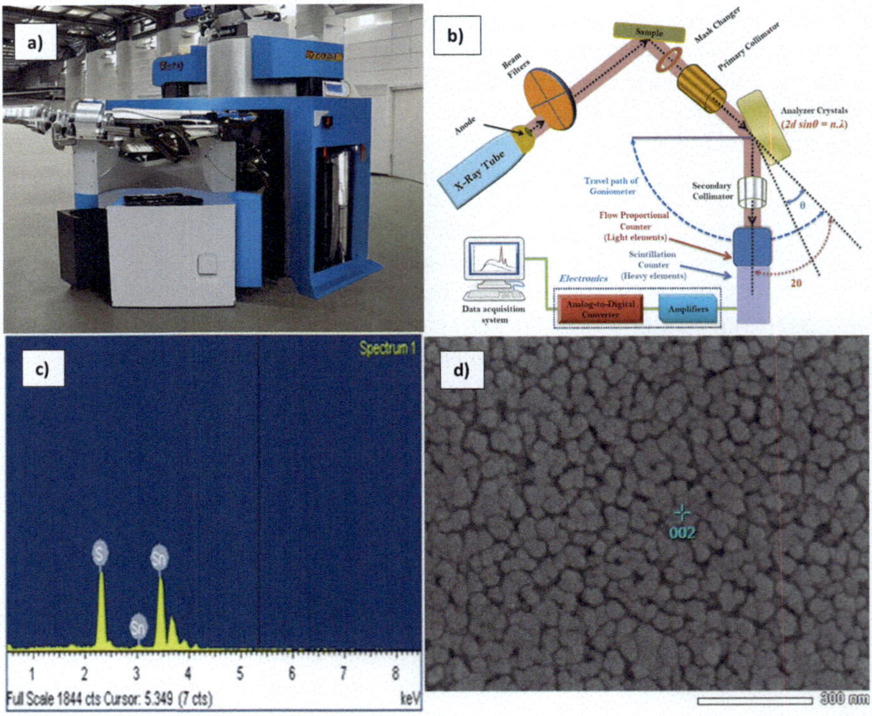

Fig. 8.9 **a** EDS machine for thin film **b** Ray-diagram of EDS **c** XRD peaks of thin film **d** microstructure of EDS thin film Combined CC BY 4.0

9.8 Energy Dispersive X-Ray Spectroscopy

EDS is a quick and non-destructive method for determining a thin film's density that is nanostructured. It operates by identifying X-rays that are released from the sample when an electron beam is focused on it. The elements and their relative quantities may be identified. Typical of the elements found in the sample ensuring the sample is applicable for various applications [42] (Fig. 8.9).

9.9 Scanning Electron Microscope

Scanning electron microscopy, or SEM, is a vital tool in nanotechnology for analyzing materials with nanostructures especially with regard to thin films by moving a concentrated electron beam across the surface of the sample, identifying secondary electrons released. Using SEM, high-resolution pictures are possible exposing the distribution and size of the particles inside the thin film's surface morphology. Understanding the structural characteristics of nanostructured materials requires such

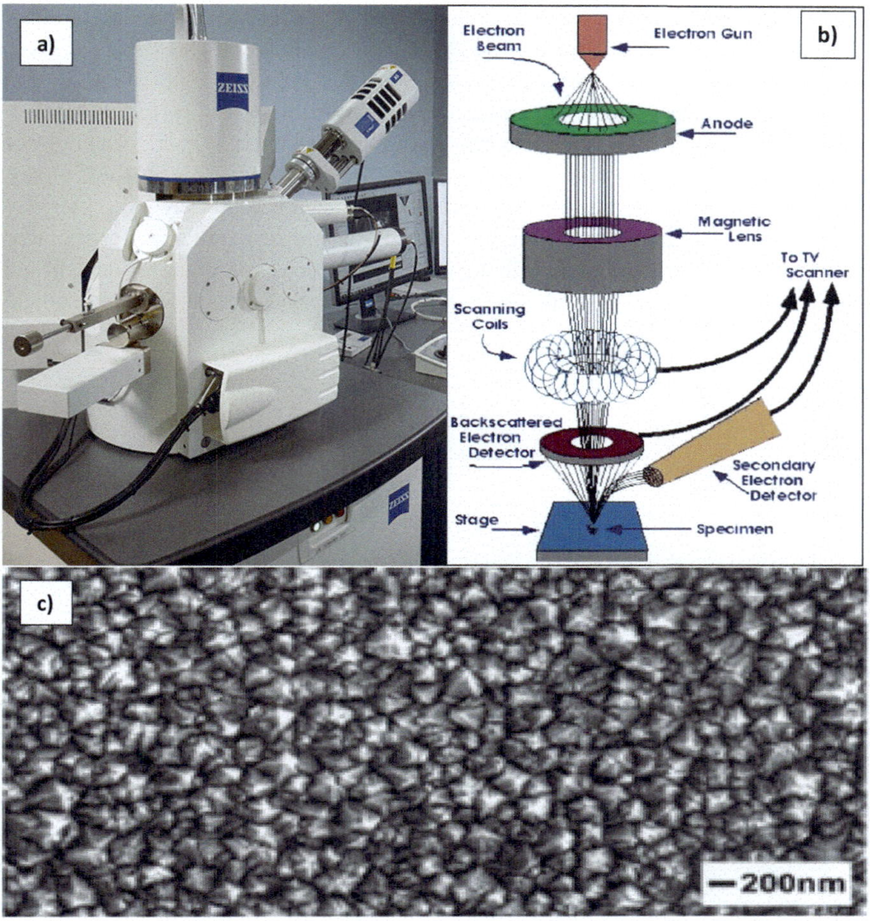

Fig. 8.10 **a** SEM machine of thin films **b** Ray-diagram schematic of scanning electron microscopy, **c** SEM image at 200 nm combined CC BY 4.0

expertise. For their effectiveness in nanotechnology applications, these characteristics are frequently essential. Applications like catalysis in electronics, and biomedical devices are also included [43] (Fig. 8.10).

9.10 Transmission Electron Microscopy

The structure and composition of nanostructured thin films may also be examined using a transmission electron microscope. This is another effective method. It creates an image of the sample's interior structure by passing an electron beam through a thin sample. TEM may be used to examine thin film cross-sectional morphology at the nanoscale. It reveals information about their composition and structure. High spatial resolution surface modification and characterization of nanostructured thin films may be achieved using sophisticated (S)TEM techniques [44] (Fig. 8.11).

These methods are crucial for comprehending distinct traits and behaviors of nanoscale thin films with nanostructures. They are essential to several disciplines. Surface engineering nanotechnology and materials science are key examples. By providing comprehensive details on the composition and structure of nanostructured thin films these methodologies will facilitate the process of material design and optimization. Particular uses include electronics, energy storage, and biomedical devices benefit as well.

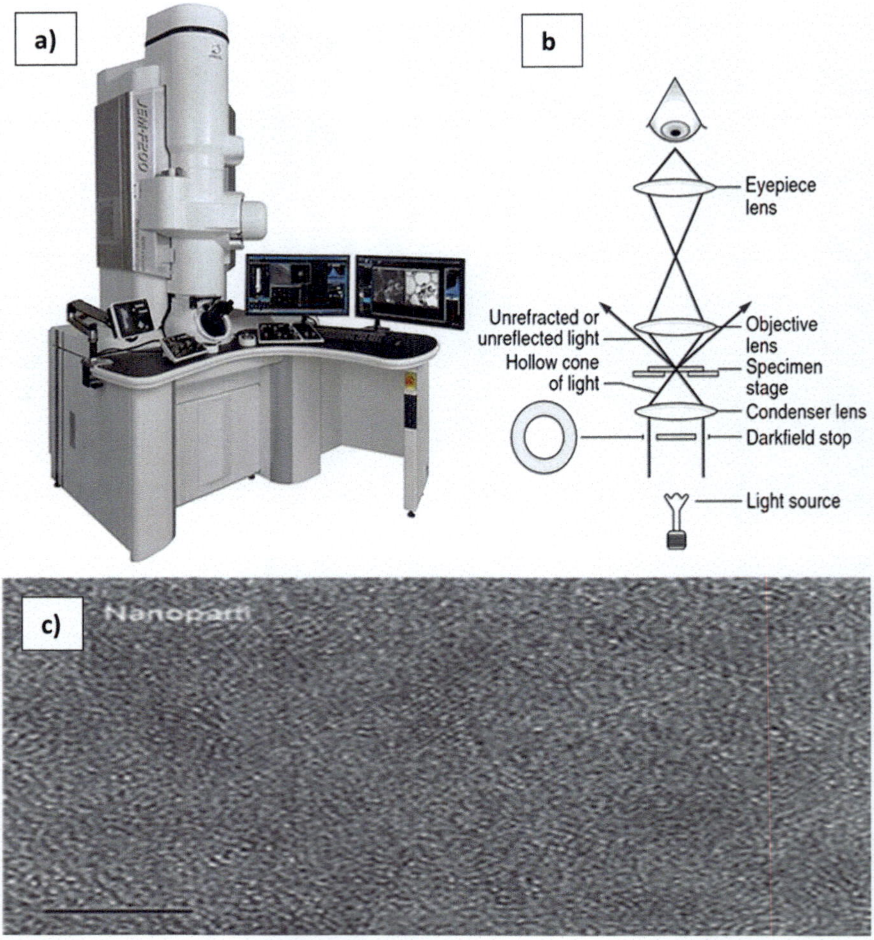

Fig. 8.11 **a** TEM microscopy machine, **b** Ray-diagram and schematic of transmission electron microscopy, **c** TEM microstructure of nanoscale thin film combined CC BY 4.0

9.11 Applications of Thin Films in Nanotechnology

Nanotechnology thin films offer a wide variety of uses in nanotechnology due to their distinct features and capabilities. These applications are used in numerous areas, including materials research, biomedical engineering, electronics, and energy.

9.12 Electronics

The development of electrical technology is greatly aided by nanostructured thin films. The production of integrated circuits, semiconductor devices, and the state of the data storage technologies like phase-change memory (PCM) and magnetic random-access memories (MRAM) is made possible in a significant way by them. Furthermore, because of their special qualities, which allow for exact control over light absorption, emission, and transmission, these films are essential parts in the production, design, and development of optoelectronic devices like organic light-emitting diodes (OLEDs) and light-emitting diodes (LEDs) [45].

9.13 Energy Sector

Nanostructured thin films act as innovative sparks in the field of renewable energy. They are widely used in solar cells and photovoltaic applications. These improve these essential technologies' performance and efficiency. These films have been significant in enhancing the solar energy industry. Due to their capacity to manufacture efficient n-type semiconductor materials, they improve light capture and conversion in dye-sensitized solar cells. The performance and lifetime of batteries and supercapacitors are enhanced by nanostructured thin films. It provides a substantial contribution to energy storage solutions [46].

9.14 Catalysis

Catalysts in the form of nanostructured thin sheets accelerate chemical reactions in systems like fuel cells, hydrogen generation, and pollution control. Because of their large surface area and distinct surface chemistry, they may interact with reactants more effectively, which increases their catalytic activity and selectivity [46].

9.15 Biomedical Engineering

Nanostructured thin films have exciting opportunities for biological applications. These films offer precise and highly biocompatible solutions for a variety of applications in the human body. It includes dental implants artificial bones and targeted drug delivery systems using hydrogels and liposomes. Their capacity to engage in nanoscale interactions with biological molecules guarantees excellent performance and integration in a range of biomedical scenarios [47].

9.16 Materials Science

Nanostructured thin films in materials science stimulate innovation in a wide range of applications. They are an essential part of ceramic coatings. They protect substrates from oxidation, wear, and corrosion. Additionally, these films are useful for sensors and detectors. Due to their increased sensitivity and response to external stimuli, instances involve biosensors and gas sensors. Additionally, surface functionalization processes depend heavily on nanostructured thin films. They boost adhesion by improving the characteristics of the substance. They also have superior biofouling resistance and lubrication [48].

9.17 Economic Considerations

The economic aspects of nanostructured thin films are complex and include factors like manufacturing costs scalability, benefits to performance, and market demand. The price of making these films comprises labor equipment and material costs. Initially, material costs are frequently greater than those of ordinary materials due to the need for special synthesis techniques. However, if production methods and economies of scale improve, there may be opportunities to cut costs in the long run. The commercial feasibility of manufacturing processes depends on scalability. Methods like chemical vapor deposition (CVD) and physical vapor deposition (PVD) offer cost-effective opportunities for production on a wide scale. Improved mechanical strength, electrical conductivity, or optical qualities are just a few of the attributes nanostructured thin films frequently possess over bulk materials. These improvements might result in noticeable performance advantages. Market demand for these films varies depending on a number of factors, such as consumer preferences, industrial requirements, and technical developments. Because of their special qualities industries including electronics, energy storage, and healthcare are progressively embracing them. The economic environment around nanostructured thin films is significantly shaped by factors such as intellectual property protection competitive landscape dynamics and regulatory concerns. The economic benefits of using nanostructured thin films include increased efficiency, lower costs, and environmental advantages. The economic feasibility and attractiveness of nanostructured thin films are anticipated to increase more as technology advances and costs come down. This will stimulate innovation and promote economic growth [49].

9.18 Conclusion

To summarize nanostructured thin films are at the forefront of nanotechnology. They represent a spectacular convergence of scientific discovery and practical application These thin films provide previously unheard-of possibilities for breakthroughs in a variety of industries. Sectors range from electronics and optics to health and renewable energy, the continuing improvement in production methods, and exact control at the nanoscale. The value of micro-inspection becomes clearer as we explore the world of nano-size thin films. It enables a thorough knowledge of their structure and characteristics. It elucidates prospective uses as we are paving the road for revolutionary discoveries. These discoveries will reshape industry and technology in the future by using the potential of nanotechnology and rising to obstacles operating at the nanoscale we are advancing. As researchers continue to push the boundaries of what is possible in nanostructured thin film technology, we stand poised. We are on the brink of a new era of scientific discovery and technological innovation.

References

1. Ghosh, M., Paul, A., Mandal, R.: Thin film fabrication for wearable electronics: journey so far (2024)
2. Grilhé, J.: Mechanical properties of thin films and coatings. In: Materials Science Forum, vol. 59, pp. 481–534. Trans Tech Publ (1990)
3. Benelmekki, M., Erbe, A.: Nanostructured thin films–background, preparation and relation to the technological revolution of the 21st century. In: Frontiers of Nanoscience, vol. 14, pp. 1–34. Elsevier (2019)
4. Li, C., Liu, X., Du, X., Yang, T., Li, Q., Jin, L.: Preparation and optical properties of nanostructure thin films. Appl. Nanosci. **11**(7), 1967–1976 (2021)
5. Fang, Z., Hou, G., Chen, C., Hu, L.: Nanocellulose-based films and their emerging applications. Curr. Opin. Solid State Mater. Sci. **23**(4), 100764 (2019)
6. Elzein, B.: "Nano Revolution:" Tiny tech, big impact: how nanotechnology is driving SDGs progress. In: Heliyon, vol. 10, no. 10 (2024)
7. Mekuye, B., Abera, B.: Nanomaterials: an overview of synthesis, classification, characterization, and applications. Nano Select **4**(8), 486–501 (2023)
8. Hiremath, A., Murthy, A.A., Thipperudrappa, S., KN, B.: Nanoparticles filled polymer nanocomposites: a technological review. Cogent Eng. **8**(1), 1991229 (2021)
9. Bayda, S., Adeel, M., Tuccinardi, T., Cordani, M., Rizzolio, F.: The history of nanoscience and nanotechnology: from chemical–physical applications to nanomedicine. Molecules **25**(1), 112 (2019)

10. Hulla, J., Sahu, S.C., Hayes, A.W.: Nanotechnology: history and future. Hum. Exp. Toxicol. **34**(12), 1318–1321 (2015)
11. Sahoo, M., Vishwakarma, S., Panigrahi, C., Kumar, J.: Nanotechnology: current applications and future scope in food. Food Front. **2**(1), 3–22 (2021)
12. Han, H., et al.: Lattice strain-enhanced exsolution of nanoparticles in thin films. Nat. Commun. **10**(1), 1471 (2019)
13. Meng, K., et al.: Superionic fluoride gate dielectrics with low diffusion barrier for two-dimensional electronics. Nat. Nanotechnol. 1–9 (2024)
14. Khan, I., Saeed, K., Khan, I.: Nanoparticles: properties, applications and toxicities. Arab. J. Chem. **12**(7), 908–931 (2019)
15. Mitchell, M.J., Billingsley, M.M., Haley, R.M., Wechsler, M.E., Peppas, N.A., Langer, R.: Engineering precision nanoparticles for drug delivery. Nat. Rev. Drug Discovery **20**(2), 101–124 (2021)
16. Memon, A.G., et al.: Ultrasensitive colorimetric aptasensor for Hg2+ detection using Exo-III assisted target recycling amplification and unmodified AuNPs as indicators. J. Hazard. Mater. **384**, 120948 (2020). https://doi.org/10.1016/j.jhazmat.2019.120948
17. Memon, A.G., et al.: Label-free colorimetric nanosensor with improved sensitivity for Pb2 + in water by using a truncated 8–17 DNAzyme. Front. Environ. Sci. Eng. **13**(1), 12 (2019). https://doi.org/10.1007/s11783-019-1094-7
18. Wang, R., Zhou, X., Liedberg, B., Zhu, X., Memon, A.G., Shi, H.: Screening criteria for qualified antibiotic targets in unmodified gold nanoparticles-based aptasensing. ACS Appl. Mater. Interfaces **9**(40), 35492–35497 (2017). https://doi.org/10.1021/acsami.7b12796
19. Memon, A.G., et al.: Citrate-capped AuNP fabrication, characterization and comparison with commercially produced nanoparticles. Crystals (Basel) **12**(12), 1747 (2022). https://doi.org/10.3390/cryst12121747
20. Ahmad, J., Memon, A.G., Shaikh, A.A., Ismail, T., Giwa, A.S., Mahmood, A.: Insight into single-element nobel metal anisotropic silver nanoparticle shape-dependent selective ROS generation and quantification. RSC Adv. **11**(14), 8314–8322 (2021). https://doi.org/10.1039/D0RA10616J
21. Memon, A.G., et al.: Utilization of unmodified gold nanoparticles for label-free detection of mercury (II): insight into rational design of mercury-specific oligonucleotides. J. Hazard. Mater. **321**, 417–423 (2017). https://doi.org/10.1016/j.jhazmat.2016.09.025
22. Srikar, S.K., Giri, D.D., Pal, D.B., Mishra, P.K., Upadhyay, S.N.: Green synthesis of silver nanoparticles: a review. Green Sustain. Chem. **06**(01) (2016). https://doi.org/10.4236/gsc.2016.61004
23. Xiong, Y., Liu, X., Xiong, H.: Aggregation modeling of the influence of pH on the aggregation of variably charged nanoparticles. Sci. Rep. **11**(1), 17386 (2021)
24. Petros, R.A., DeSimone, J.M.: Strategies in the design of nanoparticles for therapeutic applications. Nat. Rev. Drug Discovery **9**(8), 615–627 (2010)
25. Gigliobianco, M.R., Casadidio, C., Censi, R., Di Martino, P.: Nanocrystals of poorly soluble drugs: drug bioavailability and physicochemical stability. Pharmaceutics **10**(3), 134 (2018)
26. Talapin, D.V., Shevchenko, E.V.: Introduction: nanoparticle chemistry. Chem. Rev. **116**(18), 10343–10345 (2016)
27. Hishimone, P.N., Nagai, H., Sato, M.: Methods of fabricating thin films for energy materials and devices. In: Lithium-ion Batteries-Thin Film for Energy Materials and Devices. IntechOpen (2020)
28. Chopra, K.: Thin Film Device Applications. Springer Science & Business Media (2012)
29. Sampaio, W.R.V., Serra, P.L.C., Dantas, N.O., de Sousa, R.R.M., Silva, A.C.A.: Nanocomposites thin films: manufacturing and applications. In: Nanocomposite Materials for Biomedical and Energy Storage Applications. IntechOpen (2022)
30. Arifin, H.R., et al.: Corn starch-based bionanocomposite film reinforced with ZnO nanoparticles and different types of plasticizers. Front. Sustain. Food Syst. **6**, 886219 (2022)
31. Balema, V.P.: Chemical deposition techniques in materials design. Mater. Matters **1**(3) (2006)

32. Knoll, A.: Thin film physical vapor deposition (PVD) system. University of Akron. https://www.uakron.edu/soa/instruments/pvd
33. Schneider, C.W., Lippert, T.: Laser ablation and thin film deposition. In: Laser Processing of Materials: Fundamentals, Applications and Developments, pp. 89–112 (2010)
34. Sakka, S.: Handbook of Sol-gel Science and Technology. 1. Sol-gel Processing. Springer Science & Business Media (2005)
35. Oke, J.A., Jen, T.-C.: Atomic layer deposition and other thin film deposition techniques: from principles to film properties. J. Market. Res. **21**, 2481–2514 (2022)
36. Maduraiveeran, G., Jin, W.: Carbon nanomaterials: synthesis, properties and applications in electrochemical sensors and energy conversion systems. Mater. Sci. Eng., B **272**, 115341 (2021)
37. Hancharova, M., Halicka-Stępień, K., Dupla, A., Lesiak, A., Sołoducho, J., Cabaj, J.: Antimicrobial activity of metal-based nanoparticles: a mini-review. BioMetals 1–29 (2024)
38. Zeggai, F.Z., Touahra, F., Labied, R., Lerari, D., Chebout, R., Bachari, K.: Biopolymers-Clay Nanocomposites: Synthesis Pathways, Properties, and Applications (2024)
39. Guo, D., Xie, G., Luo, J.: Mechanical properties of nanoparticles: basics and applications. J. Phys. D Appl. Phys. **47**(1), 013001 (2013)
40. Hisham, F., Akmal, M.M., Ahmad, F., Ahmad, K., Samat, N.: Biopolymer chitosan: potential sources, extraction methods, and emerging applications. Ain Shams Eng. J. 102424 (2023)
41. Shimomura, S., et al.: Stabilization of polystyrene thin films by introduction of a functional end group. Polym. J. **48**(9), 949–953 (2016)
42. Ngo, P.D.: Energy dispersive spectroscopy. In: Failure Analysis of Integrated Circuits: Tools and Techniques, pp. 205–215. Springer (1999)
43. Yablon, D., Libera, M.: Microscopy 101: scanning probes or scanning electrons: a practical guide to select a method for nanoscale characterization. Microscopy Today **27**(6), 32–38 (2019)
44. Zheng, H., Claridge, S.A., Minor, A.M., Alivisatos, A.P., Dahmen, U.: Nanocrystal diffusion in a liquid thin film observed by in situ transmission electron microscopy. Nano Lett. **9**(6), 2460–2465 (2009)
45. Eda, G., Chhowalla, M.: Graphene-based composite thin films for electronics. Nano Lett. **9**(2), 814–818 (2009)
46. Moorthy, S.B.K.: Thin Film Structures in Energy Applications. Springer (2015)
47. Vendra, V.K., Wu, L., Krishnan, S.: Polymer thin films for biomedical applications. In: Nanomaterials for the Life Sciences, vol. 5 (2010)
48. Ohring, M.: Materials Science of Thin Films: Deposition and Structure. Elsevier (2001)
49. Russell, T., Baron, B., Rocheleau, R.: Economics of processing thin-film solar cells. J. Vacuum Sci. Technol. B: Microelectron. Process. Phenomena **2**(4), 840–844 (1984)

Aerospace and Thin Films

Abdulaziz Alhazaa and Muhammad Ali Shar

Abstract The chapter titled "Aerospace and Thin Films" discusses the importance of thin film coatings in the field of aerospace, mainly with regard to aircraft as well as spacecraft. In this chapter, the literature describes the types of advanced coatings that are used in the aerospace industry for the improvement of performance and for increasing the lifespan of aerospace parts. Thin films are a necessitated thermal protection system used to protect spacecraft during its re-entry and functioning in space. Also, the tribological coatings are tested to improve the way the moving parts interact to reduce the rate of friction and wear; thus, improving the durability and efficiency of the aerospace equipment. The chapter shows how these coatings assist in enhancing the fuel efficiency of aerospace vehicles, cutting down on the rate of maintenance, and making aerospace vehicles safer and have a longer life expectancy. In highlighting these advanced thin film applications in this chapter, it is established that the applications cannot be over-emphasized in the revelation of how aerospace engineering can be developed to perform under demanding circumstances. These findings of this exploration demonstrate that thin films have revolutionized the aerospace industry by improving the performance and features of airplanes and space vehicles.

1 Introduction

Modern manufacturing technology has drawn a lot of attention as industrial civilization has grown, especially in the aerospace sector. The fast advancement of information and communication technology has made the application of aircraft technology more and more significant. Aerospace science and technology are growing at a rapid rate due to industry demands. Researchers have created numerous methods to

A. Alhazaa (✉) · M. A. Shar (✉)
King Abdullah Institute for Nanotechnology, King Saud University, Riyadh 11451, Saudi Arabia
e-mail: aalhazaa@ksu.edu.sa

M. A. Shar
e-mail: m.baloch@bradford.ac.uk

A. D. Chandio and I. A. Channa (eds.), *Thin Films and Coatings*, Engineering Materials,
https://doi.org/10.1007/978-981-96-0005-2_9

produce greater performance, lower energy and resource consumption, more dependable goods, etc. Thin film integration has been a game changer in the aerospace sector, which is a large business where precision and creativity are crucial. Thin films layers of material with thicknesses on the nanoscale have several uses in aerospace technology and provide a number of advantages, including increased performance and durability [1].

The requirement of robust yet lightweight material has always been the point of concentration and desire for aeronautical engineers. Thin films are ideal for this task in the first place because they are typically characterized by lightweight. Airframes, wings, and propulsion systems are among the aerospace components on which engineers can achieve substantial weight reduction without sacrificing structural integrity by using thin film coatings. Reducing weight is significant since it makes airplanes and spacecraft perform effectively, over a long distance and with less fuel consumption, because of enhanced mobility. In addition, thin film coatings are of great importance in protecting aeronautical vehicles from the harsh environment in which they usually operate. These films are useful as they provide a layer of armor that can minimize the impact of heat stress, abrasion, and corrosion on particular parts. At the moment it can be argued that Nano thin films based on polymers could take the leading position of high-performance materials. Thus, fiber-reinforced thin films or composites that have recently been widely utilized as material in the aerospace industry possess slightly higher strength-to-weight and modulus-to-weight ratios than metallic materials. But for the last 20 years, aeronautical engineers have reflected their attention on thin films made up of nanomaterials or their reinforcements since they can improve the qualities and add utility factors also in the field of aeronautics [2].

In the economic aspect, the aerospace industry profits from thin films mostly since they contribute to the minimization of weight, optimization of performance, fastening of production, multiplicity of function, as well as provision of the competitive advantage to industries. It reduced material consumption, improved the gas mileage of vehicles, and thus extended the durability of parts which in turn reduced maintenance costs. Thin films are faster at producing functions as they allow the condensation of several layers into one layer. Also, some of the advanced attributes such as sensing and self-repair features amends to enhance dependability and efficacy. This way, through the use of thin films, aerospace firms get to position themselves to capture the market and attract customers and contracts. In light of what we have explored so far, it is imperative to recognize the importance of thin films in the future of affordable, efficient, and innovative aerospace manufacturing strategies [3].

2 Historic Background

Aerospace thin films have created immense value in the current technological world, complementing the performance, efficiency, and safety mechanisms in aircraft and space vehicle engineering, thus being a breakthrough in material science and engineering. The outlined historical background shows the transition of thin films in aerospace and their importance from prototype components to crucial elements in modern aerospace industries. With the development of thin film coating technologies, applications in the domains of aircraft have emerged [4].

Scientists began to consider the possibility of using polymers for use in aircraft and space vehicles and began to incorporate polymers into aviation and space technologies in the first decade of the twenty-first century, it is for this reason that thin films of polymers in aerospace were first used. During this period, polymeric materials like polyesters and polyimides that had properties of lightweight, flexibility, and high durability of interest for thin film coatings began to be posted for their potential use as such. The advancement of polymer-based thin films for aerospace applications has taken place in a number of fields, all of which have advanced the design and functionality of airplanes and spacecraft. In the year 2003, these thin films were originally used as components and protective coatings in composite structures. As materials science progressed, engineers designed less dense but more rigid polymer films and thereafter included them more in the design of aerospace vehicles' frameworks mainly for adequate strength and performance. In the year 2007, the use of polymer thin films found its application in thermal protection too. These films have a useful purpose as insulators, regulating the environment inside the capsule when in the air and serving as a barrier that insulates against high temperatures when entering the planet's atmosphere. They not only served as barriers but also took part in absorbing or rerouting cosmic radiation to protect humans and delicate equipment within the aircraft or spacecraft. Moreover, in between 2009 and 2010 aerodynamics and drag reduction in aircraft engineering have also been transformed by advances in polymer thin film technology. Specialized coatings and surface treatments have been developed to maximize fuel economy, minimize turbulence, and enhance flying characteristics by using the unique properties of polymer films. These films are also essential for the production of sensors and electronic parts for aircraft systems. Polymer films, which are engineered to possess specific electrical, mechanical, and thermal properties, facilitate the development of resilient and dependable sensor systems that can resist the extreme circumstances encountered in space in the year 2015 (shown in Fig. 9.1) [4, 5].

Additionally, after 2018 polymer films support aircraft industry efforts to protect the environment. They are used as anti-erosion, anti-abrasion, and anti-corrosive since they protect the surface of the aircraft to ensure that the aircraft are serviceable for as long as possible to guarantee service on the aircraft. Those polymer-based thin films, still present new learnings and improvements to aerospace engineering through the thin films that have maintained the high durability, strength, flexibility, and lightweight characteristics it needs to conquer the skies and even in

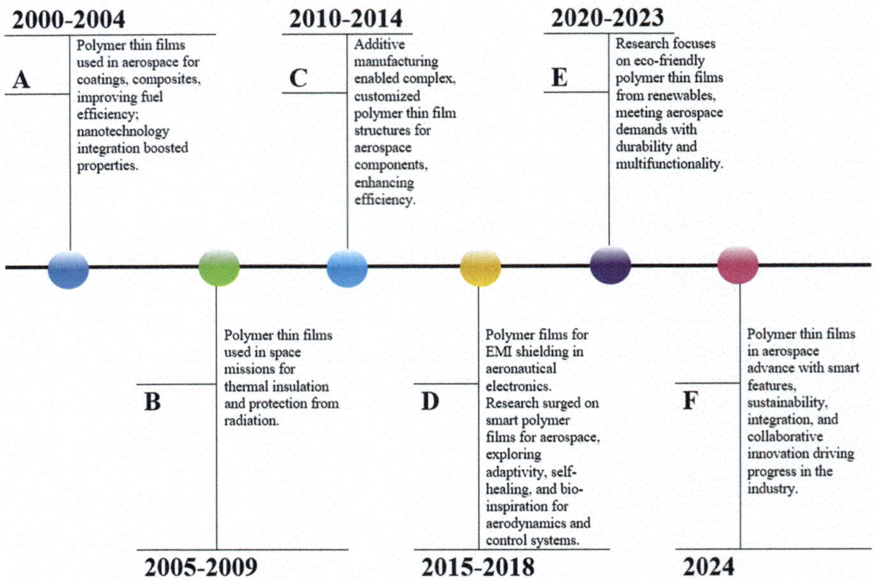

Fig. 9.1 Timeline illustrating development in the use of thin films in aerospace engineering applications

outer space [5]. At present time the thin film industry is more conscious of the multi-functional, self-healing, and sustainable polymer films that have no adverse effect on our environment and successfully meet the modern demands of aerospace industries [6].

3 Coatings for Aircraft and Spacecraft

In general, it means the outermost layer of polymer which is formed on any material (also known as substrate) for certain usages. Coating is a process that involves polymer coating placement on a substrate and permitting it to absorb on its surface [7]. A priming layer, a topcoat layer, and a conversion coating layer usually make up the coating system. These days, an electric current is employed to develop the conversion coating anodically, or a tri- and hexavalent chromium salt solution (like Alodine TM [8]) reacts with the metal to create the coating chemically on the surface. The conversion coating's main function is to facilitate the primer's adherence to the substrate. Epoxy resins crosslinked with amine or polyamide curing agents are commonly used as primers. On rare occasions, other resin systems are also employed for particular purposes [9]. Depending on the nature of the polymer that has been selected for application on the coating, the resin may be required to dry or polymerize after the coat is applied. Thin film coatings using polymer-based typically play an important role in increasing the performance, durability, and safety of the aircraft or spacecraft. It is important to consider that these coatings have other specific properties such as thermal insulating properties or even anti-fouling besides having a protective layer that shields the equipment against unfavorable environmental conditions such as high temperatures, UV radiation, corrosion, and abrasive wear. The application of polymer, polymeric composite, and nanocomposite as corrosion barriers, thermal stability, and strength is still on the rise in many fields and end-products such as in aerospace and automotive fields. In the future, these innovations should be able to provide new opportunities for UAVs and spacecraft of the subsequent series [7].

The right choice of polymer, coating technique, and processing variables lead to enhanced coatings that offer higher performance features shown in Table 9.1. Acrylics, cellulose, epoxy, polyurethane, polytetrafluoroethylene, biopolymers, and other polymers have been coated in numerous technical application sectors.

3.1 Advance Coating Techniques for Aerospace Engineering

A variety of painting, printing, and bonding procedures are used to coat thin films, outlining the requirements for the creation of high-performance materials. Generally, a particular interface adhesion between the coating and the substrate is required to produce high-quality material. For polymer coatings with thickness ranging from a few millimeters to 10 nm, numerous techniques have been developed; some of these are covered in this section. The substrate and/or surfaces have been successfully coated using a variety of coating techniques. Among them are printing processes, extrusion/dispersion coating, spin coating, dip coating, doctor blading, solution casting, spray coating, gravure coating, etc. Typically, high-volume production, cheap cost, and convenience of fabrication are taken into consideration while choosing a coating technique [17]. The most modern and widely utilized techniques used in the aerospace sector are as follows.

- **Plasma Polymerization Technique**

The process of plasma polymerization falls under the chemical vapor deposition (CVD) technology, which is widely employed in the aerospace industry to create polymer thin films. Plasma polymers are deposited as a thin film or as a powder on surfaces that come into contact with a glow discharge of organic or organometallic feed gases (shown in Fig. 9.2). One particular kind of plasma chemistry is called plasma polymerization, and it includes reactions between surface species, between plasma, and between surface species themselves. Aerospace parts including fuel tanks, avionics housings, and satellite substrates are coated with thin films with specific characteristics utilizing plasma polymerization [18].

- **Layer-by-Layer Deposition Method**

Table 9.1 Shows the Material selection and properties required for aerospace applications.

Applications	Properties required	Polymer thin films	References
Aerospace coatings systems comprise topcoats to improve the appearance and durability of aerospace vehicles	Good weather ability, UV resistance, color retention, glossy finish, and excellent adhesion	Acrylic	Visser et al. [10]
Anti-corrosive, anti-UV, and anti-erosion coatings on aircraft surfaces	Excellent abrasion resistance, chemical resistance, flexibility, and durability	Polyurethane	Yang et al. [11]
Applying a layer of protection to aircraft parts like fasteners, bearings, and seals to minimize friction, sticking, and corrosion	Outstanding chemical resistance, low friction coefficient, excellent non-stick properties	Teflon	FitzSimmons and Zisman [12]
Protecting aircraft windows, sealing components, and providing thermal insulation	Excellent weather resistance, flexibility, thermal stability, and good adhesion	Silicone	Shit and Shah [13]
Thermal protection against hot and cold environments and dielectric purposes in electric components	Exceptional thermal stability, chemical resistance, and mechanical strength	Polyimide	Sezer Hicyilmaz and Celik Bedeloglu [14]
Shielding from corrosion, wear and tear, or impact, especially on areas that are prone to the above effects	Exceptional chemical resistance, abrasion resistance, rapid curing properties, and excellent impact resistance	Polyurea	Peng [15]
Providing a layer to aircraft structures, parts, and surfaces for the purpose of protection from corrosion and strengthening	High strength, excellent adhesion, and chemical resistance	Epoxy	Song et al. [16]

LbL assembly is used in the formation of a multiple-layered coating on aerospace material for such applications as the formation of a corrosion-proof coating, fouling-resistant coating, and self-healing coating. It is also for the fine-tuning of surface characteristics such as the wettability and adhesion properties required in aerospace engineering. Layer-by-layer assembly, or LbL, is a Nanotechnological approach to negotiating a material and its surface or interface. The multilayered interfaces positively affect the stress transfer, micro-crack propagation, as well as dispersion (shown in Fig. 9.3). Layer-by-layer (LbL) self-assembly of multilayered polymer composites has received much attention over the last few decades because the sample thickness can be controlled at the nanometer scale and the applicability is versatile [19].

It is common practice to initiate the procedure with surface preparation to ensure that there is appropriate surface adhesion. To enhance adhesion between the substrate

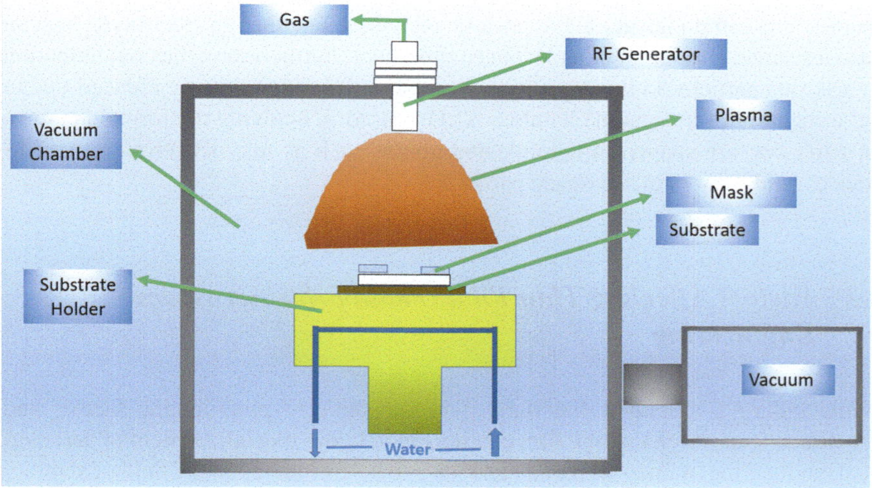

Fig. 9.2 Schematic overview of plasma polymerization method

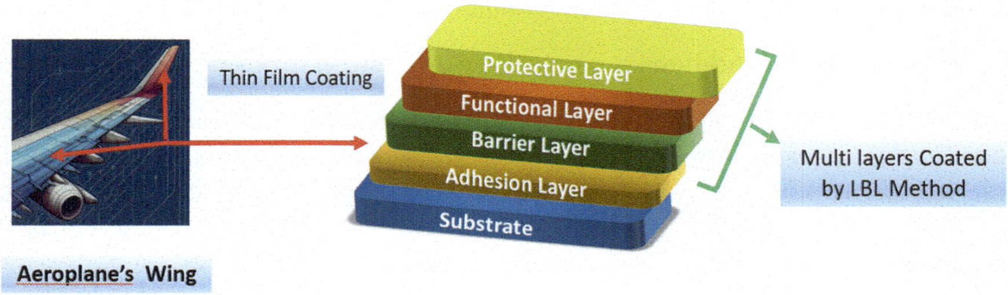

Fig. 9.3 An airplane wing coated by LBL method

and subsequent layers, there may be a primer coating added as well. Various functional coatings are then deposited and each is intended to perform a specific role such as wear protection, corrosion protection, thermal barrier, oxidation protection, and others. These coatings can be deposited using sophisticated techniques including; atomic layer deposition (ALD), chemical vapor deposition (CVD), electroplating, and physical vapor deposition (PVD) [19].

- **Atomic Layer Deposition Technique**

ALD or atomic layer deposition is one of the thin film deposition techniques in which molecules of introduced chemicals react with the substrate to grow sub-monolayers of film at a time. As the name implies, it is fundamentally atomic in its operation and it guarantees precise atomic layer build-up on a selected substrate material. Due to the above-stated properties, ALD is an ultra-thin film deposition technology that has been deployed in several capacities. They include thickness and conformal coating films with controllable thickness for depositing complex three-dimensional shapes. You can do the atomic deposition technique in either an electrolytic state, a gaseous stage, a vacuum, or a plasma environment. Furthermore, the contamination by gaseous particles in the deposition chamber will be significantly reduced due to the utilization of a vacuum chamber. ALD is used in applying conformal coatings on parts that are shaped in complicated guides such as microelectronics, MEMS devices, and sensors in aerospace parts [20].

3.2 *Factors Affecting Thin Film Coatings for Aerospace Engineering*

While talking about their functions it is inevitable to pinpoint the factors and constituents on which the thin film coatings rely. In the specialized field of aeronautical engineering, numerous parameters have to be considered in the field of stability and durability of polymer-based thin-film coatings for specific applications. They include factors involving coating selection, surface treatment, and the usage of coatings in aerospace environments to ensure the products perform at their optimal best. Hence, the detailed list of the most important components includes the following.

- **Environmental Conditions**

Aerospace vehicles operate in rather harsh environmental conditions that are often quite aggressive and change quite often, at that; temperature fluctuations, ultraviolet light impact, relative humidity, and contact with various corrosive materials. For preparing such coatings it is essential to select basic or to design such coatings that will remain performing and intact in these situations [21].

- **Weight Considerations**

Traditionally, the demand for innovation has been mainly pushed by weight-loss applications throughout the aerospace industry, fueled by safety characteristics, appropriate performance, fuel consumption, and operational range. Weight is one of the key parameters that define outright efficiency as well as the fuel consumption rate and the overall carrying capacity of the vehicle. They should also be very light but at the same time very strong to enable them to withstand the pulling forces that are exerted on them without increasing the weight of the moving parts while at the same time offering the necessary protection and functionality [21].

- **Adhesion**

Ease in the application of the coating and compatibility with the substrate are vital for more uniformization and efficiency of the coated layer. Most early studies have highlighted that adhesion is a function of the coating and the substrate chemistry, method of application, surface preparation, and age of the coating [21].

- **Flexibility and Durability**

When in operation these aerospace vehicles go through some dynamic stresses such as vibrations, expansion, and contraction forces, therefore, there is a need to incorporate aerospace vehicles with these qualities through flexibility and durability. Coatings have to be rigid enough to also hold the above movements as well so that they do not crack in the long run creating the need for frequent redecoration [21].

- **Chemical and Thermal Stability**

Matching the coating of individual components to parts in the aerospace industry, they work under the high heat of re-entering the atmosphere or the cold of space. Under these conditions, they still must remain integrated and serviceable. This also goes hand in hand with degradation especially when exposed to various things like fuels or de-icing solutions like ethylene glycol and propylene glycol among others. In order for the coatings to offer the desired protection and performance for a long time they have to be able to be chemically stable and properly shielded from various chemical attacks [22].

4 Thermal Protection Systems

In aerospace engineering, the thermal protection systems or TPS are very important subsystems that shield spacecraft and aircraft from heat during activities like re-entry, high-velocity flying, and exposure to heat events like rocket exhausts. Durable TPSs minimize thermal loads, material degradation, and damage to critical components, ensuring efficient and reliable performance in harsh aerospace environments. The formulations which have been invented, correspond to the various different hyper-thermal conditions where such material is expected to operate. The heat shield which protects the structure, the aerodynamic surfaces as well as the payload of spacecraft, missiles, warheads, or probes from heating which occurs at the entrance flight through the stratosphere of an astronomic body is made of TPS materials [21].

Thin films therefore provide adaptable solutions to various aerospace sectors involving thermal protection, heat rejection, and ablation. In fact, TPS material made from polymers is more efficient than the traditional practice of using ceramics and metallic alloys; this is because polymer-based TPS has some added advantages such as cost efficiency, availability, and enhanced manufacturing. Polymer-based thin films provide great potential to enhance heat resistant coefficient and to ensure the reliability and safety of air vehicles as material science and engineering continue to grow [22].

UV light together with high temperatures can affect the properties of polymeric materials and make them unsuitable for their intended applications. The polymer materials may suffer a chemical change over lasting exposures to high temperatures or physical changes that occur at a shorter duration. Coatings and polymers will be weakened and affected by discoloration due to the UV rays affecting their mechanical properties. It becomes very challenging for a material to endure high temperatures; every material has that point at which it starts turning soft by the influence of elevated temperature [22]. Certain polymer composite materials have already been designed and used for aerospace use for certain temperature requirements like 120–250 °C and 250–350 °C. Current work is being carried out to develop polymer-based thin films capable of application at temperatures in excess of 350 °C. These films should not change their characteristics at high temperatures; they should not experience changes in glass transition temperature, exhibit thermal stability, least weight loss, resistance to oxidation, negligible aging effects, and ability not to crack. Polymer matrix composites have been extensively investigated for applications in aerospace vehicles especially in skin and engine constituents [23].

4.1 Types of Polymer-Based Thin Films for TPS in Aerospace Engineering

These thermo-plated polymer-based thin films can be classified into various groups depending on the chemical composition of the material and characteristics where individual properties of the thin films can be obtained to meet the required necessities of aerospace engineering repairs.

1. **Organic Polymers**: Such thin films may include a long macromolecule chain, which is synthesized from hydrogen and other carbon molecules. The organic polymer is suitable for numerous TPS applications given that it is versatile, which entails that it is lightweight and can be developed simply. Some examples of industrial polymers include; polyimide commonly known as PI, polyethylene as well as polytetrafluoroethylene ordinarily referred to as PTFE [24].
2. **Nanocomposites**: A nanocomposite thin film results from the ability to introduce what is known as nanofillers to a polymer matrix in an attempt to enhance traditional parameters including mechanical, thermal, or barrier properties. That is, it may be said that dissipation energy can be further enhanced, heat deflection temperature can be improved, and thermal conductivity of ablative properties can be further improved by incorporation of nanofillers like carbon nanotubes, graphene or ceramic nanoparticles into the polymer matrix [25].

4.1.1 Specific Polymeric Materials Frequently Used in TPS

With respect to polymer composites, thermal stability, flexibility, as well as ablative characteristics are some of the signs many polymer materials use in aerospace thermal protection systems.

- **Polyimides**: Some of these films include business best polyimide films for mechanical usability under radiation and chemical aptitudes in addition to heat stability such as the Kapton®. It is expected that the TPS application of polyimides since the material used in this part must be strong, and withstand high temperatures based on the above-mentioned characteristics [26].

- **Polymers Based on Silicone**: Silicone polymer as it is referred to in its conventional surfaces of silicones/siloxanes is flexible and has low heat conductivity making it heat resistant. This means that these polymers can additionally be used to supply thin movies or coatings for high vane-speed flight for re-entry protection via heat through intensive thermal insulant and heat flux protectant [27].
- **Phenolic Resins**: These resins are convenient in TPS where there are issues in controlling thermal ablation to eliminate heat For the safety of other structural designs because they possess high thermal conductivity coupled with ablative characteristics [28].
- Copolymer types of fluoropolymers are tetrafluoroethylene (PTFE) and ethylene, propylene, and tetrafluoroethylene (FEP). Some of the qualities that can be observed in certain materials include chemical resistance which is the ability of a material when subject to chemical attack; low friction and good thermal stability. These are so because it is possible to use fluoropolymers in TPS as they can effectively resist heat and corrosive environments—replicating TPS function [29].

Thus aerospace engineers are in a position to engineer and optimize thermal protection systems and meet certain thermal control for space and aircraft vehicles for extreme environments by utilizing the flexibility and diversities of grades of these polymer materials.

4.2 Polymeric Ablative Thin Films for Thermal Protection System and Propulsion Devices

Thermal protection system (TPS) materials include Ablative polymer-based thin films among others, and they are applied in aerospace engineering for the protection of space crafts, re-entry vehicles, and other aerospace parts from heat which is produced under conditions such as re-entry into the earth's atmosphere or under very high temperatures. These thin films can be designed to dissolve, or ablate, at relatively high temperatures, thereby protecting the structure behind them and shedding heat in the process [30].

The foundation of the aerospace industry is comprised of substitute materials, such as ablative materials, which are used to protect vehicles and probes during hypersonic flight through planetary atmospheres or to control the heat shielding of propulsion systems, such as liquid and solid rocket motors (SRMs). As a result, they are sometimes referred to as materials for thermal protection systems (TPS). While some non-polymeric materials have been effectively employed as ablatives, polymeric ablatives (PAs) are the most diverse family of sacrificial thermoplastic materials because of their adaptability. Indeed, PAs have several inherent advantages over non-polymeric ablatives such as metals with high melting points, inorganic polymers, metal oxides, and metal carbides. These advantages include tunable density, reduced cost, and increased thermal shock resilience. Re-entry capsules, rocket nozzles, heat shields on spacecraft, and other aerospace components that require thermal insulation are all covered with protective layers made of ablative polymer-based thin films [30, 31].

4.2.1 Mechanism of Ablative Process

The TPS kind of ablative coating operates through a process of decomposition after exposure to extreme temperatures after which it protects the structure by offering its own life. Actually, here is how ablative coating works through the following steps (shown in Fig. 9.4). As stated before, the ablative kinds of composites work in a complicated manner for the thermal safeguard which includes eradicating the heat, the reaction of the heat that is produced from the degradation of the involved material, insulation, high heat absorption, and latent heat of the thermal degradation [31]. The three common zones in any ablative film are defined as follows:

i. **Virgin Zone**: The Virgin zone refers to part of the outer layer of an ablative material that does not undergo any change when subjected to heat for a long time. This zone ensures that no heat degradation takes place to the resin material thus keeping it with its properties as originally intended. The operation of ablation takes place when heat is applied to the material and proceeds to other zones at different progressive layers [31].
ii. **Pyrolysis Zone**: The breakdown of the ablative material occurs within the pyrolysis zone due to the high temperatures that are present. Consequently, the disintegration of molecular bonds occurs due to heat resistance, and the release of

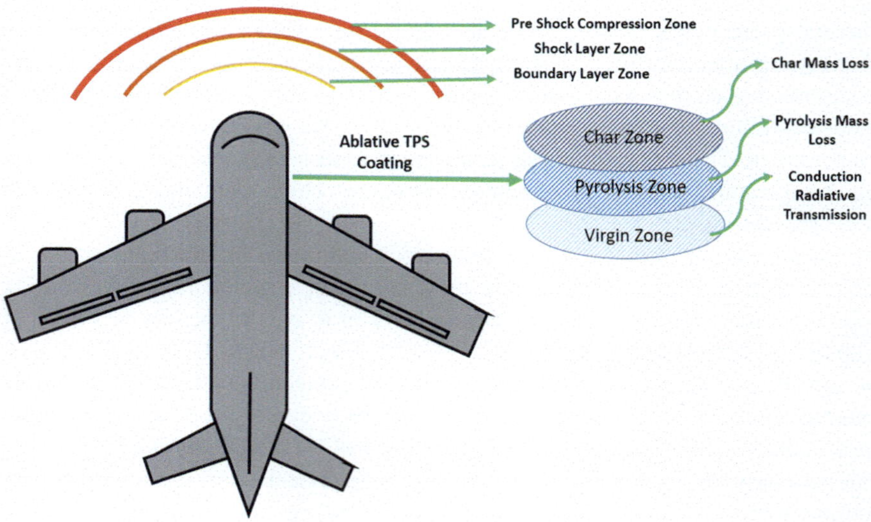

Fig. 9.4 Mechanism of an ablative thin film on an aircraft

volatile breakdown products such as gasses and vapors. These compounds aid the material in performing ablative properties and enable heat to be effectively shed through a layer formation. It generally has a company structure in the sense that the material composition for the pyrolysis zone is altered and made porous [31].

iii. **Char Zone**: The core of the layer is called the "char zone" because here the matrix of decomposition products turns into a solid carbonized residue. Another chemical transformation occurs which forms layers that are hard and dense walls of char. This layer of char acts as a shield of sorts to the rest of the material, continuing to safeguard it from further heat and its degradation effects. The design of thermal protecting systems for space suit application is dependent on the behavior and characteristics of each zone [31].

To summarize, the process of ablation entails the passage of an ablative material through three distinct zones: the Virgin region which can be clearly described as heat affected region where the material does not suffer through any kind of change. A pyrolytic region is the area in which the gases from thermal decomposition form; carbonization or char zone is the zone in which a layer attempts to form a layer of carbon with the aim of preventing the material from further heat effects.

4.2.2 Performance Evaluation and Testing of Ablative Films

Thermal Cycling Testing

Thermal shock test assesses the capacity of polymer base thin films to withstand high- and low-temperature cycles which are encountered in space missions. The films used are exposed to various temperatures, from very low to very high with a set time at every particular stage to determine the thermal reliability and durability of the films. After each cycle, the engineers go through a thorough analysis of the samples to determine the level of degradation that could have occurred, such as checking for any sign of cracking or delamination. To accomplish a proper assessment of the performance and reliability of thin films for aerospace applications, these changes need to be supervised to determine the performance of the films under thermal stress [32].

Mechanical Testing

Based on the simulated aeronautical environment, mechanical analysis examines the stiffness, toughness, and durability of the polymer-based thin films. These tests help in determining the ability of the material to withstand mechanical loads that may be likely to be exerted on the object during the takeoff, landing, and launch which include tension, compression, bending, and shear forces [33].

- **Tensile testing**: Determines the strength that the thin films undergo by stretching, elongation, and the modulus of elasticity [33].
- **Flexural testing**: Assesses the mechanical properties of the thin films with respect to how they bend or flex.
- **Shear testing**: Defines the shear strength and the measures of the thin films to the parallel forces that are applied in opposite directions [33].
- **Impact testing**: Evaluates the performance of the thin films in terms of their vulnerability to shocks or impacts in terms of hardness [33].

During mechanical testing, samples of the polymer-based thin films tested are subject to controlled loading using unique testing equipment. In order to confirm the applicability of obtained samples for aeronautical uses, their mechanical characteristics are analyzed.

Environmental Exposure Testing

Environmental exposure testing evaluates the behavior of polymer-based thin films in adverse environmental conditions that may be possible during aerospace missions. Such variables encompass ultraviolet radiation, relative humidity, sea spray or salt fog, and chemicals. The samples of the thin films are subjected to accelerated aging conditions in environmental chambers or exposure racks to determine the effects of aging by causing the films to be exposed to environmental factors for a longer period than necessary [34].

The samples' change in properties is then assessed after the exposure time; the signs of corrosion or deformation, such as color changes, rough surface, cracks, or loss in mechanical properties, are looked for. It also plays a crucial role in trying to ascertain how the material might stand up to the conditions within aircraft and how susceptible it might be to degradation from the environment [34].

5 Tribological Coatings

The word was derived from the Greek term "tribos" which means rubbing. Tribology is defined as the science and engineering of surfaces in contact that are in relative motion and it is made up of sub-disciplines such as friction, wear, and lubrication. Every wear process interferes with the weather and deforms at several scale levels at the same time, and it can also cause chemical reactions. In the best interest of the modern world, tribology plays a significant role as mechanical parts produce heat and expend most of their energy on friction. This therefore implies that something needs to be done so as to ensure that energy wastage is reduced while at the same time, more energy is used less [35]. Becoming more familiar with the tribological understanding could be useful in this regard. There are many tribological phenomena that are referred to today which has led to the development of tribological coatings

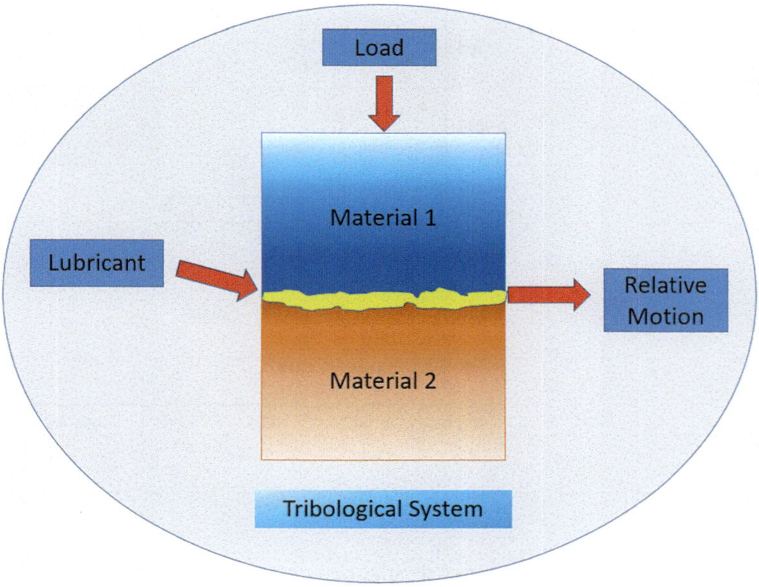

Fig. 9.5 A conventional system of tribology

in various fields that have different roles. It is crucial to understand and manage tribological effects in aeronautics because the portions utilized bear critical pressures, temperatures, and velocities. This in turn assists in enhancing efficiency, safety, and reliability as well. Part and sub-assemblies that undergo thermal, pressure, and mechanical load during aircraft operations, the parts include; engine components, bearings, gears, sealant, structural parts, etc. These conditions, if not checked, may lead to the destruction of the surface and high levels of wear together with high friction loss (shown in Fig. 9.5). Therefore, the tribological impacts should play a crucial role in the design and the lifecycle effect on the aircraft systems so that they would hold and continue to perform under the conditions imposed [36].

5.1 Levels of Tribology

Tribology is now firmly a part of the "Nano" revolution enhancing significantly from technology advancement to analyze surfaces in nanometer scales due to newly invented equipment. At the same time, owing to the development of physical and chemical instrumentation more emphasis has shifted at the molecular and atomic level for precise understanding and to control tribological factors seen at the macroscopic scale (shown in Fig. 9.6). From the preceding description, it is apparent that the tribology sub-discipline possesses five primary scales [36].

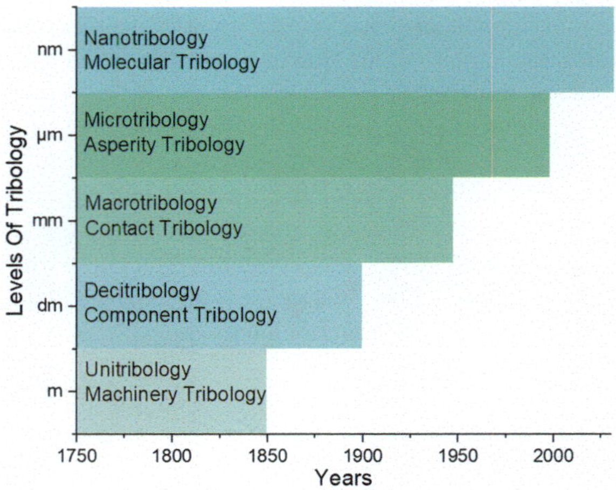

Fig. 9.6 Timeline showing different levels of tribology

i. Uni Tribology

Uni Tribology deals with the tribological factors that are involved in providing the efficiency and performance of the machines, components, or subsystems, including factors related to the statistical reliability of the systems [36].

ii. Deci Tribology

It is also known as the deterministic study is the practice of identifying and even measuring main parameters like torque forces, vibrations, clearance, and alignment that characterize the interactions of the component and their performance status [36].

iii. Macro Tribology

Macro tribology focuses on the more distant interactions that are observed in masses that impact and come into contact. It is essential to consider the overall loading response of the substrate and the additional loading response of the coating when in applications that require the bearing of high loads, for example, rollers, givers, and bearing parts. They include pitting, scoring, and scuffing which are evident in areas with intense contact stresses. Even on a macro level, there are stress values that can cause the layer adhesion to fail and result in coating separation [36].

iv. Micro Tribology

Micro tribology concerns subtopics like asperity connections. Microscopic scale phenomena including fracture, elastic and plastic deformation, debris, formation of surface layer, and the changes in the topographic features also play a very significant role [36].

v. Nano Tribology

Nano tribology which is also called molecular tribology deals with those problems that occur due to the existence of molecules and atoms. Such phenomena include the effects resulting from van der Waal forces and other intra-atomic processes that depend on the crystalline and bonding structure of the materials involved [36].

5.2 *Polymer-Based Tribological Coatings for Aerospace Engineering*

Tribological coatings based on polymers present a viable way to reduce wear, friction, and surface damage on aircraft parts. Polymers or polymer composites designed to possess particular tribological qualities, like high wear resistance, low friction, and superior lubricity, are commonly used to create these coatings (shown in Fig. 9.7). These polymer-based tribological coatings can be customized to satisfy the special needs of aerospace applications, such as compatibility with various substrate materials, chemical resistance, and temperature stability. Their benefits include cost-effectiveness, convenience of application, and the capacity to adhere to complex geometries, which makes them appropriate for a variety of aerospace components [36].

Significant Features and Metrics of Polymer-Based Tribological

Numerous critical characteristics and performance measures control the effectiveness of polymer-based tribological coatings in aircraft applications. Among them are:

i. **Friction coefficient**: The amount of resistance to a surface sliding motion [32].
ii. **Wear resistance**: The coating's capacity to withstand material loss when subjected to mechanical force [32].
iii. **Hardness**: The coating's ability to withstand scratches and indentations [32].

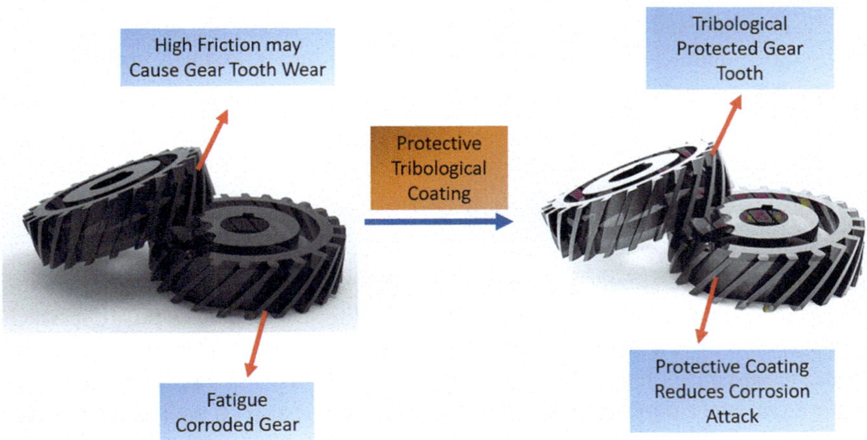

Fig. 9.7 Tribological coating application in aircrafts engine gear

iv. **Adhesion**: The degree of fusion between the substrate material and the coating [32].
v. **Chemical resistance**: The coating's capacity to endure exposure to corrosive substances or unfavorable environmental circumstances [32].
vi. **Temperature stability**: The coating's capacity to hold onto its characteristics at high temperatures, such as those found in aerospace applications [32].

In aeronautical engineering, performance parameters including adhesion strength, durability, wear rate, and coefficient of friction are essential for assessing how well polymer-based tribological coatings work [32].

Aerospace tribological coatings use a variety of polymer types because of their beneficial qualities and performance characteristics as shown in Fig. 9.8. Aerospace uses polytetrafluoroethylene (PTFE) as a preferred material to reduce wear and friction on critical components because of its exceptional lubricity, low friction coefficient, and chemical inertness. Aerospace systems working under demanding conditions can benefit from the use of Polyetheretherketone (PEEK) coatings due to their remarkable mechanical strength, resilience to high temperatures, and chemical stability. Aerospace coatings that need to have great abrasion resistance and minimal friction are especially fond of Ultra-great Molecular Weight Polyethylene (UHMWPE) due to its remarkable wear resistance, impact strength, and self-lubricating qualities. When combined, these polymers improve the endurance and performance of aeronautical components in harsh conditions. Multiple materials are blended to create synergistic tribological qualities in composite and hybrid polymer coatings. By incorporating ceramic nanoparticles into a polymer matrix, polymer-ceramic composites improve the hardness and abrasion resistance of coatings intended for use in aerospace applications. For aerospace parts like bearings and gears, polymer-metal hybrids spread metal fibers or particles in polymer matrices to improve mechanical strength and conductivity. Polymer-lubricant hybrids are lubricant-polymer blends that produce self-lubricating coatings with better wear resistance and less friction that are perfect for aerospace applications that don't require maintenance. These cutting-edge coatings maximize longevity and performance in tough aircraft settings [37].

5.3 *Fabrication of Tribological Coatings*

One of the most commonly known methods that have been used in the formation of thin films to cater for certain behaviors is the Physical Vapor Deposition which is widely referred to as PVD. Several studies in various manufacturing processes of polymeric-based tribological coatings show highly positive outcomes for employing the physical vapor deposition (PVD) method in the aerospace industry for future analysis. On the other hand, there is one more technology, plasma-enhanced chemical vapor deposition (PECVD) technology that can be used in the creation of thin coating for thermally and mechanically sensitive aerospace elements because of the

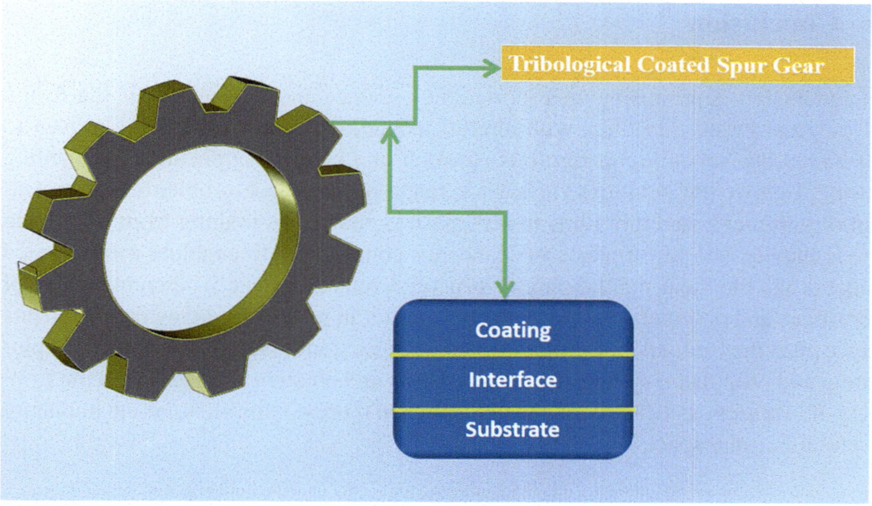

Fig. 9.8 Overview of tribological coated spur gear

possibility of fine control over the constitution and thickness of the coating along with its adherence to the base material. Thermal evaporation and magnetron sputtering tend to offer high deposition rates as well as superb wear resistance and hence these PVD techniques will respond to the airplane conditions or the demands of an airplane. When choosing which method of applying aerosol coatings is most suitable for aerospace engineering in the context of a given set of tasks, there are a lot of factors to be taken into consideration including substrate properties, the conditions for the application going to be exposed to, as well as the cost implications of the whole process [38].

In the futuristic view of Aircraft engineering, there exist chances to expand polymer tribological coating in the field of more intensive investigations and applications. Scientists have been working hard to look for green grasses as a way of achieving various formulations with better properties. Their work is exclusively vested in coming new polymers that can stick to the substrate, those that can afford high temperatures, and those that can be lubricated to order. Another promising area of interest is in nanotechnology where it is possible to seek solutions such as typical characteristics of nanocomposite coatings having an extraordinary self-numerous, tribo- coefficient that cannot be simply surpassed. These processes are the guide lamps towards the path of strengthening the avionic systems and the systems that are getting better with time [39].

6 Conclusion

In conclusion, a promising area in materials science and engineering is the fusion of polymer-based thin films with aircraft technology. These films have proven to be incredibly versatile via rigorous research and creative applications, providing strong, lightweight solutions for a wide range of aerospace problems. The potential of polymer-based thin films in aerospace is enormous, ranging from improving the longevity and performance of spacecraft components to enabling sophisticated capabilities like heat regulation and structural reinforcement. The combination of aerospace and polymer sciences will surely result in ground-breaking developments as we push the frontiers of technology and discovery, influencing future space exploration and propulsion systems. With conviction and creativity, we can push the frontiers of invention and collaborate to unleash even more possibilities, taking humanity even farther into space.

References

1. Xie, G., Bai, H., Miao, G., Feng, G., Yang, J., He, Y., Li, X., Li, Y.: The applications of ultra-thin Nano film for aerospace advanced manufacturing technology. Nanomaterials **11**(12), 3282 (2021)
2. Joshi, M., Chatterjee, U.: Polymer nanocomposite: an advanced material for aerospace applications. In: Advanced Composite Materials for Aerospace Engineering, pp. 241–264. Woodhead Publishing (2016)
3. Das, P., Banerjee, S., Das, N.C.: Polymer-graphene composite in aerospace engineering. In: Polymer Nanocomposites Containing Graphene, pp. 683–711. Woodhead Publishing (2022)
4. Papa, F.: Foreword. In: Wang, F., Wu, J. (eds.) Modern Ion Plating Technology, pp. xxv–xxvi. Elsevier (2023)
5. Kruzelecky, R.V., Haddad, E., Soltani, M., Chaker, M., Nikanpour, D.: Integrated thin-film smart coatings with dynamically-tunable thermo-optical characteristics. In: SAE Transactions, pp. 323–330 (2002)
6. Chang, T., Panhwar, F., Zhao, G.: Flourishing self-healing surface materials: recent progresses and challenges. Adv. Mater. Interfaces **7**(6), 1901959 (2020)
7. Firth, N.: Bonding over technology. Aluminium Int. Today **27**(4), 50 (2015)
8. Du, Y.J., Damron, M., Tang, G., Zheng, H., Chu, C.J., Osborne, J.H.: Inorganic/organic hybrid coatings for aircraft aluminum alloy substrates. Prog. Org. Coat. **41**(4), 226–232 (2001)
9. Umoren, S.A., Solomon, M.M.: Protective polymeric films for industrial substrates: a critical review on past and recent applications with conducting polymers and polymer composites/nanocomposites. Prog. Mater Sci. **104**, 380–450 (2019)
10. Visser, P., Terryn, H., Mol, J. M.: Aerospace coatings. In: Active Protective Coatings: New-Generation Coatings for Metals, pp. 315–372 (2016)
11. Yang, X.F., Li, J., Croll, S.G., Tallman, D.E., Bierwagen, G.P.: Degradation of low gloss polyurethane aircraft coatings under UV and prohesion alternating exposures. Polym. Degrad. Stab. **80**(1), 51–58 (2003)
12. FitzSimmons, V.G., Zisman, W.A.: Thin films of polytetrafluoroethylene resin as lubricants and preservative coatings for metals. Ind. Eng. Chem. **50**(5), 781–784 (1958)
13. Shit, S.C., Shah, P.: A review on silicone rubber. Natl. Acad. Sci. Lett. **36**(4), 355–365 (2013)
14. Sezer Hicyilmaz, A., Celik Bedeloglu, A.: Applications of polyimide coatings: a review. SN Appl. Sci. **3**(3), 363 (2021)

15. Peng, H.X.: Polyurethane nanocomposite coatings for aeronautical applications. In: Multifunctional Polymer Nanocomposites. CRC Press, Boca Raton, pp. 337–387 (2011)
16. Song, S., Yan, H., Cai, M., Huang, Y., Fan, X., Zhu, M.: Multilayer structural epoxy composite coating towards long-term corrosion/wear protection. Carbon **183**, 42–52 (2021)
17. Butt, M.A.: Thin-film coating methods: a successful marriage of high-quality and cost-effectiveness—a brief exploration. Coatings **12**(8), 1115 (2022)
18. Shi, F.F.: Recent advances in polymer thin films prepared by plasma polymerization synthesis, structural characterization, properties and applications. Surf. Coat. Technol. **82**(1–2), 1–15 (1996)
19. Zhang, X., Xu, Y., Zhang, X., Wu, H., Shen, J., Chen, R., Xiong, Y., Li, J., Guo, S.: Progress on the layer-by-layer assembly of multilayered polymer composites: strategy, structural control and applications. Progr. Polym. Sci. **89**, 76–107 (2019)
20. Oviroh, P.O., Akbarzadeh, R., Pan, D., Coetzee, R.A.M., Jen, T.C.: New development of atomic layer deposition: processes, methods and applications. Sci. Technol. Adv. Mater. **20**(1), 465–496 (2019)
21. Boyer, R.R., Cotton, J.D., Mohaghegh, M., Schafrik, R.E.: Materials considerations for aerospace applications. MRS Bull. **40**(12), 1055–1066 (2015)
22. Kausar, A.: Polymer coating technology for high performance applications: fundamentals and advances. J. Macromol. Sci., Part A **55**(5), 440–448 (2018)
23. Hołyńska, M., Tighe, A., Semprimoschnig, C.: Coatings and thin films for spacecraft thermo-optical and related functional applications. Adv. Mater. Interfaces **5**(11), 1701644 (2018)
24. Ching, Y.C., Gunathilake, T.U., Ching, K.Y., Chuah, C.H., Sandu, V., Singh, R., Liou, N.S.: Effects of high temperature and ultraviolet radiation on polymer composites. In: Durability and Life Prediction in Biocomposites, Fibre-Reinforced Composites and Hybrid Composites, pp. 407–426. Woodhead Publishing (2019)
25. Grilli, M.L., Valerini, D., Slobozeanu, A.E., Postolnyi, B.O., Balos, S., Rizzo, A., Piticescu, R.R.: Critical raw materials saving by protective coatings under extreme conditions: a review of last trends in alloys and coatings for aerospace engine applications. Materials **14**(7), 1656 (2021)
26. Chen, J., Ding, N., Li, Z., Wang, W.: Organic polymer materials in the space environment. Prog. Aerosp. Sci. **83**, 37–56 (2016)
27. Kausar, A., Rafique, I., Muhammad, B.: Aerospace application of polymer nanocomposite with carbon nanotube, graphite, graphene oxide, and nanoclay. Polym.-Plast. Technol. Eng. **56**(13), 1438–1456 (2017)
28. Sezer Hicyilmaz, A., Celik Bedeloglu, A.: Applications of polyimide coatings: a review. SN Appl. Sci. **3**, 1–22 (2021)
29. Shi, S., Lei, B., Li, M., Cui, X., Wang, X., Fan, X., Tang, S., Shen, J.: Thermal decomposition behavior of a thermal protection coating composite with silicone rubber: experiment and modeling. Progr. Organic Coat. **143**, 105609 (2020)
30. Ding, J., Qin, Z., Luo, H., Yang, W., Wang, Y., Huang, Z.: Nano-silica modified phenolic resin film: manufacturing and properties. Nanotechnol. Rev. **9**(1), 209–218 (2020)
31. Améduri, B.: The promising future of fluoropolymers. Macromol. Chem. Phys. **221**(8), 1900573 (2020)
32. Ellahi, R.: Recent trends in coatings and thin film: modeling and application. Coatings **10**(8), 777 (2020)
33. Natali, M., Kenny, J.M., Torre, L.: Science and technology of polymeric ablative materials for thermal protection systems and propulsion devices: a review. Prog. Mater Sci. **84**, 192–275 (2016)
34. Maaza, M., Bouziane, K., Maritz, J., McLachlan, D.S., Swanepool, R., Frigerio, J.M., Every, M.: Direct production of thermochromic VO_2 thin film coatings by pulsed laser ablation. Opt. Mater. **15**(1), 41–45 (2000)
35. Brotzen, F.R.: Mechanical testing of thin films. Int. Mater. Rev. **39**(1), 24–45 (1994)
36. Schmidt, D.L.: Ablative polymers in aerospace technology. J. Macromol. Sci.—Chem. **3**(3), 327–365 (1969)

37. Matthews, A., Franklin, S., Holmberg, K.: Tribological coatings: contact mechanisms and selection. J. Phys. D Appl. Phys. **40**(18), 5463 (2007)
38. Brostow, W., Kovačevic, V., Vrsaljko, D., Whitworth, J.: Tribology of polymers and polymer-based composites. J. Mater. Educ. **32**(5), 273 (2010)
39. Friedrich, K.: Polymer composites for tribological applications. Adv. Industrial Eng. Polym. Res. **1**(1), 3–39 (2018)

Emerging Trends and Innovations

Zarmeen Suhail, Ali Dad Chandio, and Syed Sarim Ahmed

Abstract The chapter also reveals an understanding of the global advancement in thin film technology especially the smart coatings' positive aspects of self-healing thin films and convenient deposition methods and materials. In this section, a detailed description of flat device technology in relation to smart coatings, self-decorating films, and deposition techniques is outlined. Responsive films are coatings that undergo changes in properties depending on the stimulation received by the film in terms of temperature, lighting, and chemical impact to name but a few; smart films are applied in the automotive industry, aerospace industry, and consumer electronics among others. The invention of self-healing thin films is another advancement; with the thin films possessing the capability to heal themselves and consequently, the coated material, minimizing the frequency of maintenance and lowering the maintenance price of the material. Also, this chapter explicates novel methods of depositions that offer more accuracy, speed, and modularity in thin film formation including atomic layer deposition and pulsed laser deposition. These methods are further useful in generating films of higher uniformity, purity, and receptive characteristics for any application. Thus, exploring these future trends and novelties, the chapter shows that the advancements in next-generation thin films can contribute to the further development of technology and generate fresh possibilities in various fields. The detailed evaluation therefore emphasizes the essence of constant innovation as a central theme in improving thin film science.

Z. Suhail (✉)
Centre for Regenerative Medicine and Stem Cell Research, Aga Khan University, National Stadium Rd, Karachi, Sindh, Pakistan
e-mail: zarmeen.ssoomro@gmail.com

A. D. Chandio
Thin-Film Lab, Advanced Materials and Sustainable Environment (AMSE) Group, Department of Metallurgical Engineering, Main Campus, NED University of Engineering and Technology, University Road, Karachi, Sindh 75270, Pakistan
e-mail: alidad@neduet.edu.pk

S. S. Ahmed
Department of Chemistry, Main Campus, NED University of Engineering and Technology, University Road, Karachi, Sindh 75270, Pakistan

A. D. Chandio and I. A. Channa (eds.), *Thin Films and Coatings*, Engineering Materials,
https://doi.org/10.1007/978-981-96-0005-2_10

List of Abbreviations

UV	Ultraviolet
3D	3 Dimensional
PVA	Polyvinyl Alcohol
FT	Freeze Thawing
PAA	Poly acrylic acid
SDS	Sodium dodecyl sulfate
DOPA	3,4-Dihydroxyphenyl-L-Alanine
HA	Hyaluronic Acid
SEM	Scanning Electron Microscopy
PEG	Polyethylene Glycol

1 Emerging Trends and Innovation

Numerous new developments and trends are propelling the rapid progress of the coatings industry. There has been a noticeable trend toward sustainability, as seen by the rising demand for eco-friendly formulations that use bio-based components and water-based solutions. The use of smart coatings is growing because they include features like the ability to mend themselves and react to outside stimulation The application of nanoparticles to increase durability, UV protection, and scratch resistance is one of the key ways that nanotechnology is improving coating performance. Antimicrobial coatings are becoming indispensable, particularly in public and clinical settings, as hygiene becomes increasingly important. The development of 3D printing and digital printing technology is opening up possibilities for more elaborate and customized products. Additionally, the industry is seeing advancements in robotic application systems, UV-curable coatings, and functional coatings designed to meet certain requirements.

2 Smart Materials and Coatings

Smart materials, also referred to as intelligent or advanced materials, cannot be explained by a single particular definition. They can be described as materials that can regain their shape in response to particular stimuli or can be defined as smart materials that can react intelligently to changes in their surroundings [1]. Smart materials are divided into different categories based on their properties like active and passive. Passive smart materials can transport some kinds of energy; such as optical fibers can transfer electromagnetic waves. Active smart materials are categorized into two groups. The first group consists of materials that, when exposed to external stimuli, cannot change. For example, photochromatic spectacles only change color when exposed to sunlight. The other kinds can change one type of energy (e.g., electrical, optical, chemical, thermal, and mechanical) to another type such as piezoelectric materials are able to create an electric charge when subjected to external strain. These materials have some remarkable qualities that set them apart from other materials like transiency that can respond to different external stimuli, self-diagnostic properties that can diagnose cracks on materials automatically, self-actuation that can create mechanical actuation in response to external stimuli, and self-healing qualities that are able to automatically repair damages [2].

When it comes to coatings, in the past few years, there has been an increase in research and development as well as the commercial production of coatings with new functions has increased. These products not only serve conventional protective and decorative purposes; they also perceive and interact with their surroundings. A common term for these coatings is "smart coatings." More precisely, a smart coating is one that, while preserving compositional integrity, senses changes in its surroundings and interacts with them. It may react to variations in several factors such as pH, light, temperature, pressure, polarity, biological variables, and other conditions. Therefore, a smart coating is designed so that, contingent upon the kind and intensity of external stimuli, one or more of the functionalities mentioned can be altered.

A variety of multifunctional coatings have been created that sense pressure, temperature, and corrosion. Moreover, decorative and protective coatings possessing self-healing properties are currently in commercial use. Smart coatings are also having a significant impact on the medical industry by providing long-lasting antibacterial and anti-inflammatory medical devices, such as implants and drugs that can be released when needed. Piezoelectric paint, reversible thermochromic, hydrophilic/hydrophobic switching, pH-responsive, self-cleaning, and self-erasing inks are a few more examples of smart coatings.

Smart Coatings can be developed in various ways such as by integrating stimuli-responsive materials like pH, light, temperature, pressure, etc. nanoparticles, sensitive molecules, and antimicrobial substances as additives, or by effectively creating polymer frameworks and coatings that react to either external or internal stimuli.

In general, two activities need to occur to achieve responsiveness [3]:

- The material needs to receive the signal after a stimulus has been introduced.
- A physical or chemical process is initiated by this signal.

For the response to be effective and predictable, the stimulus must be precise and clear. This also prevents the subsequent responses from interfering with or canceling each other out completely. Internal stimuli, like those found in self-healing or corrosion-sensing coatings, are signals that work by inducing reactions within the coating itself with the intention of changing the bulk properties. External stimuli are responses that modify the surface properties in relation to the surroundings, such as in self-cleaning coatings.

Since the signal needs to function on some kind of chemical entity such as polymer, the stimulus can be physical including temperature, light, electrical field, electromagnetic and acoustic waves, solubility, ionic strength, pH, electrical and surface tension gradients, and pressure or chemical in nature such as acid–base, biochemical, electrochemical, redox, and photochemical reactions as well as formation and breakings of chemical bonds. Comparing the physical signals to their chemical counterparts, the former is more prevalent, more adjustable, and possibly simpler. Although a variety of chemical reactions can be employed as stimuli, it is significantly more challenging and complex to monitor the extent, intensity, and limits of reactions.

Inspired by systems in biology, such as skin and bones, where damage prompts reorganization, reconstruction, and healing, polymeric materials have been created that possess self-healing properties.

3 Self-Healing Thin Films

Self-healing films are a special kind of intelligent/smart films that can recover their internal damage generated in the matrix and recover the mechanical properties like tensile strength of the damaged area through a self-healing mechanism, similar to the human skin when it is wounded. Figure 10.1 describes the basic mechanism of how a self-healing material recovers following damage. These smart films have attracted great attention in the past years because they can prolong a material's lifespan, drop the cost of replacement, as well as foster product safety [4]. The characteristic of self-healing is especially important in materials that are meant to work in a particular manner for an extended duration when a repair is not possible.

Although many materials can be employed to create self-healing systems, polymers have received great attention due to their tunability (chemical and mechanical) and potential to synthesize dynamic materials [5]. A great deal of earlier research has looked into the mechanisms of healing in polymeric systems, such as epoxy coatings [6, 7], elastomers [8], and more delicate systems like thin films [9] with self-healing properties that can repair mechanical damage. Self-healing materials have several applications in multiple areas, such as construction, aerospace, automobile, defense, and biomedical engineering.

We will go into further detail about various kinds of self-healing mechanisms, and some interesting studies in the next part.

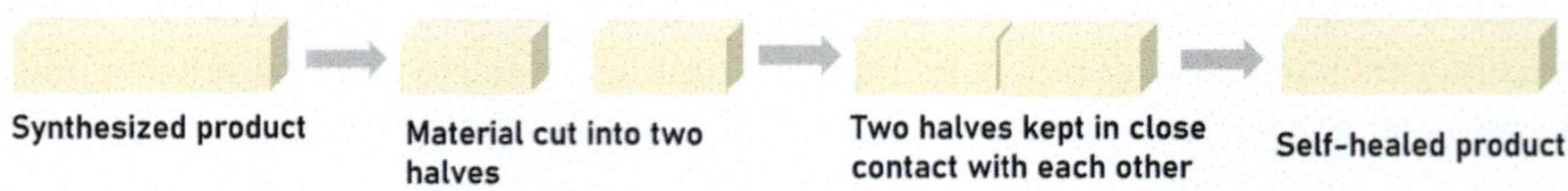

Fig. 10.1 Basic demonstration of self-healing materials

3.1 Self-Healing Mechanisms

Major advancement has been achieved in the design and production of novel self-healing polymers employing various chemical approaches. These polymeric materials can heal damage while improving performance in real-world applications.

There are two categories of self-healing polymers: intrinsic materials and extrinsic materials. In the case of extrinsic mechanisms, external healing agents such as organic monomers or crosslinkers are applied as encapsulating elements to the area that has been scraped, cracked, or damaged in order to repair the product's problematic areas [10]. In contrast, intrinsic self-healing takes place in the product's fractured or damaged area as a result of intermolecular or intramolecular forces such as hydrogen bonds, strong covalent linkage bonds, physical/chemical crosslinking networks, and supramolecular interactions [11, 12].

The process of polymer phase separation may occasionally encourage self-healing. Polymers' dynamic and intricate chemical structure makes them great options for self-healing applications. The self-healing process can be incorporated into polymers by applying particular chemistries or by exposing them to thermal, chemical, or physical stimuli [12]. A polymeric material's healing efficiency is often described as the toughness ratio between the original sample and the healed sample [4].

The self-healing process of different polymeric films can be obtained through various methods, including non-covalent linkages and covalent bonds (Fig. 10.2). The polymeric films synthesized from non-covalent approaches are typically more flexible and self-recover because of their capability of easy breaking and rebuilding of crosslinks, while those made from covalent linkages are highly stable [13, 14]. The mechanisms and strategies in detail are listed below.

3.1.1 Non-covalent Interactions

Non-covalent linkages, including ionic bonds, hydrophobic bonds, and hydrogen bonds can be employed to develop self-healing films (Fig. 10.3). Non-covalent bonds are less stable and more susceptible to environmental factors (e.g., pH and temperature) in comparison to covalent interactions. Nevertheless, strong self-healing polymeric films or other structures can still be produced based on these approaches by employing particular manufacturing methods or nanostructures [15].

Hydrogen Bonding

Hydrogen bonding is an interaction between two kinds of atoms, namely hydrogen and electronegative. The crosslinked polymeric films based on hydrogen bonds are obtained through the reversible interactions of polymers in which H-atoms interact

Fig. 10.2 Schematics of different bonds involved in the self-healing mechanism

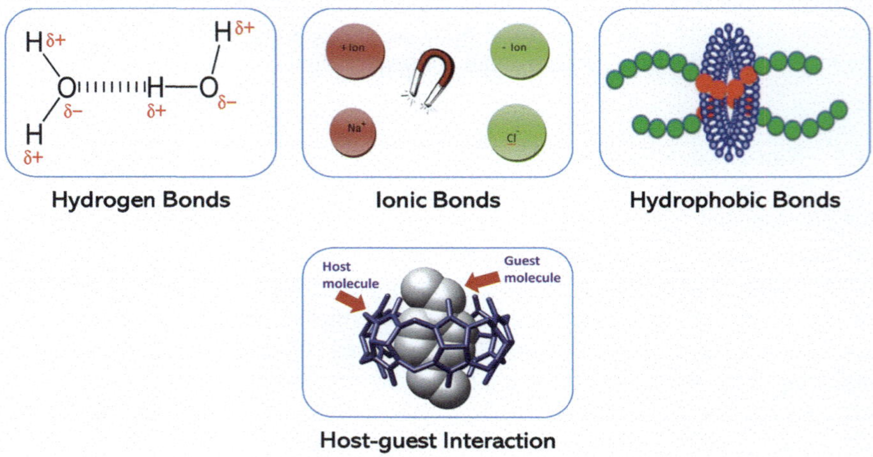

Fig. 10.3 Different types of non-covalent self-healing mechanisms

with electronegative atoms including fluorine, nitrogen, and oxygen. These structures possess improved bond strength and self-healing capacity [13].

The freeze-thawing approach has been used to produce polyvinyl alcohol (PVA)-based hydrogels with self-healing properties that make multiple H-bonds at the fractured area due to the presence of ample free hydroxyl groups PVA chain [16, 17]. PVA, 3,4-dihydroxyphenyl-L-alanine (DOPA), iron complex showed rapid healing efficiency after disintegration and a continuous strain sweep test has been employed to analyze it. The catechol groups present in DOPA play an important part in self-healing [18]. Samadi et al. have produced tough, self-healing triple network nanocomposite polymeric material which was composed of agar, PVA, and graphene nanoplatelets as the first, second, and third networks, respectively. Remarkably, physical crosslinking between all three networks through hydrogen bonding connections produced concurrent high strength, toughness, and autonomous self-healing [19].

In another research by the Luo group [20], agar and PVA were utilized to create strong, self-healable, and conductive hydrogels employing the freeze–thaw (FT) approach followed by immersion in the solution containing ammonium sulfate, which led to the formation of stronger H-bond interaction between PVA and agar. Thereby, the developed polymeric system offers >80% self-healing efficiency. The toughness and tensile strength increased from 0.2 MPa and 0.3 MJ/m^3 to 18.0 MPa and 42.3 MJ/m^3, respectively. A microgel with the aim of enhancing self-healing properties and mechanical strength was synthesized with guar gum, PVA, and borax, incorporated with hyaluronic acid for potential drug release applications. Inverse emulsion polymerization and cross-reaction with PVA and borax have been used to develop self-healable hydrogels [21]. Similarly, a conductive hydrogel composed of galactomannan, guar gum, together with acidic poly(3,4-ethylenedioxythiophene): poly(styrenesulfonate) was reported to foster wound closure and rearrangement of tissue in order to promote wound healing process [22]. Moreover, Wang's group [23] created Zn- functionalized chitosan or polypyrrole, which were crosslinked with PVA to produce a hydrogel through hydrogen bonding, di-diol complexations, and zinc-based coordination bonds. These interactions provide the material with autonomous ability and exceptional stretchability.

Ionic Bonds

The contact between polymers of opposing charges facilitates crosslinking in polymers during the development of films. The movement of free ions to the broken or non-crosslinked part of the polymeric chain leads to self-healing, giving rise to reversible reactions brought on by ionic interaction between polymers and free ions [24]. For instance, employing Fe ions for the synthesis of self-healing material appertaining to ionic bonding. An ionic bond obtained when the carboxylic group in polyacrylic acid (PAA) interacts with Fe^{3+} ions results in the self-healing capability of the polymeric material. The self-healing ability in synthesized material is due to the presence of ionic bonds while an enhanced mechanical strength is because of the covalent interaction by the carboxylic acid ions that exist in PAA. It showed a tensile strength of 860 and 560 kPa for the original product and the self-healed material, respectively. Their toughness, self-repair, and stretchability make them promising candidates for various tissue engineering applications [25].

In research conducted, a carboxybetaine acrylamide-based self-healing hydrogel consisting of zwitterionic material that was healed through ionic interaction restored almost 90% of its compressive capacities after repair and demonstrated time-independent healing characteristics [26]. Catechol has been used for generating ionic bond-based self-healing hydrogels. By electrostatically crosslinking catechol-grafted chitosan and Fe^{3+} ions, the hydrogels with self-repairing capacity and good mechanical properties regained their strength within 1 m 40 s in cyclic time sweep experiments [27]. In another study, the synthesis of self-healing material via reversible alginate polyelectrolyte complexes and 2-hydroxypropyl trimethyl ammonium chloride chitosan was reported. The polymers were combined to create hydrogels at charge neutrality, which were then precipitated for 12 h. The hydrogel also has strong adhesive behavior, cytocompatibility, and shear-thinning characteristics in addition to its capacity for self-healing [28].

Hydrophobic Bonds

Hydrophobic interaction is a reversible non-covalent bond that describes the connection between hydrophobic monomers and water. It occurs as a result of the aggregation of hydrophobes in aqueous solution. Hydrophobic interaction can be used to build physical self-healing polymeric structures via the monomer's self-assembling process in an aqueous medium. A widely used technique to produce hydrophobic interaction-based self-healing material is micellar polymerization. This method occurs with the help of several important elements, including hydrophobic units, surfactants, hydrophilic units, and electrolytes. In this method, a surfactant is utilized to solubilize the hydrophobic component, while hydrophilic compounds and electrolyte-aqueous solutions help in copolymerization [29].

In many studies of hydrophobic bond-based self-healing hydrogels, liposomes [30, 31] or surfactant micelles [32, 33] are used as crosslinking points to make the polymer chains consisting of both hydrophobic and hydrophilic monomers. For instance, self-healing hydrogel could be generated through micellar copolymerization of hydrophilic monomer acrylamide and hydrophobic monomer stearyl methacrylate in an aqueous solution containing micelles of sodium dodecyl sulfate (SDS) [34–36]. When salt was introduced to aqueous sodium dodecyl sulfate solutions, micellar growth along with the solubilization of hydrophobes occurred within SDS micelles. This hydrogel with the time-dependent dynamic moduli showed good self-healing capacity as well as a high elongation ratio whereas following SDS extraction, the same hydrogel with time-independent dynamic moduli displayed excellent mechanical strength, and no self-healing activities [36].

Host–Guest Interactions

Host–guest interactions are a kind of non-covalent bond that occurs when two chemical species interact via non-covalent bond linkages including hydrogen bond, Van der Waals force, hydrophobic interaction, and electrostatic interaction. These interactions involve host and guest moieties. The host moieties are usually the macrocyclic molecules that are introduced within the guest moieties in order to create unique inclusion complexation through hydrophobic interactions and π-π stacking [37]. A number of guest groups, including adamantane, azobenzene, ferrocene, cholesterol, and cholic acid are currently being utilized in biomedical research demonstrating a promising future [38].

Self-healing films or hydrogels are frequently created by host–guest interactions, and many of these materials depend on external stimuli like light [39], temperature [40], redox potentials [41], and pH [42] to initiate the healing process. Some hydrogels have also been produced based on host–guest interactions that could recover themselves without any kind of external stimuli [43, 44]. Rodell et al. developed a hyaluronic acid (HA) self-healing hydrogel based on host–guest interactions using β-cyclodextrin-modified HA as a host macromer while adamantane-modified HA as a guest macromer. The resultant hydrogels showed rapid self-healing at 25 °C and shear-thinning properties [44]. β-cyclodextrin based hydrogels possess good self-healing properties and due to their ability to rapidly repair, these hydrogels could be used in 3D printing applications [45]. Shear thinning and self-healing hydrogels based on methacrylated gelatin and cyclodextrin-modified alginate were formed through a host–guest interaction between cyclodextrin cavity which is hydrophobic in nature and aromatic gelatin residues with potential in various biomedical applications such as scaffolds, and cell or drug delivery carriers [46]. Furthermore, in another research conducted, a shape-memorable and self-healing polymeric system with dual host–guest was described. These hydrogels assembled two complexes namely β-cyclodextrin—ferrocene and β-cyclodextrin—adamantine into a supramolecular network in order to develop stimuli-responsive hydrogels [47].

3.1.2 Dynamic Covalent Bonding

Dynamic covalent interactions, such as imine bonds, disulfide bonds, boronate ester complexation, and Diels–Alder reaction (Fig. 10.4), are broadly employed in the synthesis of self-healing products, including thin films and hydrogels. Comparing dynamic covalent bonds to non-covalent interactions, dynamic covalent bonds show a stronger, however slower dynamic equilibrium.

Imine Bonds

The imine, also called Schiff base, is a promising group of compounds with a double bond between carbon and nitrogen which is used to create hydrogels by forming

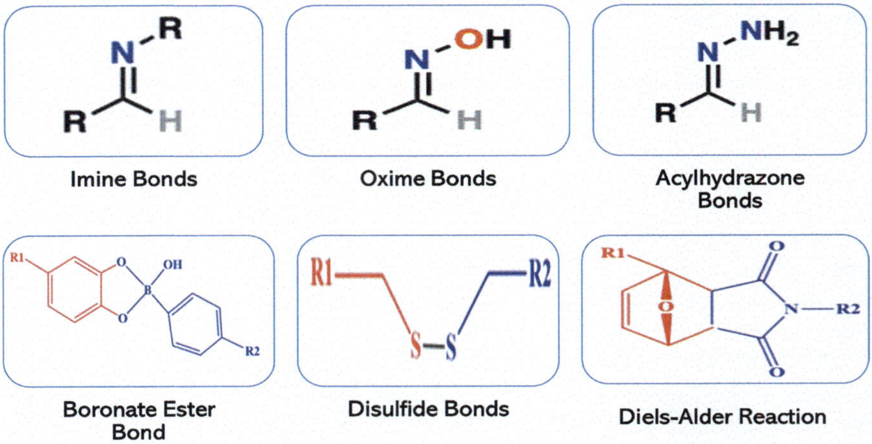

Fig. 10.4 Dynamic covalent bonds involved in self-healing mechanisms

crosslinks between amine groups and polymer's ketone or aldehyde groups under physiological circumstances [48, 49]. Numerous self-healing materials have been synthesized by aromatic Schiff bases [50, 51] or aliphatic Schiff bases [50, 52, 53]. Unlike aliphatic Schiff bases, aromatic Schiff bases display higher stability to preserve mechanical properties unlike the latter [54].

A natural injectable material was fabricated with dialdehyde bacterial cellulose and chitosan for wound dressing. The synthesized hydrogels demonstrated injectability, good mechanical properties, antibacterial properties, drug delivery, and dynamic self-healing capability because of the formation of imine bonds, making them a suitable material for applications involving tissue engineering [55]. In another study conducted, the naturally occurring polymer chitin, in its slightly altered form known as acrylamide-modified chitin containing amino, hydroxyl, and carboxyl groups, was mixed with oxidized alginate to make a hydrogel with suitable biocompatibility and biodegradability. This polysaccharide-based material possesses self-healing properties because of dynamic imine bonds and has the potential application in a variety of biomedical fields [56].

Furthermore, chitosan-based self-healing hydrogels have been developed by crosslinking chitosan or its derivatives with a benzaldehyde-terminated poly (ethylene glycol) (DF-PEG). The aldehyde groups present in DF-PEG could make an imine bond (Schiff base) with the amine group present in chitosan, leading to the synthesis of self-healing hydrogel formed by a dynamic network. This reversible Schiff base interaction maintains the balance between fracture/rupture and regeneration in the environment of hydrogel, resulting in remarkable self-healing properties [54, 57].

Oxime and acylhydrazone are derivatives of Schiff base (imine) that have excellent stability and have also been used to create self-healing hydrogels [58–60]. Oxime bonds are formed by condensation of hydroxylamine with ketones or aldehydes, whereas arylhydrazone bonds are generated when ketone or aldehyde groups interact with hydrazine groups to create a dynamic covalent linkage at an acid pH environment [61, 62]. To give an idea, the self-healing materials were synthesized by combining oxidized sodium alginate with adipic acid dihydrazide and N-carboxyethyl chitosan through dynamic acyl hydrazone and imine bonds [63].

Boronate Ester Complexation

The combination of boronic acid with diol results in the formation of the reversible boronate ester bond**.** The stability of this bond is dependent on glucose concentration and pH value. Different hydrogels with self-healing ability have been developed using boronic acid and its derivatives, including phenylboronic acid or polymers that contain phenylboronic acid.

Synthesizing self-healing hydrogels using other types of bonds/linkages usually involves the polymer's chemical modification in order to achieve the required functional groups. However, polyol/borax self-healing hydrogels can be produced simply by combining the borax with commonly available polyols in an aqueous medium. Meanwhile, these hydrogels have demonstrated weak mechanical characteristics [64].

Over the past five years, self-healing hydrogels made of polyol (polysaccharides, glycopolymers, and PVA) and borax (as a crosslinking agent) have received a lot of interest. Borax is non-toxic that, in aqueous media, separates into boric acid and borate. The resultant borate can form dynamic borate-diol bonds with polyols containing 1,2- or 1,3-diol groups, such as PVA and different types of glycopolymers and polysaccharides in order to construct self-healing hydrogels [65]. Figure 10.5 shows the general process for crosslinking polyols with borax.

Disulfide Bonds

Disulfide bonds are another kind of covalent linkage that is reversible, enabling cleaved bonds to repair. They provide dynamic covalent interactions to develop self-healing products, which are sensitive to redox potential and pH [66].

A composite hydrogel containing silver nitrate (AgNO3) and thiolated polyethylene glycol for diabetic wound repair application was produced which exhibited antibacterial properties and resisted external mechanical resistance. The existence of Ag^+ and disulfide linkage of Ag–S strengthened the bactericidal function and the self-healing capability of the hydrogels, respectively, making it a suitable choice for diabetic skin wounds [67].

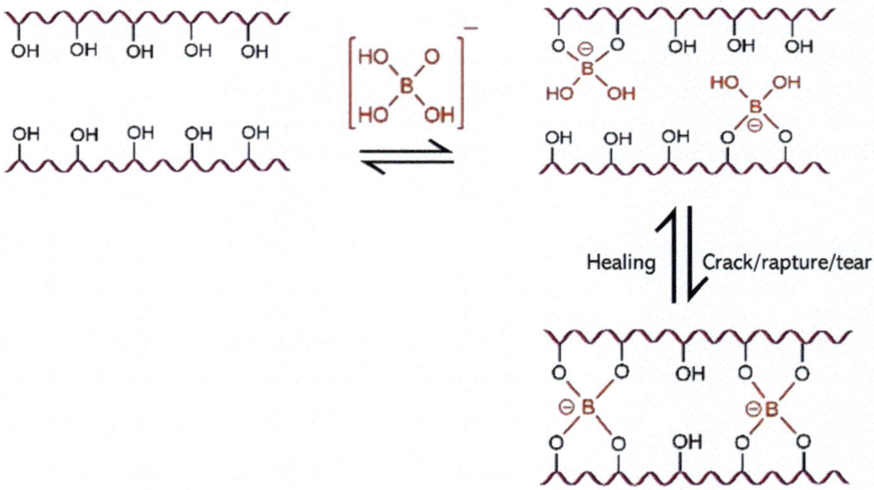

Fig. 10.5 General process for crosslinking of polyols with borax

Diel's-Alder Reaction

Diels–Alder reaction is another dynamic covalent interaction in self-healing hydrogels, which is thermally reversible. This click reaction takes place between a dienophile, typically an alkene or an alkyne, and a conjugated diene [61, 68]. It does not require any coupling reagent or catalysts and generates no side products [69], which makes it an appropriate candidate for biomedical applications in self-healing and shear-thinning hydrogels [70].

Diels–alder reaction was employed to create self-healing hydrogels along with some other reversible linkages, including coordination bond [71], electrostatic interaction [72], imine bond [73], and acylhydrazone bond [74]. Recently, Diels–alder chemistry was examined in the interaction between maleimide-end-functionalized PEG and furyl-modified cellulose nanocrystals that exhibited approximately 78% healing efficiency because of the high concentration of Diels–alder bonds [75].

4 Conclusion

The analysis of the updates of the modern trends and novelties in the area of smart coatings, self-healing thin films, and advanced deposition techniques define a well-developed and dynamically growing industry with high potential to change various fields. Smart coatings are coatings that can exhibit a response to one or many stimulations and have applications in the field of corrosion protection to energy conservation. Self-healing thin films are a major advancement in the field of material science as they drastically cut the need for maintenance and act as a surefire way of ensuring that components last longer than ever before. On the other hand, developments in deposition technologies are expanding the ability to deposit these RSSs with high complexity and superior performance and functionality.

Such purchasing of these advanced technologies brings a deal of hope to several industries such as automotive, aerospace, electronics, and construction industries. Various investigations and innovations are constantly held in the field of RD all over the world, these will subsequently lead to improvement as well in the functions and essential attributes of smart coatings and self-healing thin films. Finally, all these innovations are expected to accelerate the production of better and socially friendly materials that will support advancements in material science and engineering.

References

1. Bogue, R.: Smart materials: a review of capabilities and applications. Assem. Autom. **34**(1), 16–22 (2014)
2. Qader, I.N., Mediha, K.Ö.K., Dagdelen, F., AYDOĞDU, Y.: A review of smart materials: researches and applications. El-Cezeri **6**(3), 755–788 (2019)
3. Mertoglu, M., Garnier, S., Laschewsky, A., Skrabania, K., Storsberg, J.: Stimuli responsive amphiphilic block copolymers for aqueous media synthesised via reversible addition fragmentation chain transfer polymerisation (RAFT). Polymer (Guildf) **46**(18), 7726–7740 (2005)
4. Reddy, K.R., El-Zein, A., Airey, D.W., Alonso-Marroquin, F., Schubel, P., Manalo, A.: Self-healing polymers: synthesis methods and applications. Nano-Struct. Nano-Objects **23**, 100500 (2020)
5. Urban, M.W.: Stratification, stimuli-responsiveness, self-healing, and signaling in polymer networks. Prog. Polym. Sci. **34**(8), 679–687 (2009)
6. White, S.R., et al.: Autonomic healing of polymer composites. Nature **409**(6822), 794–797 (2001)
7. Cho, S.H., White, S.R., Braun, P.V.: Self-healing polymer coatings. Adv. Mater. **21**(6), 645–649 (2009)
8. Cordier, P., Tournilhac, F., Soulié-Ziakovic, C., Leibler, L.: Self-healing and thermoreversible rubber from supramolecular assembly. Nature **451**(7181), 977–980 (2008)
9. South, A.B., Lyon, L.A.: Autonomic self-healing of hydrogel thin films. Angew. Chem. **122**(4), 779–783 (2010)
10. White, S.R., Moore, J.S., Sottos, N.R., Krull, B.P., Santa Cruz, W.A., Gergely, R.C.R.: Restoration of large damage volumes in polymers. Science (1979) **344**(6184), 620–623 (2014)
11. Rao, Y.-L., et al.: Stretchable self-healing polymeric dielectrics cross-linked through metal–ligand coordination. J. Am. Chem. Soc. **138**(18), 6020–6027 (2016)
12. Zhang, M.Q., Rong, M.Z.: Theoretical consideration and modeling of self-healing polymers. J. Polym. Sci. B Polym. Phys. **50**(4), 229–241 (2012)
13. Jeon, I., Cui, J., Illeperuma, W.R.K., Aizenberg, J., Vlassak, J.J.: Extremely stretchable and fast self-healing hydrogels. Adv. Mater. **28**(23), 4678–4683 (2016)
14. Saunders, L., Ma, P.X.: Self-healing supramolecular hydrogels for tissue engineering applications. Macromol. Biosci. **19**(1), 1800313 (2019)
15. Liu, Y., Hsu, S.: Synthesis and biomedical applications of self-healing hydrogels. Front. Chem. **6**, 449 (2018)
16. Zhang, H., Xia, H., Zhao, Y.: Poly (vinyl alcohol) hydrogel can autonomously self-heal. ACS Macro Lett. **1**(11), 1233–1236 (2012)

17. Zhang, Z., Li, T., Chen, B., Wang, S., Guo, Z.: Self-healing supramolecular hydrogel of poly (vinyl alcohol)/chitosan carbon dots. J. Mater. Sci. **52**, 10614–10623 (2017)
18. Shi, D., et al.: PH-dependent and self-healing properties of mussel modified poly (vinyl alcohol) hydrogels in a metal-free environment. RSC Adv. **5**(100), 82252–82258 (2015)
19. Samadi, N., Sabzi, M., Babaahmadi, M.: Self-healing and tough hydrogels with physically cross-linked triple networks based on Agar/PVA/Graphene. Int. J. Biol. Macromol. **107**, 2291–2297 (2018)
20. Sun, X., Luo, C., Luo, F.: Preparation and properties of self-healable and conductive PVA-agar hydrogel with ultra-high mechanical strength. Eur. Polym. J. **124**, 109465 (2020)
21. Chang, Y.-A., Chou, Y.-N., Lin, Y.-J., Chen, W.-Y., Chen, C.-Y., Lin, H.-R.: Microgel-reinforced PVA hydrogel with self-healing and hyaluronic acid drug-releasing properties. Int. J. Polym. Mater. Polym. Biomater. **70**(17), 1224–1235 (2021)
22. Li, S., Wang, L., Zheng, W., Yang, G., Jiang, X.: Rapid fabrication of self-healing, conductive, and injectable gel as dressings for healing wounds in stretchable parts of the body. Adv. Funct. Mater. **30**(31), 2002370 (2020)
23. Zhang, J., et al.: Highly stretchable and conductive self-healing hydrogels for temperature and strain sensing and chronic wound treatment. ACS Appl. Mater. Interfaces **12**(37), 40990–40999 (2020)
24. Zhang, W., et al.: Recent achievements in self-healing materials based on ionic liquids: a review. J. Mater. Sci. **55**, 13543–13558 (2020)
25. Zhong, M., et al.: Self-healable, tough and highly stretchable ionic nanocomposite physical hydrogels. Soft Matter **11**(21), 4235–4241 (2015)
26. Bai, T., et al.: Zwitterionic fusion in hydrogels and spontaneous and time-independent self-healing under physiological conditions. Biomaterials **35**(13), 3926–3933 (2014)
27. Yavvari, P.S., Srivastava, A.: Robust, self-healing hydrogels synthesised from catechol rich polymers. J. Mater. Chem. B **3**(5), 899–910 (2015)
28. Ren, Y., et al.: A self-healing hydrogel formation strategy via exploiting endothermic interactions between polyelectrolytes. Chem. Commun. **52**(37), 6273–6276 (2016)
29. Okay, O.: Self-healing hydrogels formed via hydrophobic interactions. Supramolecular Polym. Netw. Gels 101–142 (2015)
30. Rao, Z., Inoue, M., Matsuda, M., Taguchi, T.: Quick self-healing and thermo-reversible liposome gel. Colloids Surf. B Biointerfaces **82**(1), 196–202 (2011)
31. Hao, X., Liu, H., Xie, Y., Fang, C., Yang, H.: Thermal-responsive self-healing hydrogel based on hydrophobically modified chitosan and vesicle. Colloid Polym. Sci. **291**, 1749–1758 (2013)
32. Gulyuz, U., Okay, O.: Self-healing poly (acrylic acid) hydrogels: effect of surfactant. In: Macromolecular Symposia, Wiley Online Library, pp. 232–238 (2015)
33. Liu, Y., Li, Z., Niu, N., Zou, J., Liu, F.: A simple coordination strategy for preparing a complex hydrophobic association hydrogel. J. Appl. Polym. Sci. **135**(25), 46400 (2018)
34. Tuncaboylu, D.C., Sari, M., Oppermann, W., Okay, O.: Tough and self-healing hydrogels formed via hydrophobic interactions. Macromolecules **44**(12), 4997–5005 (2011)
35. Tuncaboylu, D.C., Argun, A., Sahin, M., Sari, M., Okay, O.: Structure optimization of self-healing hydrogels formed via hydrophobic interactions. Polymer (Guildf) **53**(24), 5513–5522 (2012)
36. Tuncaboylu, D.C., Sahin, M., Argun, A., Oppermann, W., Okay, O.: Dynamics and large strain behavior of self-healing hydrogels with and without surfactants. Macromolecules **45**(4), 1991–2000 (2012)
37. Szejtli, J.: "Bender, ML, and M. Komiyama: Cyclodextrin Chemistry-Reactivity and Structure, Concepts in Organic Chemistry Vol. 6 (Editors: K. Hafner, J.-M. Lehn, CW Rees, P. v. R. Schleyer, BM Trost and R, Zahradnik)(Cyclodextrin Chemistry. Reaktionsfähigkeit und Struktur, Konzepte der Organischen Chemie Bd. 6). Springer-Verlag, Berlin-Heidelberg-New York 1978. X, 96 pages, with 14 figs, and 37 tables, cloth DM 44,-; US $22.00." Wiley Online Library (1979)
38. Jin, J., Cai, L., Jia, Y.-G., Liu, S., Chen, Y., Ren, L.: Progress in self-healing hydrogels assembled by host–guest interactions: preparation and biomedical applications. J. Mater. Chem. B **7**(10), 1637–1651 (2019)

39. Yamaguchi, H., Kobayashi, Y., Kobayashi, R., Takashima, Y., Hashidzume, A., Harada, A.: Photoswitchable gel assembly based on molecular recognition. Nat. Commun. **3**(1), 603 (2012)
40. Zheng, Y., Hashidzume, A., Takashima, Y., Yamaguchi, H., Harada, A.: Temperature-sensitive macroscopic assembly based on molecular recognition. ACS Macro Lett. **1**(8), 1083–1085 (2012)
41. Nakahata, M., Takashima, Y., Yamaguchi, H., Harada, A.: Redox-responsive self-healing materials formed from host–guest polymers. Nat. Commun. **2**(1), 511 (2011)
42. Zheng, Y., Hashidzume, A., Harada, A.: PH-responsive self-assembly by molecular recognition on a macroscopic scale. Macromol. Rapid Commun. **34**(13), 1062–1066 (2013)
43. McKee, J.R., Appel, E.A., Seitsonen, J., Kontturi, E., Scherman, O.A., Ikkala, O.: Healable, stable and stiff hydrogels: combining conflicting properties using dynamic and selective three-component recognition with reinforcing cellulose nanorods. Adv. Funct. Mater. **24**(18), 2706–2713 (2014)
44. Rodell, C.B., Kaminski, A.L., Burdick, J.A.: Rational design of network properties in guest–host assembled and shear-thinning hyaluronic acid hydrogels. Biomacromol **14**(11), 4125–4134 (2013)
45. Highley, C.B., Rodell, C.B., Burdick, J.A.: Direct 3D printing of shear-thinning hydrogels into self-healing hydrogels. Adv. Mater. **27**(34), 5075–5079 (2015)
46. Jalalvandi, E., Shavandi, A.: Shear thinning/self-healing hydrogel based on natural polymers with secondary photocrosslinking for biomedical applications. J. Mech. Behav. Biomed. Mater. **90**, 191–201 (2019)
47. Miyamae, K., Nakahata, M., Takashima, Y., Harada, A.: Self-healing, expansion–contraction, and shape-memory properties of a preorganized supramolecular hydrogel through host–guest interactions. Angew. Chem. Int. Ed. **54**(31), 8984–8987 (2015)
48. Trask, R.S., Williams, H.R., Bond, I.P.: Self-healing polymer composites: mimicking nature to enhance performance. Bioinspir. Biomim. **2**(1), P1 (2007)
49. McKay, C.S., Finn, M.G.: Click chemistry in complex mixtures: bioorthogonal bioconjugation. Chem. Biol. **21**(9), 1075–1101 (2014)
50. Qu, J., Zhao, X., Ma, P.X., Guo, B.: PH-responsive self-healing injectable hydrogel based on N-carboxyethyl chitosan for hepatocellular carcinoma therapy. Acta Biomater. **58**, 168–180 (2017)
51. Raza-Karimi, A., Khodadadi, A.: Mechanically robust 3D nanostructure chitosan-based hydrogels with autonomic self-healing properties. ACS Appl. Mater. Interfaces **8**(40), 27254–27263 (2016)
52. Luü, S., et al.: Injectable and self-healing carbohydrate-based hydrogel for cell encapsulation. ACS Appl. Mater. Interfaces (Providence) **7**, 13029–13037 (2015)
53. Huang, J., et al.: Novel in situ forming hydrogel based on xanthan and chitosan re-gelifying in liquids for local drug delivery. Carbohydr. Polym. **186**, 54–63 (2018)
54. Zhang, Y., Tao, L., Li, S., Wei, Y.: Synthesis of multiresponsive and dynamic chitosan-based hydrogels for controlled release of bioactive molecules. Biomacromol **12**(8), 2894–2901 (2011)
55. Li, W., et al.: All-natural injectable hydrogel with self-healing and antibacterial properties for wound dressing. Cellulose **27**, 2637–2650 (2020)
56. Ding, F., et al.: A dynamic and self-crosslinked polysaccharide hydrogel with autonomous self-healing ability. Soft Matter **11**(20), 3971–3976 (2015)
57. Li, Y., Wang, X., Wei, Y., Tao, L.: Chitosan-based self-healing hydrogel for bioapplications. Chin. Chem. Lett. **28**(11), 2053–2057 (2017)
58. Lin, F., et al.: Peptide-functionalized oxime hydrogels with tunable mechanical properties and gelation behavior. Biomacromol **14**(10), 3749–3758 (2013)
59. Grover, G.N., Lam, J., Nguyen, T.H., Segura, T., Maynard, H.D.: Biocompatible hydrogels by oxime click chemistry. Biomacromol **13**(10), 3013–3017 (2012)
60. Mukherjee, S., Hill, M.R., Sumerlin, B.S.: Self-healing hydrogels containing reversible oxime crosslinks. Soft Matter **11**(30), 6152–6161 (2015)
61. Li, Q., Liu, C., Wen, J., Wu, Y., Shan, Y., Liao, J.: The design, mechanism and biomedical application of self-healing hydrogels. Chin. Chem. Lett. **28**(9), 1857–1874 (2017)

62. Tu, Y., et al.: Advances in injectable self-healing biomedical hydrogels. Acta Biomater. **90**, 1–20 (2019)
63. Wei, Z., et al.: Novel biocompatible polysaccharide-based self-healing hydrogel. Adv. Funct. Mater. **25**(9), 1352–1359 (2015)
64. Seidi, F., et al.: Self-healing Polyol/borax hydrogels: fabrications, properties and applications. Chem. Rec. **20**(10), 1142–1162 (2020)
65. Schultz, R.K., Myers, R.R.: The chemorheology of poly (vinyl alcohol)-borate gels. Macromolecules **2**(3), 281–285 (1969)
66. Wei, Z., et al.: Self-healing gels based on constitutional dynamic chemistry and their potential applications. Chem. Soc. Rev. **43**(23), 8114–8131 (2014)
67. Chen, H., et al.: An injectable self-healing coordinative hydrogel with antibacterial and angiogenic properties for diabetic skin wound repair. NPG Asia Mater. **11**(1), 3 (2019)
68. Talebian, S., et al.: Self-healing hydrogels: the next paradigm shift in tissue engineering? Adv. Sci. **6**(16), 1801664 (2019)
69. Cadamuro, F., Russo, L., Nicotra, F.: Biomedical hydrogels fabricated using Diels-Alder crosslinking. Eur. J. Org. Chem. **2021**(3), 374–382 (2021)
70. Uman, S., Dhand, A., Burdick, J.A.: Recent advances in shear-thinning and self-healing hydrogels for biomedical applications. J. Appl. Polym. Sci. **137**(25), 48668 (2020)
71. Li, S., Wang, L., Yu, X., Wang, C., Wang, Z.: Synthesis and characterization of a novel double cross-linked hydrogel based on Diels-Alder click reaction and coordination bonding. Mater. Sci. Eng., C **82**, 299–309 (2018)
72. Ghanian, M.H., Mirzadeh, H., Baharvand, H.: In situ forming, cytocompatible, and self-recoverable tough hydrogels based on dual ionic and click cross-linked alginate. Biomacromol **19**(5), 1646–1662 (2018)
73. Li, S., Yi, J., Yu, X., Shi, H., Zhu, J., Wang, L.: Preparation and characterization of acid resistant double cross-linked hydrogel for potential biomedical applications. ACS Biomater. Sci. Eng. **4**(3), 872–883 (2018)
74. Yu, F., Cao, X., Du, J., Wang, G., Chen, X.: Multifunctional hydrogel with good structure integrity, self-healing, and tissue-adhesive property formed by combining Diels-Alder click reaction and acylhydrazone bond. ACS Appl. Mater. Interfaces **7**(43), 24023–24031 (2015)
75. Shao, C., Wang, M., Chang, H., Xu, F., Yang, J.: A self-healing cellulose nanocrystal-poly (ethylene glycol) nanocomposite hydrogel via Diels-Alder click reaction. ACS Sustain. Chem. Eng. **5**(7), 6167–6174 (2017)

Environment Consideration and Sustainability—Thin Film and Environment Impact—Sustainability Coating Technologies—Green Engineering Approaches

Tasneem Pervez, Farooq Khalfan Saif Al Jahwari, and Moosa Salim Moosa Al Kharusi

Abstract This chapter discusses on environmental issues of thin films, green coating techniques, and green engineering. The synthesis also presents the life cycle of thin films and stresses the necessity of decreasing ecological impacts. Options of applying environmentally friendly materials and energy-conserving processes are mentioned as resolutions that lead to environmental degradation as the main goals of the coating technology. The chapter also looks into green engineering solutions that can be applied in thin film design and fabrication to uphold sustainability in all sectors. Thus, the outlined aspects of the chapter stress the need for environmentally friendly approaches in thin film technology to achieve sustainable development and avoid negative environmental effects. A general understanding of the analysis reveals that thin film applications are not invulnerable to change and that through the introduction of new ideas and engineering approaches to the manufacturing process, sustainable use of technological resources and protection of the environment is fully achievable.

Abbreviations

SDG	Sustainable Development Goals
TF	Thin Film
CdTe	Cadmium telluride

T. Pervez (✉) · F. K. S. Al Jahwari · M. S. M. Al Kharusi
Department of Mechanical and Industrial Engineering, College of Engineering, Sultan Qaboos University, Muscat, Sultanate of Oman
e-mail: tasneem@squ.edu.om

F. K. S. Al Jahwari
e-mail: farooq@squ.edu.om

M. S. M. Al Kharusi
e-mail: m.alkharusi@squ.edu.om

A. D. Chandio and I. A. Channa (eds.), *Thin Films and Coatings*, Engineering Materials,
https://doi.org/10.1007/978-981-96-0005-2_11

OPV	Organic Photovoltaics
QDPD	Thin-Film Quantum Dot Photodiode
VOCs	Volatile organic compounds
WEEE	Waste Electrical and Electronic Equipment
EMS	Environmental Management Systems
SMEs	Small and Medium-sized Enterprises
PV	Photovoltaic

1 Introduction

This chapter is centered on the "Environment Consideration and Sustainability in Thin Film Manufacturing". In this part of the discussion, the symbiotic relationship between "maturity in thin film technologies" and sustainability will be analyzed. It begins with a general understanding of thin film production and a brief detail of how such thin layers are produced. From the following point, we will proceed with the analysis of the environmental impact of thin film production with special emphasis on such indicators as the resources consumed, CO2 emission, and waste generated. However, this chapter goes beyond the simple staking out of problems. It will disclose the area of sustainability coatings or thin films with green engineering inclusion. We shall also give some illustrations of thin films in clever uses such as in the manufacture of solar cells and daily widely used products for instance biodegradable food packs.

The journey would only be complete by going through some of the regulations set by the different authorities. According to this, while analyzing the subject under consideration it will be efficient to concentrate on the basic set of rules and standards, thus indicating the directions for the further evolution of the thin film industry, thus adhering to the logic of technological progress. Therefore, it is progressive and our focus remains constant as we move forward, on making these ideas simple so that they can be reasoned by everyone. This means, no matter if the reader is a fan of updated technologies, a future scientist, or the ordinary person who always searches for the unknown to gain new knowledge about the world, this chapter will introduce him to the interesting topic of thin films and its significance in the creation of the new high-tech society. Well, then, fasten your seatbelts while we drive to the realm of innovation and environmental focus, of thin films for a cleaner world.

2 SGDs of Thin Film

In this regard, the concept of the SDGs emerges as a significant concept that guides the behavior of companies toward the right approach. Sustainability goals, ranging from one to seventeen, issued by the UN under the name of Sustainable Development Goals (SDGs), act as a beacon for companies, pointing out the future direction on which a company's operations should be aligned. The following objectives are a little more sophisticated than mere commercial objectives, and outline how crucial it is for industry to somehow rise up as the problem solver for the world's environmental challenges. The general movement of the SDGs becomes both strategically and morally obligatory with the rise of thin film technology. In this regard, it performs the function of an invitation, showing how the conjunction of the perspective field of thin film and the broader context of sustainability, sustainability coating technologies, and green engineering methodologies exists. It sets the context for a consideration of these aspects that make up this composite whole in the context of reconceptualizing the field of ethical business practices with thin film applications.

I. Affordable and Clean Energy (SDG 7)

In line with Sustainable Development Goal 7: In the thin film sector, Affordable and Clean Energy is a perfect example that demonstrates this new age vanguard that is altering the structure of green energy sources. This contribution is in the area of Energy efficient thin film technology, particularly concerning solar energy. This alliance of clean energy with thin films is indeed revolutionary which was expanding to a tomorrow when energy supply is both clean and affordable. Of course, consistently, thin film technology is associated with energy efficiency, especially in the sphere of the solar industry. These films make inexpensive and versatile substitutes for conventional solar panels because they are produced from thin film materials. Due to their flexibility, it can easily be installed in different surfaces enhancing the efficiency of collecting solar energy within the developed and the rural settings. Materials and processes that are used in creating thin film solar cells also experience changes over time, and these changes make the conversion of energy increase as thin film solar cells are developed more.

Moreover, concerning SDG 7, the process of thin film solar technologies manufacturing typically consumes less energy compared to the conventional photovoltaic approaches. Due to these technologies' lowered environmental footprint, they are crucial for realizing the goals established for the availability, affordability, and sustainability of the energy supply worldwide. However, it remains evident that these developments do not only alter our energy environment but also reflect commitment toward sustainability and the attainment of cleaner and more accessible energy as we navigate through the multiple facets of the thin film industry toward SDG 7 attainments.

II. Responsible Consumptions (SDG 12)

In the criterion of the seventh sub-goal of the 2030 Sustainable Development Goals (SDGs)—"Responsible Consumption and Production," the thin film industry plays a crucial role in the transformation of production and consumption patterns. Thus, this objective reflects the importance of the efficient usage of resources, which is a concept closely tied to that of thin film. I would note that in addition to innovation, which is intrinsic to the actual development of thin film deposition technology, the industry's commitment to sustainable thin film technology is parasitic to reduce the amount of waste produced, sparing resources, and supporting green manufacturing processes. Thus, the thin film business contributes to SDG 12 more through material efficiency improvements and recycling programs. This is something that is inevitable and is heading toward us, ethical procurement and manufacturing practices will be key players in today 's technological development.

III. Climate Action (SDG 13)

Regarding Climate Action or SDG 13, the thin film industry automatically becomes one of the key actors in mitigating the adverse effects of climate change. The industry is pledging to environmentally friendly manufacturing and production by orienting the efforts of sustainable thin films toward the eradication of greenhouse gases in the production process. This, together with the shift of energy sources toward environmentally friendlier ones, the enhancement of the production techniques, and the application and adherence to strict emission standards makes the thin film industry not only a supporter of the concept of SDG 13 but also a critical actor toward positive change in the battle against climate change. However, the advancement of technology and social responsibility for the climate change issue gives a revolutionary implication that thin films are products of environmental awareness.

IV. Industry, Innovation and Infrastructure (SDG 9)

To induce technological innovation for better infrastructure, or SDG 9 ("Industry, Innovation and Infrastructure"), the thin film industry is one of the best examples of technical innovation that correlates highly with sustainability. In the case of the mobile phone industry and SDG 9 on industry innovation, their requirement of innovation in production indicates similar concerns in the development of sustainable thin film technology. On the same note, the industry's ongoing R&D programs aim to enhance the performance of thin films regarding the environment aside from stretching the technological frontier. The industry remains a main driver of green innovation to provide effective, accessible, and sustainable systems for improvement in the development of infrastructure worldwide as it continues to advance new products and methods of creation.

V. Decent Work and Economic Growth (SDG 8)

Sustainable Development Goals (SDG 8) entitled "Decent Work and Economic Growth" can also be inferred from the thin film industry function which is the creation of job opportunities and social impact. In addition to innovations in technology, the industry helps in the advancement of the economy by providing employment opportunities and improvement of skills. Not only that, the thin film industry creates decent work and also assists communities through employment practices which is an aspect of Sustainable Development Goal 8 (shown in Fig. 11.1). It evolves into a structural foundation that contributes to the concept of sustainable economic inclusion and openness to visitors who adhere to the principles of responsible tourism.

Fig. 11.1 Sustainable development goals

3 Identification of Environmental Impact in Thin Film Production

Resource Consumption

The technique of creating thin films requires exact proportions in the usage of resources and this is just like a delicate juggling of ingredients in a recipe. You are in a kitchen and want just sufficient flour, sugar, and eggs to prepare a good cake but not more. In the same way, the production of thin films requires such materials as thin films themselves, metals, plastics, semiconductor materials, and insulators. These are inputs or our "baking ingredients" and we cannot afford to be wasteful with any of these.

Now this is a problem because not all of those materials are as freely available as flour or sugar are today for instance. For instance, thin film technology utilizes certain materials such as rare earth elements, however, their extraction impacts the environment. As we know, everyone who is involved in thin film research is trying to save as many resources as possible. You know when you look for a substitute for the ingredients needed in a recipe—the goal is similar—get the same taste but with a lesser impact on the environment.

Emission of Greenhouse Gases

Like a car produces exhaust gases when it is using fuel, thin film production can also produce greenhouse gases. Constitute it as the environmental equivalent of the carbon footprint per driver-to-passenger transport. Depending on the specific manufacturing process there may be emissions of gases that normally are not present outside factories, such as nitrogen oxides or fluorinated compounds in the case of high temperature or chemical reactions in manufacturing processes.

Minimizing these emission gases is an issue of interest to thin film scientists. We are like earth-conscious drivers seeking more environmentally friendly fuel or hybrid cars. Thus, using new technologies for producing thin films, it is focused on reducing the company's effect on the greenhouse effect and creating lighter thin films.

Waste Generation

Waste generation in thin film production is the last stage process before cleaning just like leftovers in a kitchen. We desire to reduce waste similar to when you would not wish to discard unwanted leftover food. Manufacturing thin films may generate by-products or unused material and managing these wastes properly is essential.

It is very essential to note that scholars work hard and come up with processes that minimize the production of waste and also come up with the best techniques which shall enhance the reusing and recycling of waste. The accomplishment could indeed be compared to turning scraps from the kitchen into compost or finding ways to use all leftovers to the maximum in a single meal. In other words, by observing a strict procedure on how waste is produced during the engineering of thin films, the goal is to achieve economical production of thin films while adhering to ecological principles of ensuring a sustainable technological culture.

4 Sustainable Coating Technologies

Regarding thin film coating technologies, they can be broadly described as sustainable coats for environmental protection. It should be interesting to think of them as a layer of defense for our technical advancements, carefully constructed with a passion for sustainability while working at the highest level of efficiency possible.

Sustainability coatings are, therefore, a layer that ranges thin and that is put on a variety of surfaces and has a dual function. First of all, they make their underlying material more useful/advantageous and also more durable; we get the idea of a superhero enhancing his armor to make it tougher and sturdier. Indeed, environmental harmony is another area that gets priority from these coatings which might be even more important. Their formulation is more focused on reducing the environmental effects of the typical methods of coating the paints.

Sustainable coatings are relevant because it is possible to achieve both efficient ecology and technology in some of them. They liberalize the organization from conventional approaches which could even compound adverse effects on the environment. Successful integration of the thin film sector is using sustainable coatings that help create a balance between creativity and preservation. These coatings are signals of a fresher world in thin film technological advancement, a world where the natural environment is not compromised for advancement. Science continues to develop such coatings and scientists envision a day when technology enhancement is not just about creating push but also about integrating with the environment. Sustainability coatings rise to the level of being major players as new characters in the thin film novel, signifying both the responsibly engineered and the progressive.

Fig. 11.2 Illustrating the sustainable characteristics of thin films

5 What Are the Characteristics of Sustainable Thin Film Coatings?

It is essential to consider at least three ways that justify setting apart the scientific factors of the foundations of sustainable thin film coatings and their ecological tendencies (shown in Fig. 11.2).

Eco-Friendly Composition: The concept of sustainability coatings is one of the eco-friendly compositions that has come to be associated with it. ReIly, these coatings can be imagined as components in which each of them is an environmentally sensitive area and all of them were chosen in order to have less impact on the ecosystem. It is also crucial to know that the approach was employed in defining material to fix the sustainability specification of the composition.

Low Environmental Impact: As can be seen, almost all types of sustainable thin film coatings have the concept of low environmental impact built in from their design. Regarding the optical coatings, they are per design and there is no inclusion of pollution that may affect the environment and hence classified as wastes. This characteristic is at the stage of manufacturing the clear coatings, during application of the coatings, and it will be so even at the disposal of the clear coatings in that they are environmentally friendly.

Energy-Efficient Application: Maintaining thin film coatings can be likened to an incredibly complex dance where each of the movements pointed at ensuring lower energy utilization. These coatings are designed for effectiveness and are the complete opposite of the coatings that need heat or some type of intense energy in order to be used. Just like switching from a general lamp which always requires more energy, to a laser which will light up all the intended area as required with remarkable precision within the shortest time possible.

6 Real-Life Applications Using Thin Films

Specifically, in the field of thin films, sustainability coatings have not only been merely theoretical concepts but also more practically oriented containing a number of successful types that in several applications altered industries and initiated possibilities for the future. Now I would like to classify the developments of sustainability coatings in different technical fields and illustrate this by several examples.

Fig. 11.3 Showing the biodegradable thin film packaging for daily usage items

1. Biodegradable Food Packaging

Thus, sustainable coatings applied to packaging, have revolutionized biodegradable food packaging which is the world's most important weapon against uncontrolled plastic pollution (shown in Fig. 11.3). These coatings often consist of materials of plant origin, and therefore, the protective layer created around the package is biodegradable and has an exclusively beneficial effect on the environment. The thin film coating enhances the usability of the package by adding the much-needed strength and moisture barrier.

Since thin film technology puts a layer over packaged foods that enhances the shelf life of the foods, the biodegradable food packaging, therefore, operates better. In one aspect this invention provides the food industry with sustainable solutions in an environmentally friendly way by eradicating the conventional package.

2. Thin Films and Intelligent Packaging

These thin films can be incorporated into the sensors to detect different conditions of the packaged product including temperature, humidity, or freshness (shown in Fig. 11.4). For instance, in food packaging, the integration of thin film sensors changes their sensitivity regarding their surroundings and is capable of signaling back the status of the interior packing. Some thin films can have the functionality of actively preserving the quality and freshness of the product that is to be packaged as well. For example, thin films with waterproofing characteristics can prevent the formation of a layer of microorganisms and thus increase the shelf life of food products.

Some of the thin films' responsive properties can change to the external stimulus. For example, in the packages of pharmaceuticals, thin films could alter shades according to the temperature or humidity to sustain the enclosed drugs' stability.

Fig. 11.4 Showing the intelligent packaging solution, that became possible by utilizing thin film technology CC by 3.0

3. Semiconductors

Thin films of semiconductors encompass films of semiconductor material that are deposited or developed on a substrate and may normally vary in thickness between a few nanometers to microns. This procedure utilizes several kinds of depositions such as; electron beam (e-beam) evaporation and chemical vapor deposition (CVD). These thin films, having enhanced conductivity, flexibility, and Controllability of the semiconductor properties, are components of electronic devices as shown in Fig. 11.5.

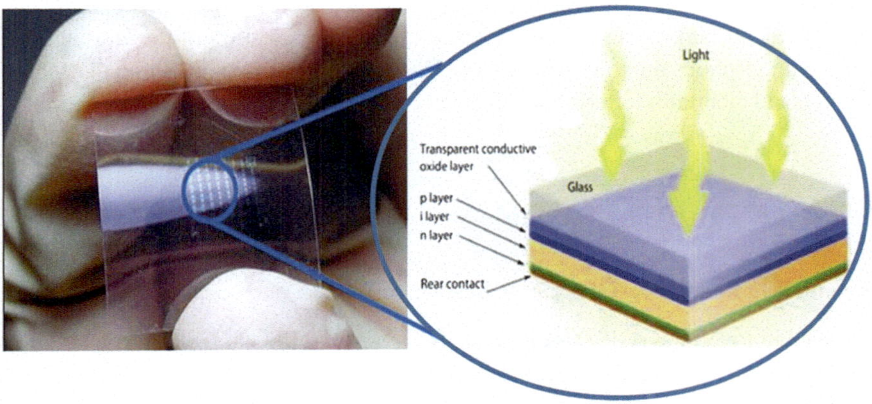

Fig. 11.5 Showing the thin film semiconductors and their specific layers

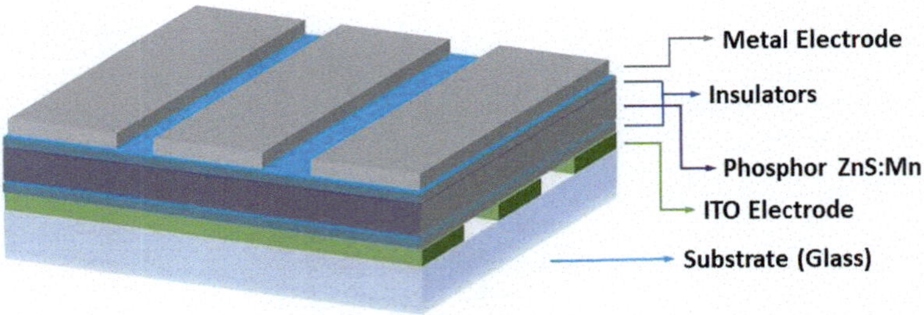

Fig. 11.6 Showing the thin film solar cell (creative common under licence 4.0)

The evolutions of electronic technology are well enhanced by the usage of Semiconductor Thin Films as they provide a platform for the fabrication of small efficient and flexible devices. Small semiconductor structures with improved performance characteristics can now be manufactured because of their thinness sizes and controlled geometries—the possibilities of developing new circuitry, sensors, and electronic products. Semiconductor thin films provide a new technique for preparing semiconductors and improving and miniaturizing the sizes of electronic apparatus.

4. Solar Cells

The thin film technique occurs when one or a few thin layers of photovoltaic material are deposited on a substrate which could be glass, plastic, or metal; thin film solar cells revolutionize the collection of solar energy. These solar cells are generally of a few nanometer nm to a few micron μm in thickness. Thin film solar cells have less thickness therefore different from the crystalline silicon (c-Si) based solar cells which have a thickness between 100 and 200 μm (shown in Fig. 11.6).

Employing thin film solar cell technology is a better and more efficient way of converting power as compared to traditional technologies. This change in the thickness of the material lays the ground to enhance the elasticity besides having the potentiality to blend into other surfaces. The commercial application of these solar cells includes amorphous thin film silicon (a-Si, TF-Si), copper indium gallium diselenide (CIGS), and cadmium telluride (CdTe) that enhances solar energy solutions.

7 Green Thin Film Technologies

Organic Photovoltaics (OPV)

Organic Photovoltaics (OPV) is an emerging green thin film photovoltaic technology that incorporates organic molecules as the working layer of the photo voltaic cell shown in Fig. 11.7. These organic materials can be lightweight and flexible and are potential replacements for traditional rigid inorganic semiconductors.

Anti-reflective coatings that can be made using OPV have low energy consumption and are easier to recycle because the use of toxic materials is eliminated. The flexible

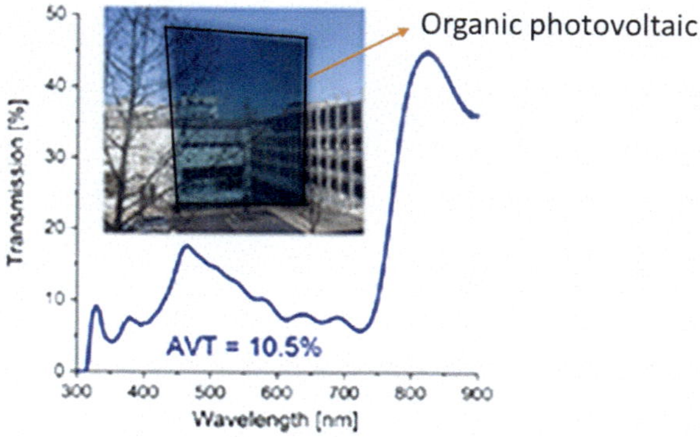

Fig. 11.7 Showing flexible organic photovoltaic with transparency CC by 4.0

nature of the organic materials also increases the versatility of utilized applications, which is why OPV is considered a viable option for obtaining solar energy.

Quantum Dot Thin Films

Quantum dots are semiconductor particles containing on the order of 10 to the power of 3–10 to the 6 atoms that possess size-dependent optical and electronic characteristics shown in Fig. 11.8. These properties are applied in Thin Films of quantum dots in display devices, sensors, and solar cells. They are designed with an efficiency of energy utilization and provide an improved performance in several electronic gadgets.

Thin films of QDTs help to reduce the power consumption in operating electronics hence energy efficiency. Further, their incorporation in the solar cells boosts the ability of light trapping hence increasing the efficiency of the energy conversion from solar

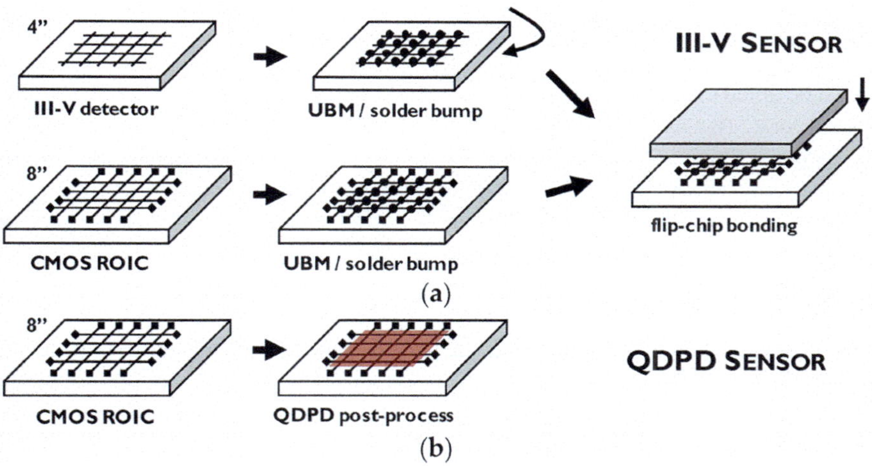

Fig. 11.8 Showing QDPD (thin film quantum dot photodiode) process ***CC by 4.0***

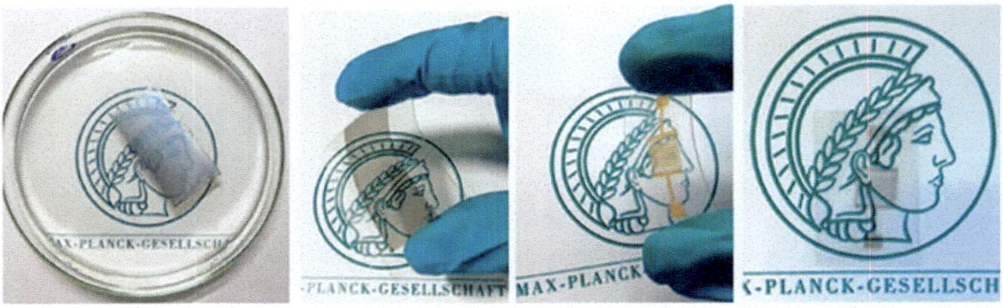

Fig. 11.9 Showing graphene-based thin film supercapacitors ***CC by 4.0***

into electricity. The process of synthesizing quantum dots is well controlled and this aspect reduces the use of resources hence conforming to the green technology.

Graphene-Based Thin Films

Graphene is the basic material of graphene-based thin films (shown in Fig. 11.9); it constitutes a resembling one atom thickness honeycomb lattice. These films are used in applications like flexible electronics; sensors, and energy storage devices owing to their high electrical, as well as mechanical, properties.

The processes for creating thin films of graphene are as a rule ecological, and the great variability of the graphene nature contributes to creating eco-friendly electronic devices. Their light and high-strength features make them appropriate for numerous uses, which in turn lowers the environmental footprint of products with electronics.

8 Environmental Benefits of Sustainable Coatings

Reduced Carbon Footprint

It must be noted that many of the coatings with sustainability as a factor are conceived to incorporate two aspects, namely, environmentally friendly materials and production methods. Such a decision-making process results in improvements in the environmental impact by lowering the level of carbon emissions tied to the production of coatings. It is a greener cycle in manufacturing because it reduces the energy consumed and employs harmless materials.

Minimized Emissions

Some conventional coatings can emit dangerous substances both in the process of their production and during the application. Sustainable coatings focus on using low or even no VOCs during the formulation of coatings. Such a reduction has benefits to the emissions and the environment concerning the application of coatings on the surface under consideration.

Resource Conservation

Eco-friendly coatings often contain materials of renewable resource origin or use recycled material. It reduces the need for scarce resources as well as has a positive impact on the concept of the circular economy. Sustainable coatings provide more resource-efficient developments as the usage of virgin materials declines in the present production.

Extended Lifecycle and Reduced Waste

The fact is that sustainable coatings usually mean greater durability, which in turn makes the coated products have a longer useful life span. This makes the option have a long-lasting solution that does not necessarily require regular recoating or replacement which in turn reduces the consumption and utilization of the material in the long run and the generation of waste products. Sustainable coatings can therefore be said to support a circular economy and or a reduction of waste.

Enhanced Indoor Air Quality

Most conventional coatings emit VOCs which affect the indoor environment in the form of air quality. Sustainable coatings prefer those that have low VOC to enhance the quality of the interior environment. This benefit is especially important in the cases of application in architecture where coated surfaces should not detract from the quality of indoor air.

9 Green Engineering Approaches

Even more crucial is the fact that there are more and more thin film companies that choose sustainable measures—this is their stern pledge to a better future and the protection of the environment. The main endeavors in this regard are the non-toxic content initiative, renewable content initiative, and detailed strategies concerning recycling and circular economy principles.

1. Use of Renewable Resources

Therefore, the use of renewals as sources is impactful in green engineering in the thin film sector. This method entails that some of the components in the processes of creating the thin films for instance bio polymers be of renewable origin. The sector is also capable of decreasing its reliance on the few sources of resources and shifting to renewable resources thus relieving resource sustainability in the longer period and organic issues.

The following are advantages of the integration of renewable resources: Helps to decrease the level of carbon emissions This type of resistance is closely related to the great experience of the industry, where the problem of resource scarcity resistance is felt most keenly. Exploitation of thin films from renewable materials implies that the impact of production is comparatively small; hence fingering the atmosphere in the manufacturing of the products as environmentally friendly.

2. Non-toxic Materials

Typically, some information about general green engineering concepts and their implementation in thin film manufacture reveals that throughout the process, non-toxic materials should be used. This entails the evaluation of the compounds and materials that are incorporated into encapsulation, and deposition among the other processes that are involved in manufacturing. This is also significant considering that substitution also facilitates the exclusion of effects on the environment coupled with the laborer in the production processes since the substation releases chemical products.

The realization of the non-toxic aspect encourages thin film products to be aligned with no toxic content emitting into ecosystems while at the same time, it minimizes risks that industry people meet in their practice. This goal of non-toxic solar films is combinable with general sustainability objectives and allows the thin film segment to present itself as an environmentally friendly protective shield.

3. Recycling and Circular Economy Initiatives

Green engineering concepts are to be adopted by the thin film sector about recycling and circular economy options. This includes research and analysis of innovative recycling methods of post-consumer products, designing with recyclability at heart, and being involved in circular economy business models of recycling, reusing, and recovering materials.

The impact of disposing of thin films on the environment is minimized when recycling and circular economy solutions are used in parallel. Also, to reduce the waste generation and the environmental footprint the industry is keen on developing closed-loop solutions. This strategy focuses on the sustainable development of the economic aspect of the sector together with the environmental challenges.

10 Environmental Regulations and Standards

A. Overview of Existing Environmental Regulations

In the context of the current technologically developing world, environmental laws play a crucial role in pursuance, the thin film sector is no exception, as it offers guidelines for enterprise's accountability and modulation of the adverse effects on the environment and ecosystems. This section provides an overview of the current environmental laws currently implemented in governing the thin film sector toward compliance and sustainable operations.

1. Restrictions on Hazardous Substances, or RoHS Directive

The second law affecting the thin film is the "RoHS" which stands for restriction on the use of certain hazardous substances in electrical and electronic equipment, it is a directive of the European Union. It reduces the risk of some compounds such as lead, mercury, cadmium, and some flame retardants. It was passed to demystify the utilization of hazardous chemicals in electronic and electrical products. Such a directive aims to eliminate the effect of pollutants on the environment and the human body.

Flavors and/or compositions used in thin film products should be scrutinized for endorsement with the RoHS Directive. Despite having to meet the specific requirements for the material used, there are broader goals that need to be met by the industry—reducing pollution and encouraging sustainable practices.

2. Registration, Evaluation, Authorization, and Restriction of Chemicals (REACH) Regulation

Also, there is another significant piece of legislation under the European Union law that exists in the context of chemicals known as the REACH Regulation. REACH is a regulation that aims to provide updated data and measures on chemical registration, examination, and restriction, as well as their authorization to minimize hazards to the environment and people. Its general principles encourage the manufacture of chemicals and the right utilization of those manufactured.

The thin film sector performs the risk evaluations of chemicals under REACH and creates the communication. Apart from ensuring the safe handling of chemicals, this rule boasts of the commitment to these chemistries throughout the life cycle of thin film products within the industry.

3. Waste Electrical and Electronic Equipment (WEEE) Directive

The WEEE Directive gives guidelines on the treatment of end-of-use or waste electrical and electronic equipment due to increasing worry about the menace posed by electronic waste. Similarly, this regulation ensures that manufacturers are held accountable for properly disposing of their products thus promoting the circulation of a circular economy and proper dispositions of waste. So, strong waste management and recent regulations implemented by the WEEEE Directive provided the thin film sector. It is therefore important that the end-of-life thin film goods are collected, treated, and recycled in a responsible manner that will not have any degenerative effects on the environment for the sector to deliver its sustainability and circular economy mandate.

4. ISO 14001: Environmental Management Systems (EMS)

The general framework of EMS development and implementation is supported by the international normative document ISO 14001. The thin film firms that have embraced the ISO 14001 standards vow to institute appropriate environmental management systems that can be used to enhance environmental conservation and or enhancement.

This kind of accreditation gives the thin film industry the face of a responsible and accountable organization. However, this standard has made it possible to enhance environmental sustainability, construct and improve a culture of development, and a commitment to the negative effects that are posed on the environment different from legal compliance.

5. Green Chemistry Principles

The Green Chemistry principles offer structures that guide businesses, including thin film production, to develop goods and processes that do not require dangerous chemicals and whose formation does not pose a danger, despite their absence of legal statutes. All these guidelines promote the use of safer chemicals, reduction of waste, and optimization of the usage of resources and energy. Thus, the use of Green Chemistry principles in the thin film industry encourages an innovative culture. It leads the industry to a post-PC paradigm that embraces increased environmental conservation and the coming up of environmentally friendly products as well as procedures. This methodology does not only deal with these questions of ecological significance but also contributes to the development of long-lasting sustainable solutions.

There are numerous regulations that one has to come across regarding the surroundings of the thin film industry and such have to be acquaintances as well as controlled. The laws that I am referring to are not mere laws that only regulate but laws that help encourage more innovation, create responsible actions, and place the sector on the frontline toward making the technology world a better and moral one.

11 Compliance Challenges and Solutions Faced in the Thin Film Industry

As for the environmental standards, analyzing what has appeared in the tin film industry with the several difficulties mentioned above requires innovations to address them as follows. The following are the threats that characterize the industry that undertakes to meet a level of compliance in a process that entails some measure of strategizing and coordination: This part clears confusion as well as offers operational recommendations to the topicality concerning the thin film industry's compliance questions.

Challenge: Varied International Regulations

Problem description	Solution
The issue of environmental regulation varies across different jurisdiction that deals with the thin film industry	The recommendation that should be made is the creation of specific global regulatory compliance teams within the organizations with a regional focus. These teams follow and analyze shifts in regulations across the global environment

Challenge: Continuous Updates and Amendments

Problem description	Solution
Requirements of the environmental rules and regulations for businesses change from time to time and therefore need to be closely monitored	Put into action effective compliance management systems that are complemented by tracking solutions in real time. The management of change and the implementation of new requirements should be automated to some extent

Challenge: Resource Intensive Compliance

Problem description	Solution
The registration for, and implementation of environmental standards are time-consuming, costly, and require human resources, a challenge to SMEs	Encourage industry associations whereby the SMEs can combine efforts, share common compliance information, and share specialized knowledge

Challenge: Complexity of Chemical Management

Problem description	Solution
Adherence to environmental standards requires a lot of time, money, and people which are constraints to SMEs due to environmental legislation	Adopt a Multi-Chemical Tracking System for timely tracking of chemicals and management documentation. This focalized platform helps in compliance and regulation-related documentation

Challenge: Stringent Waste Management Requirements

Problem description	Solution
In this regard, waste management policies, more so, on end-of-life products, hold a complex planning and execution process	Develop a recycling relationship with recycling centers of the product life cycle or a closed-loop recycling agreement. Engage with the ideas for take-back programs and guarantee the management of poly thin film end of its life cycle

12 Sustainability Approach in the Thin Film Industry

Having compared the market structure of the thin film industry the following organizations reveal themselves as sustainable businesses. The thin film organizations depicted in the two cases show some glimpses of the innovative strategies that the firms are employing to augment the effects on the environment and make the world green.

1. FirstSolar: Pioneering Sustainable Photovoltaics

Intensive work by the company due to several activities it has embarked on in its production of thin film FirstSolar has placed the company among some of the best thin film solar firms across the whole world. The business relates to thin film solar modules developed from CdTe that present stiff competition to crystalline silicon solar cells.

Sustainability Initiatives

It has sustainable technologies operations methods and processes, and the company's logo has imprints of sustainability throughout (shown in Fig. 11.10). Analyzing the CdTe thin film technology as equally as with the silicon-based option, the utilization of energy and other materials is comparatively low. Thus, there is a decreased effect on the environment and its negative carbon influence. The business offers its care also in the life cycle of the products through engagement in the recycling of the products.

2. SolarWindow Technologies: Advancing Transparent Solar Films

Among different transparent solar films that create various levels of interest among the generating stakeholders, SolarWindow Technologies stands out as the company that leads in creating innovative films that can be applied to windows and other types of glasses. Its innovation seeks to capture light without necessarily spoiling the view and light coming into the house.

Fig. 11.10 Showing the first solar advance thin film PV modules ***CC BY 3.0***

Fig. 11.11 Showing the transparent solar films that are to be used on Windows CC by 3.0 ***[by solarwindows website]***

Sustainability Initiatives

In this regard, the translucent solar films that the firm helps buildings embrace assist in making them power efficient by converting light to power. In this manner, Solar-Window (shown in Fig. 11.11) solves the issues of sustainability in the construction sector by applying the aspects of clean power generation in design features. This strategy aligns in conformation with the overall maxims of reducing the carbon intensity as well as utilizing the generation of renewable energy in the urban fabric.

3. NatureFlex™ by Futamura: Sustainable Packaging Solutions

A market leader in cellulose-based films, NatureFlex™ is the company's portfolio of bio-based and fully biodegradable super-thin films for flexible packaging (shown in Fig. 11.12). Films employed in the packaging of these products which are in reality made from renewable material such as wood pulp serve as evidence of the company's green attitude toward the selection of the material used in the packaging processes.

Sustainability Initiatives

Renewable Resources: The films produced with NatureFlexTM are obtained from wood pulp and as a result, the utilization of fossils which are non-renewable natural resources is minimized.

Compostability: These thin films are the epitome of simple and environmentally friendly films that can serve as optimistic mechanisms to promote circular acts since they are capable of degrading under industrial composting.

4. Heliatek: Organic Photovoltaics for Flexible Applications

Fig. 11.12 Showing the transparent thin film compostable packaging CC by 3.0

Commencing from strategy 1 as discussed above, OPV technology user Heliatek develops thin and light organic solar films as seen in the following Fig. 11.13. From the perspective of the structure, it can be downright arrogant to say that their subject elicits flexibility, which makes them ideal for use in abnormal situations, and environments.

Sustainability Initiatives

However, versatile ultrathin solar cells of OPV technology offered by Heliatek's competitors are much more sustainable in production as they use organic materials in the construction; and have a significantly less detrimental effect on the environment

Fig. 11.13 Showing the flexible PV thin film CC by 3.0

as compared to other well-known solar technologies. Speaking of their solar films, clients can be potentially installed on different kinds of surfaces or construction that are curved or even out-of-the-ordinary. They increase the given probability of producing renewable energy in the open-ended context.

13 Future Trends and Innovations

As for different sectors, Thin film technology was applied, and depending on the new technologies created it is expected to increase even more because of greater sustainability, efficacy, and performance. The following emerging technologies haven't clearly defined their impacts on the trajectory of thin film applications, as they are still under development.

Quantum Dot Thin Films

In thin film technology, the products based on novel material-Q dots i.e., nanoscale semiconductor particles are becoming popular day by day. Thin films from quantum dots exhibit certain properties, namely, optical and electrical properties, which thus enable very fine control of lighting and the absorption of light. These films are employed in Solar cells, Sensors, and Displays.

Products normally associated with a revolutionary capability include the quantum dot thin films and bright quantum dot panels with enhanced energy efficiency to fit this slot. They improve the light trapping in the solar cell hence its efficiency is enhanced. The requirement for high-performance economical green solutions matches with the quantum dot principle generally used in the business.

2D Materials and Graphene-Based Thin Films

Thin Films Based on 2D Materials and Graphene.

The general field of thin film is a relatively new branch that includes the technological study of thin layers of 2D material like graphene. Graphene-based thin films are used incoherently with high mechanical properties and great electrical conductivity. This makes them suitable for applications including energy storage and flexible electronics.

The thin film particularly graphene will be able to bring out the flexible electronics using 2D thin films to get lighter bendable electronics. Their life span and efficiency can therefore be improved when integrated into energy storage technologies making energy solutions cheap and sustainable.

Organic Photovoltaic (OPV) Innovations

This leads to the general objective of OPV to enhance the material stringer as well as its functionality. The new developments in the OPV increasing the efficiency of the power conversion and the operational life of OPV devices are the new organic materials in the development of tandem structures, and conductive coats.

Opportunities in the given field demonstrated the fact that the advances are beneficial for the development of applications for solar energy harvesting. Higher stability and efficiency make OPV applicable to be incorporated into wearable electronics and building skins among others for conservation of the energy cause.

14 Developments in Sustainable Thin Film Coatings

Sustainability as one of the important parameters of thin film coatings is one of the important criteria for its development. Future research on thin film coatings with improved environmental PSA performance will provide functionality while working considerately. Finally, the prospects of the work carried out in this field are studied in this section.

Biodegradable Thin Film Packaging

There is growing research on the development of biodegradable thin film packaging materials. These coatings are designed to biodegrade after specific years of application and eliminate the adverse impacts of plastics on the environment and certain challenges with plastic packaging.

Biodegradable thin film packaging provides a solution to the efforts of lesser use of plastic that is globally being employed. The packaging industry's anticipated wide application of these coatings holds the promise of reducing environmental impact very significantly and promoting a circular economy.

Self-Healing Thin Film Coatings

Self-healing thin film coatings have functionalities such as novel materials that can recover from the inflicted damage. Products coated with these polymeric materials have longer durability and need less maintenance because such coatings' surface is abrasion and scratch-resistant.

Humans believe that there will be changes in the service life of products, and recoating or replacing the thin films will no longer be frequently needed. This invention helps individuals engage in practices that will enable the conservation of scarce resources.

Nanotechnology-Enhanced Thin Films

Nanotechnology as a field of technology has the potential to enhance thin film barrier characteristics, electrical conductivity, and outright optical properties. These developments may lead to thin film applications of diverse types in various industries which will be sustainable and efficient.

Based on the forecast of favorable characteristics when merged with thin film coatings, nanotechnology is expected to improve the performance across various applications, including biomedical and electronics. They assist in Resource utilization thereby helping in the development of green technology.

Smart and Responsive Coatings

Smart coatings or coatings that can react to evoked stimuli or changes in the environment are coatings designed this way. Such coatings exhibit dynamic functioning and probable energy conservation as it is likely to change their characteristics concerning temperature, light, or humidity.

The integration of the smart and sensing coating layers means that a more flexible and energy-efficient future will be implemented within the thin film plans. As such, these coatings can be used to enhance energy applications and embrace the use of eco-friendly activities in different fields.

15 Conclusion

The chapter titled Environmental Considerations and Sustainability points to the significance of understanding the environmental implications of thin films and promoting the development of novel eco-friendly coatings as well as establishing the principles of green integrated solutions to coatings. With the increased rate of the use of high-performance materials in industries, there should be an analysis of the life cycle impacts of these materials.

Thin films present a myriad of advantages but are causing havoc with the environment when not well dealt with. The positive development and establishment of practical coat technologies are crucial to the reduction of these effects. Initiatives like using recycled or biodegradable materials, using energy-efficient processes in manufacturing, recycling the material used in coating, etc. are of prime importance in decreasing the burden on earth.

Today various green engineering practices have further increased the sustainability of thin film technologies. Due to the focus of green engineering on resource efficiency, reduction of waste, and utilization of renewable resources, it can serve as a prototype for creating non-harmful coating solutions. The use of such practices ensures that adverse impacts on the environment are kept at a minimal level and is also adjudged to or close to sustainable standards set by the international community as well as meeting legal requirements.

Conclusively, the invoking of environmentalism in the creation and use of thin film and coat applications is essential for sustainable development. They shrugged off sustainable technologies as well as green engineering to give good results of technology advancement with special recognition to material science without damaging the health of the planet.

References

1. Dobrucka, R., Cierpiszewski, R.: Active and intelligent packaging food-research and development-a review. Polish J. Food Nutr. Sci. **64**(1), 7–15 (2014). https://doi.org/10.2478/v10222-012-0091-3
2. Fortunato, E., Barquinha, P., Martins, R.: Oxide semiconductor thin-film transistors: a review of recent advances. In: Advanced Materials, vol. 24, no. 22, pp. 2945–2986 (2012). https://doi.org/10.1002/adma.201103228
3. Kalpana, S., Priyadarshini, S.R., Maria Leena, M., Moses, J.A., Anandharamakrishnan, C.: Intelligent packaging: trends and applications in food systems. In: Trends in Food Science and Technology, vol. 93, pp. 145–157. Elsevier Ltd. (2019). https://doi.org/10.1016/j.tifs.2019.09.008
4. Kamalisarvestani, M., Saidur, R., Mekhilef, S., Javadi, F.S.: Performance, materials and coating technologies of thermochromic thin films on smart windows. In: Renewable and Sustainable Energy Reviews, vol. 26, pp. 353–364. Elsevier Ltd. (2013). https://doi.org/10.1016/j.rser.2013.05.038
5. Mersagh Dezfuli, S., Sabzi, M.: Deposition of self-healing thin films by the sol–gel method: a review of layer-deposition mechanisms and activation of self-healing mechanisms. Appl. Phys. A: Mater. Sci. Process. **125**(8) (2019). https://doi.org/10.1007/s00339-019-2854-8
6. Pedersen, H., Barry, S.T., Sundqvist, J.: Green CVD—toward a sustainable philosophy for thin film deposition by chemical vapor deposition. J. Vacuum Sci. Technol. A: Vacuum, Surf., Films **39**(5) (2021). https://doi.org/10.1116/6.0001125
7. Sheng, J., Jeong, H.J., Han, K.L., Hong, T.H., Park, J.S.: Review of recent advances in flexible oxide semiconductor thin-film transistors. J. Inf. Display **18**(4), 159–172 (2017). https://doi.org/10.1080/15980316.2017.1385544
8. Tyagi, P., Salem, K.S., Hubbe, M.A., Pal, L.: Advances in barrier coatings and film technologies for achieving sustainable packaging of food products—a review. In: Trends in Food Science and Technology, vol. 115, pp. 461–485. Elsevier Ltd. (2021). https://doi.org/10.1016/j.tifs.2021.06.036

Image References

https://www.linkedin.com/pulse/difference-between-capacitive-touch-screen-resistive-chi-ho/
https://www.solarreviews.com/blog/organic-solar-cells
https://www.mdpi.com/1424-8220/17/12/2867
https://research.qut.edu.au/surface/research/graphene-based-thin-film-supercapacitors/
https://www.firstsolar.com/en-Emea/
https://www.solarwindow.com/
https://www.natureflex.com/uk/
https://www.heliatek.com/de/

Case Studies and Practical Examples

Abdulmoseen Segun Giwa and Emmanuel Alepu Odey

Abstract This chapter explores detailed consideration and examples of thin films and coatings based on case studies and examples of their usage in engineering fields. This way, the chapter aims to illustrate the capacity and potential application of advanced materials when dealing with sophisticated engineering problems. The case studies range from using nanoparticles to boost the performance of an electronic device to using nano-enhanced concrete, designing better building materials, and increasing the effectiveness of working renewable energy systems among others. Therefore, this chapter builds up the arguments about how thin film and coating technologies are being enforced to set newer fronts in performance, sustainability, and cost-effectiveness in diverse sectors.

1 Case Study # 1

Case Study: Pratt and Whitney—Enhancing Jet Engine Performance with DLC Coating [1].

Overview

The well-known aerospace firm Pratt and Whitney was struggling with several issues typical for the industry—minimizing the level of friction and wear of the risky aircraft components like bearings and jet engine lubrication. To tackle this problem, the firm resorted to thin film coatings known as Diamond-Like Carbon or DLC which is appreciated for its high hardness, low friction rates, and resistance to wear.

A. S. Giwa (✉)
School of Civil and Environmental Engineering, Nanchang Institute of Science and Technology, Nanchang 330000, China
e-mail: giwaabdulmoseensegun@ncpu.edu.cn

E. A. Odey
Ireland Knowledge Centre for Carbon, Climate and Community Action (IKC3), Munster Technological University, Tralee, Ireland
e-mail: emmanuel.odey@mtu.ie

A. D. Chandio and I. A. Channa (eds.), *Thin Films and Coatings*, Engineering Materials,
https://doi.org/10.1007/978-981-96-0005-2_12

Objective

The overall purpose of this process was to introduce DLC coating to the jet engine bearings in order to reduce the wearing out of the bearings, increase the rate of friction as well as improve the performance of the engines.

Implementation

This layer however was still not as good as Pratt & Whitney required hence a teaming up with other specialists in thin film coatings in order to design their own DLC coating appropriate for the jet engine bearings. The coating process may include applying a thin layer of DLC onto the surface of the bearings by a method including but not limited to Physical Vapor Deposition (PVD) or Chemical Vapor Deposition (CVD).

Results

As for the DLC coatings, the experimental results showed that DLC coatings can produce the expected results efficiently. High levels of friction cause high levels of wear and tear, low energy efficiency, and increased fuel consumption: these issues were improved through the usage of coated bearings. Furthermore, the wear levels of the coatings were increased and the bearing's maintenance and replacement period were seen to have improved.

Benefits

- Extended Bearing Life: This was quite a considerable achievement by DLC coatings, as it dramatically improved the durability of the jet engine bearings, which had to be replaced much less often and, therefore, the extent of downtime.
- Reduced Friction: Due to the ability of DLC coatings to possess low friction elements, there were benefits such as better engine competence, less fuel use, and reduced expense of operation.
- Cost Savings: The overall benefits obtained as a result of applying the proactive approach consisted in the realization of long-term cost savings thanks to the decreased expenditures on the maintenance of the components and their replacement.
- Enhanced Performance: These measures improved not only the general performance of the jet engines but also increased their reliability, as well as made them much more operational.

Conclusion

Indeed, the application of thin film coatings, particularly DLC on Pratt & Whitney shows how those coatings can enhance the performance of aerospace equipment. As has been seen by Pratt & Whitney which coated its jet engine bearings with DLC with the purpose of increasing bearing life, reducing friction, and hence improving the performance of the engine with a view of achieving a balance between the cost of manufacturing as well as the cost of upheld maintenance.

Reference

Pratt and Whitney Website (https://www.prattwhitney.com

2 Case Study # 2

Case Study: Advancements in Indoor Photovoltaic Technologies [2].

Introduction

Indoor photovoltaics (PV) is a newer approach that has been proposed to deliver electricity from indoor lights to low-power devices. A recent advance of the third generation of thin film solar cells including dye-sensitized solar cells (DSSCs), perovskite solar cells (PSCs), and organic solar cells (OSCs) played an important role in the development of indoor PV technologies. This paper presents a broad understanding of essential developments in indoor PV devices and systems made from the third generation of solar cells including the substrates, cell performance, and principles for encapsulation of PV devices and systems for indoor applications.

Materials Design and Performance

The materials used in thin film solar cells at the third generation are developed in such a way that there is a tremendous impact on photovoltaic activity. With DSSC, PSC, and OSC, great improvement has been made in the efficiency and stability levels. Other attributes like high light absorption coefficients and tailored bandgap values for these materials make them ideal for use in indoor PV systems.

Visual Characteristics and Other Important Principles of Indoor PV

For an efficient and long-lasting PV system, the design of indoor PV systems becomes a complex proposition. The design criterion for new-generation interior thin film PV systems is predicated on the inherent attributes of interior light and the distinctive features of the third-generation photovoltaic cells. This involves the design, light management, and arrangement of the cell, carrier, and transportation system in an attempt to achieve higher efficiency mainly under indoor lights.

Current Research Progress

Specifically indoor thin film PV systems have been said to undergo high efficiencies in laboratories based all over the world. However, the issue of establishing sustainable stability has remained a domestic one up to this present time altogether. Ongoing studies are aimed at furthering understanding and improving the performance of indoor PV systems that can be made available in the market today.

Perspective

It can also be seen that forms of indoor PV generated from third-generation photovoltaic cells are promising for the future. More research is required to tackle the efficiency, stability, and scalability problems plaguing socialbots. To enhance and optimize the efficiency of indoor PV devices, as well as ensure they are appropriately implemented to their full capacity with regard to powering the numerous low-power devices indoors, attribute cooperation among researchers, industry players, and policymakers is crucial.

Conclusion

In sum, indoor PV technologies have developed rapidly with the help of the third generation of solar cells. Thus, the efficiency of indoor PV systems has been achieved through advanced MDA that enhances both the performance of the materials and the optimization of photovoltaic efficiency and the prevalent concepts of design. More studies are therefore needed to address other existing challenges in order to unravel the true potential of indoor PV technology for broader sustainable energy use.

3 Case Study # 3

Case Study: Advancements in Thin Film Transistor (TFT) Technologies [3].

Introduction

Another special device that has played a great role in the development of the digital world, influencing present-day culture, is the field effect transistor (FET). It is TFT that played the most essential part among all the other FET types that facilitate the modern flat-panel displays. TFTs are used devices with different electrical and mechanical parameters produced by a few types of materials for fabrication. In this case study the relevance of a number of parameters for bodies for the objective of appropriately coordinating TFT technologies to new applications prone to replace flexible, stretchable and transparent large-area microelectronic devices.

Metrics for TFT Technologies

1. Channel Miniaturization and large-Scale Integration: Channel Miniaturization and Large-Scale Integration

Therefore, the downscaling of the channel and integration of TFTs into larger circuits are two significant elements that determine the improvement of device functionality as TFTs are applied to more sophisticated usages.

2. Carrier Mobility

Alternative objectives, among which the foremost is the achievement of high carrier mobility, directly influence the velocity of devices based on TFT technology. But they are not without their drawbacks and it is here that other measures come into play in scenarios that are actual.

3. Channel Length, Contact Resistance, and Threshold Voltage Fluctuations

In systems like display, such parameters are as valuable if not more than factors like carrier mobility. Adjusting these parameters provides for stable, accurate, and reliable operation of the device.

4. Operating Instability

For OTFTs more specifically, operating instability is an issue that has been identified and that requires sort of attention. Stability is a critical characteristic for a long-term business, which would define its commercial sustainability in the long run.

5. Large RC and Manufacturing Uniformity

These values are not stated frequently but boost consistency in large-area microelectronics performance.

Moving Performance Metrics into the Right Perspective

Although it is clear that high carrier mobility is one of the important issues highlighted in the TFT description, it is also crucial to consider other parameters in the framework of the actual use of the technology. Techniques like operating instability, large RC, and manufacturing uniformity are some parameters that hold major significance in the commercial potential of the TFT technologies. For emerging TFT technologies to successfully transition into commercial applications, it is crucial to contextualize performance metrics. Understanding and evaluating these metrics within the broader scope of practical use and market demands is essential for the technology's commercial viability.

4 Case Study # 4

Case Study: Advancements in Flexible Organic Electronic Devices (FEDs) [4].

Introduction

FEDs represent multipurpose, organic electronic devices that have attracted interest across the globe thanks to their prospective applicability to various facets of life. The present paper aims at showcasing the methods of manufacturing FEDs, the materials that can be used in the process, developments in the field, and uses for these devices.

Materials Advancements

Flexible Polymeric Substrate: This implies that devices based on FPs can be lightweight and flexible because the flexible polymeric substrates allow this. Slight enhancements in this area have resulted in the availability of versatile and more robust substrates.

Hybrid Barrier Layers: The barrier layers are essential while encapsulating the FHEDs and preventing them from any damaging influence by the surroundings. Organic/inorganic barrier layers are a relatively new development that provides better barrier properties and thermal stability.

Inorganic and Organic Materials: New breakthroughs have been achieved in designing new classes of inorganic and organic materials that boast semiconducting, conducting, or insulating characteristics, as well as enhanced stability. These materials are vital in the production of efficient FEDs due to their desirable qualities such as high transmittance.

1. Nanomaterials: The use of nanomaterials is very important in FED devices due to their contributions to improved performance. They allow for enhancing the efficiency and functioning of devices used in other scientific fields.
2. Production Processes Roll-to-Roll Production: Technological advancements such as easily scalable and inexpensive large-area roll-to-roll fabrication techniques are facilitating the practical realization of FEDs in applications of the market. This approach makes it possible to produce future RO-based large-area and large-volume systems such as displays, lighting, photovoltaics, and sensors.
3. Real-time Optical Monitoring: Receiving real-time optical information and extending the capability of film thickness characterization of growing films on flexible polymeric substrates, SE methods include spectroscopic ellipsometry. It gives increased control over the fabrication process with the result of enhancing the functions of the device.

Conclusion

The flexible forms of organic electronics have been found to have a lot of potential for development in the areas of application such as display, light emitting, photovoltaic, and sensing. The future development of FEDs is expected to be driven by new material technologies, more advanced manufacturing techniques, and novel characterization methods, all of which are rapidly coming of age in the high-technology sector. There is a need for more work and innovation with human satisfaction as the goal in this area for more improved science and technology achievements.

5 Case Study # 5

Case Study: Apple Inc. - Enhancing Electronic Product Durability with Thin Film Coatings [5].

Overview

Apple Inc., one of the most recognized technological companies in the world, is famous for its dissertation on the aspects of product design and manufacture. In this regard, as part of Apple's drive to pursue sustainability and improve its products, the surfaces of many Apple products, such as the iPhone and iPad, are coated with thin film coatings, contributing to improved sturdiness and appearance.

Objective

As applied to Apple's electronics, the primary purpose of thin film coatings address the issue of durability and effectiveness in shielding against scuffs and fingerprints.

Implementation

Apple collaborates with its suppliers and partners to create and apply thin layers of metal oxides to the external part of the electronics. These coatings require careful application by using techniques that will result in uniform coverage on the iPhone manufactured to meet Apple's stringent quality standards.

Results

Apple's use of thin film coatings has been found effective in the full accomplishment of its goals. The coatings enable the devices to remain well protected from scratching, scuffing, and even fingerprints which are likely to spoil the outlooks of the devices. Also, these coatings are transparent which makes it ideal since the screens of the devices are not clouded making their display bright and clear.

Benefits

1. Enhanced Durability: This is because the thin layers that are applied on the electronics prevent them from being damaged by daily use, and therefore make them to remain looking as new even after a relatively short time.
2. Improved Aesthetics: Being protective layers over the devices, the coatings assist to prevent the devices from being marred by scuffs and fingerprints, which compromises their stylish look.
3. Customer Satisfaction: One of the major advantages of using thin film coatings is improving general usability for Apple's products since they continue to look and feel like new.

Conclusion

The use of thin film coatings in Apple Inc.'s products case clearly illustrates that thin film coatings play a vital role in boosting the resilience and the look of electronic gadgets. The following are some of the benefits that Apple has been able to offer its customers through using these coatings on its iPhone and iPad devices; The devices from Apple's iPhone to iPad are long lasting and are easy on the eye.

6 Case Study # 6

Case Study: Medtronic—Enhancing Biocompatibility of Implantable Devices with Thin Film Coatings [6].

Overview

The case you will be working with focuses on Medtronic, a premier medical device company that is dedicated to using its technologies to enhance patient care. In exercising this policy, Medtronic employs thin film coatings on its implantable cardioversions that include defibrillators and pacemakers to improve their biocompatibility and safety to patients.

Objective

The main purpose of applying thin film coatings on implantable devices manufactured by Medtronic is to minimize complications that are associated with rejection and infection of these devices which will ultimately improve the safety of the patient and the functionality of the device.

Implementation

As an industry leader in implantable equipment manufacturing, Medtronic participates in shared thin film coatings with other professionals who specialize in thin film coatings. These layers can be sprayed on or sputter coated onto the surface of the devices in such a manner as to provide homogeneity and biocompatibility.

Results

Medtronic has achieved its goals with the help of thin film coatings as it is described below. These coatings have also been proven to lower the possibility of the development of complications in machines that are implanted into the body thus enhancing the performance of the machines thereby making the patient safer.

Benefits

1. Enhanced Biocompatibility: These thin film coatings benefit the implantable devices from Medtronic in that they lessen the chances of being rejected by the body or lead to infections.
2. Improved Patient Safety: Due to a decrease in the likelihood of occurrences, the coatings improve the safety of treatment output and minimize the demand for supplementary procedures.
3. Enhanced Device Performance: These coatings play a significant role in the performance and durability of Medtronic implantable medical devices and their ability to meet the intended performance of the product continuously.

Conclusion

Looking at the specific actions that have been taken in this business, the role of thin film coatings in increasing the biocompatibility and safety of implantable medical devices is apparent in the case of Medtronic. With these coatings for its defibrillators and pacemakers, Medtronic has been able to enhance its clients' experiences while at the same time improving patient care and thus guaranteeing the durability of its equipment.

7 Case Study # 7

Case Study: Ford Motor Company—Improving Automotive Performance with Thin Film Coatings [7].

Overview

Under automotive engineering, the durability of the automotive part and its whole performance are key factors in the general driving experience and costs of operations. The reduction of friction, improvement of fuel outcomes, as well as the increase of the useful life of some of the parts have turned thin film coatings to be useful solutions for the following aims.

Objective

A popular application of thin film coatings came into light when the Ford Motor Company, a renowned automotive manufacturing company, planned to improve the lifespan and efficiency of its new-generation Turbocharged engines by applying thin film coatings directly on the piston. The initial purpose was to minimize the contact with the surfaces and thus minimize energy loss to achieve better efficiency of the combustion and longer lifespan of the engine.

Implementation

Due to this, specialists in thin film coatings partnered with Ford in order to create a more fitting overlay for its Turbocharged engines. The chosen coating was then deposited on the pistons' surfaces using such techniques as PE-CVD and MPCVD, enhancing the conformality and coating adhesion to Ford's quality standards.

Results

Thin film coatings were applied to the pistons of Ford's Turbocharged engines and there were great benefits that resulted from the process. It took me a while to assess the coatings but they did reduce friction and restraining wear hence increasing fuel efficiency and the life expectancy of the engine. Furthermore, the coated pistons offered improvements in terms of strength that would allow the product to last longer before it would need to be replaced, and frequently, less maintenance work was necessary.

Benefits

1. Improved Fuel Economy: Through the easing of contact between parts which have the thin film coatings on the pistons, the fuel consumption was also minimized hence cutting down the expenses on the fuel by the owners of vehicles and also helping to minimize environmental degradations.
2. Extended Engine Life: These coatings were used to prevent wear and corrosion in the pistons which in turn extended the life of the Turbocharged engines and increased their durability.

3. Enhanced Performance: This way, the decrease in friction and prolonged durability as a result of the coated piston made it easier for clients to drive with improved Engine performance.

Conclusion

The use of thin film coatings can be illustrated by the example of Ford Motor Company and its positive impact on the vehicles' efficiency and durability. This was evident in its application of thin film coatings on the pistons in its Turbocharged engines, which was instrumental in the realization of enhanced aspects such as improved fuel economy, durability, and enhanced performance which are core reasons why thin film coatings still remain relevant in the automotive industry to date.

8 Case Study # 8

Case Study: Nikon Corporation—Enhancing Optical Clarity with Thin Film Coatings [8].

Overview

In the optical industry, vision plays a vital role in cameras, microscopes, and even telescopes to name a few in ensuring they provide the best possible image quality. Thin film coatings are also used to prevent most of the unwanted reflections to occur and thus ensure that the quality of the images produced by these devices is impressive.

Objective

A Japanese company that has specialized in the production of cameras and microscopes namely the Nikon Corporation intended to enhance the existing images of the cameras and microscope through applying thin film to the lenses. The main goal was maximum reflection and transmission so that the lens could produce sharper and less blurred images.

Implementation

To achieve the above functions, Nikon worked with specialists in thin film coatings to create dedicated layers on the lenses. These coatings were applied using exacting deposition techniques in order to facilitate complete and adequate coverage and to meet and/or surpass Nikon's quality specifications.

Results

Decades of experience enabled Nikon to apply thin film coatings to its lenses and significantly enhance optical margins and picture quality. There were also minor reflections cut by the coatings thereby enabling the revealed images to be even clearer than before with superior contrasts. Further, it improved the durability of lenses and made them longer lasting besides improving the reliability aspect of the lenses.

Benefits

1. Improved Clarity: To minimize these reflections, Nikon utilized thin film optical coatings, enabling better light transmission and greater clarity in images by equipment such as Nikon's camera and microscopes.
2. Enhanced Image Quality: The coatings provided some benefits to the quality and sharpness of images from traditional LCD displays: it eliminated glare and enhanced contrast so that users could end up seeing more detailed pictures.
3. Increased Durability: As for the coatings that have been used on the lenses, they shielded them from scratches and other forms of degradation, thus enabling it to maintain optimal performance in the long run.

Conclusion

This study on Nikon Corporation explains the extent to which thin films are valuable within the applications of optics. Using thin film coatings for its cameras and microscopes, Nikon enhanced the ability of lenses to produce a brighter image as a result of cutting down off the light that did not converge on the focal point by magnifying the image quality, this made thin film coatings a central component in any modern optical technology that desired high quality and endurance.

9 Case Study # 9

Case Study: Anti-Reflective Coatings for Crystal Clear Vision [9].

Overview

Thin layers referred to as anti-reflective (AR) coatings are applied on the screen of smartphones, the lens covers for cameras, and eyewear among others to minimize glare and light reflection to increase the clarity of images. Usually, silicon oxide or magnesium fluoride is utilized in the creation of these coatings.

Objective

The application of the AR coatings' core goal is optimizing the visual pleasure for the consumers as more images become clearer and less glare appears particularly under the bright or reflecting circumstances.

Implementation

Deposition techniques are employed to deposit AR coatings on the lenses' or screens' surfaces using either PVD or CVD techniques. These methods ensure that the coatings are uniform, thin, and adhere to the substrate of the article to be coated.

Results

The applications of anti-reflective layers in the lenses have enhanced the quality of images and the comfort of users tremendously. Apart from enhancing the aesthetics, the AR coatings enable the consumers to cut on glare and light reflections hence they have a better view of what is around them.

Benefits

Improved Image Clarity

Reducing the level of light reflection that is returned by AR coatings enhances image quality and offers consumers more detailed images.

Decreased Glare

AR coatings eliminate any glare, thus improving vision in the areas that may be too bright or where there are reflections such as outside, or under artificial lighting.

Improved Visual Comfort: Due to the specific anti-glare and anti-reflective properties of the coatings, visual comfort is improved; eye fatigue and stress are minimized.

Conclusion

The usage of thin films in day-by-day life can be described through antireflection coatings Arab is the best example of this. The manufacturers enhance consumers' visual outcomes through applying these coatings on the smartphone screens, camera lenses, and eyewear; illustrating the beneficial applications of thin films in enhancing run-of-the-mill products.

10 Case Study # 10

Case Study: Touchscreens - A Symphony of Thin Films [10].

Overview

Thus, high-precision touch sensing used in modern touch screens common in tablets, smartphones, and other gadgets necessitate multi-layer thin films. ITO (indium tin oxide) is a necessity in the construction of the screen since it is both a conductor and transparency agent. Subsequently, dielectric films are employed to insulate the ITO layer from the circuitry beneath it.

Objective

These thin films' chief purpose is to ensure that electronic gadgets are sensitive to touch in an accurate manner so as to afford people accessibility for easy and natural operations.

Implementation

ITO and dielectric layers are some of the thin film coatings placed on glass or plastic bases via chemical vapor deposition or sputtering. Procedures must produce films that are uniformly thin and have low surface energy; the above methods ensure this for touchscreens' operation through stickiness.

Results

In this case, thin films used in touchscreens are some of the innovations that have changed the manner in which people engage with electronic gadgets. Thus, the use of such films lets users input into apps, type in messages, and manage interfaces with no mistouch thanks to accurate touch identification.

Benefits

- Improved User Experience

Thin films in touchscreens add to the general usability of electronic devices since they provide users with a slippery and easy manner of interacting with the devices.

- Enhanced Responsiveness

They feel smoother and more responsive to inputs due to improved detection by thin films that are incorporated in touch screen devices.

- Durability

Thin films also help protect the life span of screens by increasing their resistance to scratches and other sorts of wear.

Conclusion

Thus, thin films prove to be of great significance to modern technologies and their flagship products like touchscreens. Thin films have changed the way we interface with electronic devices; besides, thin films can detect touch well.

11 Case Study # 11

Case Study: Solar Cells—Capturing Sunlight with Thin Films [11].

Overview

Well, light is used to give power to solar cells especially thin film solar cells to produce electricity. These are relatively thin layers of the substance, which may be silicon, cadmium telluride, or CIGS—copper indium gallium selenide—depending on the efficiency of the cell in receiving solar light and energizing electricity.

Objective

Thin film solar cells are another solution to use to capture and convert the sun's rays effectively and in a way that does not diminish or run out; the cells are meant to offer an effective level of renewable power in many ways.

Implementation

Thin film solar cells are made through layering on a substrate of the thin layer of semiconductor material. These thin films are developed by chemical vapor deposition (CVD) and physical vapor deposition (PVD) which makes the unitary aspect of the thin films and the photovoltaic element transform photon light energy to electricity.

Results

Thin film solar cells have been central in the commercialization process of solar-generated electricity which is well known as a renewable energy source. It is versatile, economic, and efficient which is why these cells can be used in different non-luxury, compact buildings beginning from the living units, and offices, and culminating with industrial constructions.

Benefits

- Efficiency

The thin film solar cell can convert a reasonably good quantity of watts of light into electricity due to various levels.

- Cost-Effectiveness

As thin film-based solar cells are concerned, they are cheaper than normal silicon-based solar cells in terms of conversion of solar energy into electronic energy resulting from low material fabrication cost.

- Versatility

That is because thin film solar cells are flexible and can not only be embedded in different forms but also in contact with different surfaces and structures.

Conclusion

The thin film solar cell instance describes how thin film solar cells could be used in solar energy collecting as a way of improving the present means of power generation. Photovoltaic thin film solar cells have therefore incorporated their solar conversion and thus play a progressively rising position in the renewable energy sector.

12 Case Study # 12

Case Study: LED Revolution—Thin Films Illuminate the World [9].

Overview

Deposition of thin films is an essential part of LED functioning; these have revolutionized the lighting industry. LEDs emit light by having thin layers of semiconductors like indium gallium nitride or gallium nitride in thin sheets.

Objective

The use of thin films in LEDs aims at developing lightings that are energy efficient with a long life span and have a lesser degree of drawback on the environment compared to the ordinary types of lightings.

Implementation

III-nitride thin films such as gallium nitride or indium gallium nitride thin films are grown on a substrate by molecular beam epitaxy (MBE) or metalorganic chemical vapor deposition (MOCVD) method. The LED structures that emit light are then defined and developed from thin films.

Results

The use of thin films has helped in the advancement of technology for lighting, particularly in the manufacture of LEDs. LED lights are applied at the present moment in numerous settings, as these are long-lasting and energy-saving light sources that can be used in various spheres of human life, including offices, houses, stadiums, streets, etc.

Benefits

– Energy Efficiency

Thin film LEDs are far more energy efficient compared to other forms of lighting technologies and therefore they use less power and entail lower energy costs.

– Extended Lifespan

Another advantage of LED lighting is that it is durable compared to other conventional lighting systems; hence it incurs less cost in terms of maintenance and frequent replacements.

– Environmental Impact

Thin film LEDs are cooler when compared with traditional lighting sources and have no mercury in them as well, therefore they are more environment-friendly.

Conclusion

This paper shows how the thin films used in the formation of LEDs have affected the sector, through the use of LED thin films. Thin films have contributed to the production of long-lasting and energy-conservation technologies, leading to the existence of LED and much better lighting for everyone.

13 Case Study # 13

Case Study: Computer Chips—The Brains of Electronics, Powered by Thin Films [12].

Overview

An integrated circuit commonly known as a chip is a critical component of modern electronics. The complex working of these chips and their size are attributed to the several thin films that these chips are built on. These layers serve different purposes and classify them as semiconductors, insulators, or conductors.

Objective

Hence, the primary use of thin films in computer chips is to enable the efficient creation of portable electronics with new features in computers, mobile phones, and any other gadgets.

Implementation

In chemical vapor deposition or physical vapor deposition processes, thin layers required to make computer chips are deposited onto a substrate like silicon. The thin films are then patterned and processed in order to form the interconnects and devices that result in the system structures of integrated circuits (ICs).

Results

Due to the use of thin films in computer chips, compact, quicker, and more potent electronic devices are manufactured. These chips enable the expansion of technology in many sectors since they facilitate the processing of complex data and the performing of numerous calculations.

Benefits

- Miniaturization

Thin film technology makes it possible to have improved functions and miniaturization of gadgets in terms of size.

- Improved Performance

Thin films in CPU and other chips contribute by ensuring that electrical equipment offers desired performance standards by operating more efficiently and with enhanced processing rates.

- Complex Functionality

Thin film technology brings about the aspect of designing complex circuits and structures in computer chips which enhances the functions and power of the devices.

Conclusion

Computing chips, in fact, provide an instance of thin films which Anon underlined how such material are vital for introducing new technologies. Individual thin films have brought a drastic change in the electronics industry due to the complex functioning and minuteness of electronic devices and therefore the creation of more challenging gadgets.

14 Case Study # 14

Case Study: Hard Drives—Thin Films for Data Storage [10].

Overview

Hard disk drives (HDDs) are distinguishable computer and digital devices to deliver and sustain an array of data. Due to the fact that magnetic thin films help in storing data magnetically as well as allow the retrieval of data to be efficient, HDDs depend on them.

Objective

The main purpose of incorporating thin films in hard drives is to ensure the apparatus achieves good performance by offering reliable and fast storage and access to data in electronic devices.

Implementation

A common example of thin film materials is the formation of hard drives, which are mainly composed of iron, cobalt, or nickel. Film formats After these kinds of the film have been placed onto a substrate, these are magnetically formed into small areas of magnetization known as bits corresponding to the data to be recorded. The magnetic coefficients of thin films can be tuned to enable high writing as well as the reading of data.

Results

Based on the usage of thin films in hard drives, the technologies for data storage have been improved significantly. Due to high storage ability, fast data access, and good storage data, the HDDs are used in numerous ways such as data centers, and PCs.

Benefits

- High Storage Capacity

Thin films in hard drives also promote efficient storage because many papers can be stored in a single entity.

- Fast Data Access

Thin film characteristics proffer swift data access speeds along with the data search from the hard drive.

- Reliable Data Storage

There is backing up of data on hard drive thin films to ensure that data is safe when not in use or in the process of being used.

Conclusion

Thin films are very essential in data storage or computers for instance the case with hard drive thin films. Thin films, the basis of the digital age, have revolutionized how data is stored and retrieved efficiently.

15 Case Study # 15

Case Study: Architectural Glass—A Balancing Act with Thin Films [13].

Overview

Architecture glass has become a very significant part of contemporary building construction and thin films play an important role in its functionality. Thin film-coated architectural glasses have the following advantages; low maintenance; energy saving, and thermal control.

Objective

The main reasons for applying thin films on architectural glass in order to improve the performance of buildings are as follows: improving the thermal comfort, reducing the energy consumption, and at the same time reducing the maintenance of the buildings.

Implementation

Sputter-coated architectural glasses have films like self-cleaning and low emissivity also known as low-e. Hence, depending on the season, low-e coatings modules restrict the amount of heat gained from entering or lost by reflecting the infrared rays. Self-cleaning coatings also do not require frequent cleaning because they are slim and thus they do not allow for the accumulation of dust and filth.

Results

Current achievements demonstrate that the integration of thin films onto architectural glass has produced tremendous enhancements in building performance. Because thin film coatings are again self-cleaning; buildings with thin film-coated glass have got better thermal insulation, consume less energy for heating/cooling, and are easier to maintain.

Benefits

- Better Thermal Comfort

Reducing heat to be supplied or taken away through the glass thin coats on architectural glass aid in maintaining a favorable temperature indoors.

- Energy Efficiency

Thin films, therefore, reduce energy consumption thereby automatically reducing the amount of carbon that is released in the atmosphere.

– Less Maintenance

Seemingly, self-cleaning coatings reduce the quantity of grime and other particles that accumulate on glass material hence reducing the frequency of cleaning required.

Conclusion

The sample of the thin film's application and importance in the modern architecture of buildings: architectural glass. Thin films help in the making of buildings with enhanced comfort and sustainability due to reduced maintenance needs, enhanced thermal conditions, and efficiency in energy use.

16 Case Study # 16

Case Study: Medical Devices—Thin Films for Improved Healthcare [13].

Overview

Thin films are instrumental components that increase the utility and effectiveness of numerous medical products. Of the various thin films, thin film coatings are critical to endoscopy and catheterization treatments since flexibility and maneuverability are critical in minimally invasive surgical procedures.

Objective

Optimization of employing thin films in medical instruments to boost performance, functionality, and safety with the aim of improving overall patients' well-being are the main benefits.

Results

Specialized deposition processes are used in applying thin film coatings to medical facilities such as catheters and endoscopes. These coatings make procedures easier and more accurate and make the flexibility of the device better, reducing friction by lubrication.

Implementation

This has led to advancements in minimally invasive surgery whose foundation is based on the use of thin film in medical devices. Thus, these technologies are far from traditional surgical techniques in terms of dexterity, in terms of affecting the patient, recovery time, and side effects.

Benefits

– Increased Flexibility

The medical equipment made from thin film coatings is more flexible as it becomes easier to maneuver through the human body for procedures.

– Improved Maneuverability

Thin coatings decrease potential friction that exists between the surface of a piece of equipment and an operator's hands, this means that thin coatings make it easier for a doctor, nurse, or technician to maneuver equipment with better control.

– Minimally Invasive Surgery

The key advantage of thin films is the ability to have minimal invasions, which is highly beneficial because they entail one or more of the following benefits; reduced cuts, minimized pain, and short recovery time.

Conclusion

Thin films in medical devices also show how crucial they are in improving the lives of patients. Thin films play a signification role in the specialization of MIM in surgery since they improve the functionality and the performance of medical equipment which is useful to patients and surgeons.

17 Summary

The case studies and practical examples considered in the present chapter provide quite a depiction of the aspects concerning the revolutionary prospects of thin films and coatings in various branches of engineering. Such examples clearly summarize a rather vast use of advanced materials and the considerable improvement of the performance and sustainability of the products. Logically, every case study is focused on particular problems and thin film technologies solutions for them, proving the opportunities of thin film applications to solve various engineering challenges effectively.

The practical examples also covered here stress the fact that thin film applications are constantly growing with time and thereby, there is a constant need for research and development in this field. The strengths of these case studies are in the fact they highlight practical aspects of the utilization of advanced coating technologies through translating methodological research **conducted** in this field into practice.

Summing up, the discussed actual technologies and applications of thin films and coatings in this chapter prove their indispensability in the development of engineering technologies. Chemical industries today still strive to obtain solutions to improving these objectives, and with these innovative thin film systems that were discussed in these case studies, one can see that thin film technologies are not only evolving but is also becoming increasingly essential in the current engineering world.

References

1. Pratt and Whitney: https://www.prattwhitney.com/
2. Yan, N., Zhao, C., You, S., Zhang, Y., Li, W.: Recent progress of thin-film photovoltaics for indoor application. Chinese Chem. Lett. **31**(3), 643–653 (2020). https://doi.org/10.1016/j.cclet.2019.08.022
3. Paterson, A.F., Anthopoulos, T.D.: Enabling thin-film transistor technologies and the device metrics that matter. Nat. Commun. **9**(1), 1–4 (2018). https://doi.org/10.1038/s41467-018-07424-2
4. Logothetidis, S.: Flexible organic electronic devices: materials, process and applications. Mater. Sci. Eng. B **152**(1), 96–104 (2008). https://doi.org/10.1016/j.mseb.2008.06.009
5. Apple: https://www.apple.com/
6. Medtronic: https://www.medtronic.com/me-en/index.html
7. Ford: https://www.ford.com/
8. Nikon: https://www.nikon-mea.com/mirrorless-z-f?utm_source=google&utm_medium=cpc&utm_campaign=google-search&gad_source=1&gclid=CjwKCAjwko21BhAPEiwAwfaQCIW1KjWzpnXlst0WtBY4ECzCyOnxNUp3yFcKlu3Lffkg24dsPQPUcRoCodwQAvD_BwE
9. Alca Technology: https://www.alcatechnology.com/en/blog/thin-films-in-everyday-life/#:~:text=Thin films are used for, fingerprints%2C and even from corrosion.
10. Korvus Technology: https://korvustech.com/
11. Yang, D. (ed.): Thin Films. IntechOpen, Rijeka (2023). https://doi.org/10.5772/intechopen.100701
12. Thin film nad Coating: https://vaccoat.com/blog/thin-film-and-thin-films-types/#:~:text=Electroplating%2Csol-gel%2Cdip, are popular chemical deposition methods
13. Mindstick - 8 Applications of thin film in modern technology: https://www.mindstick.com/articles/332415/8-applications-of-thin-film-in-modern-technology

The manufacturer's authorised representative in the EU is Springer Nature Customer Service Centre GmbH, Europaplatz 3, 69115 Heidelberg, Germany. If you have any concerns regarding our products, please contact ProductSafety@springernature.com

Printed and bound by CPI Group (UK) Ltd, Croydon, CR0 4YY

30/09/2026

02225298-0008